Green's Functions and Finite Elements

Friedel Hartmann

Green's Functions and Finite Elements

Springer

Friedel Hartmann
Structural Mechanics
University of Kassel
Baunatal
Germany

ISBN 978-3-642-29522-5 ISBN 978-3-642-29523-2 (eBook)
DOI 10.1007/978-3-642-29523-2
Springer Heidelberg New York Dordrecht London

Library of Congress Control Number: 2012938233

Printed on acid-free paper

Springer is part of Springer Science+Business Media (www.springer.com)

Preface

This book is about finite elements and Green's functions, two seemingly very different topics, one representing modern numerical analysis and the other representing the old way of doing it, slowly converging infinite sums and strange looking integrals—none of which will appear in this book—and limited in its scope to linear problems. But the opposite is true, finite elements and Green's functions have very many things in common, may we just mention the fact that the columns of the inverse stiffness matrix are the discrete Green's functions of the nodal values. To a young engineer who is easily fascinated by all the powerful tools and machinery he has available today it may come as a surprise that the finite element method basically can be seen as a Green's function method and that when it is applied to nonlinear problems many of these features shine through as the success of goal-oriented adaptive refinement proves conclusively. So in some sense a relic of the old past is the driving force in the computer programs we use today.

The Green's functions are so to speak the *physical basis functions*, the "true" basis functions of a problem and finite element analysis is all about approximating these function with nodal basis functions. These discrete Green's functions form the machinery behind the finite element code, they produce the output the engineer sees on the screen.

In 1999 the author attended an IUTAM-symposium on boundary elements in Cracow, Poland. At lunch Wolfgang Wendland mentioned that Fehmi Çirak in Stuttgart had applied "Rannacher's method" to an engineering problem in shell theory [1], that Çirak had traveled for one month twice a week to Heidelberg for private lessons by Rannacher on goal-oriented adaptive refinement with Green's functions.

Later Ekkehard Ramm, Çiraks thesis adviser, discovered that the key equation had already been published by Hugh Tottenham in Southampton as early as 1970 [2], though hardly anyone noticed this at that time or seemingly understood the central message this equation contained—otherwise we could have had goal-oriented adaptive refinement as early as 1970.

As an aside, probably there are also other precursors because in mathematics Green's functions are the tool of choice when it comes to pointwise estimates [3]. The intention of this (unproven) remark is not to diminish the pioneering insight of Hugh Tottenham.

The conference in Cracow was the first time the author heard about goal-oriented refinement. After returning home and reading Çirak's thesis it suddenly dawned on him that Green's functions were not just a tool for pointwise estimates but that they were the gist of the matter: the finite element method itself was a Green's function method—from the start.

Seemingly, there is no need to draw circles around the point loads, to soften their impact, to avoid the infinite energy associated with most Green's functions. The algorithm itself does not care with the consequence that the machinery in an finite element program, so to speak, consists of kernel functions which are the solutions of ill-posed problems. One is reminded of the success of the X-FEM which also does everything "wrong".

A first tentative paper was published [4], and in a later book on Structural Analysis with Finite Elements we and our co-author Casimir Katz chartered the new found territory in all directions [5]. Suddenly we could explain things from a new perspective, point at possible difficulties due to the complex nature of the influence functions. With each day the two authors learnt something new. Now after more than 10 years have elapsed since the conference in Cracow the current author has the feeling that it is time to treat the subject more systematically, to summarize the main points, and to trace out the main features. This is why we have written this book.

It is the attempt of an engineer to come to terms with the subject of Green's functions and finite elements. We are well aware that a more elaborate exposition of this topic would have been desirable but we hope that what the book misses in rigor and exactness is made up by an ample provision of engineering examples and applications.

Kassel, Germany, March 2012 Friedel Hartmann

References

1. Çirak F, Ramm E (2000) A posteriori error estimation and adaptivity for elastoplasticity using the reciprocal theorem. Int J Numer Method Eng 47:379–393
2. Tottenham H (1970) Basic Principles. In: Tottenham H, Brebbia C (eds) Finite Element Techniques in StructuralMechanics. Southampton University Press, Southampton
3. Babuška I, Strouboulis T (2001) The finite element method and its reliability. OxfordUniversity Press, Oxford

4. Grätsch T, Hartmann F (2004) Duality and finite elements. Finite Elements AnalDes 40:1005–1020
5. Hartmann F, Katz C (2007) Structural analysis with finite elements, 2nd edn. Springer, New York

Contents

Chapter 1
Introduction

A Green's function is the response of a medium to a point source and because any load or charge can be considered a sum of infinitely many such point sources the Green's functions play a fundamental role in linear systems. Green's functions can be considered "*physically based basis functions adapted to a particular geometry and particular constraints*" [1], and what we do in FE-analysis is that we approximate these physical basis functions with piecewise polynomials. This is all there is to FE-analysis of linear problems—of course put aside the delicate question of how to choose the best approximate. Testimony to this tight connection between finite elements and potential theory is the fact that the columns of the inverse stiffness matrix are the discrete Green's functions of the nodal values.

Any (or nearly any) value an FE-program outputs has been processed by these numerical influence functions and the kernel in these influence functions are the approximate Green's functions pieced together from the nodal basis functions.

The better an FE-mesh can react to the point loads which generate the Green's functions the better the accuracy of the FE-solution. That is the shapes a mesh can assume, its kinematics, the quality of the Green's functions which can be generated on the mesh, ultimately determines the quality of an FE-solution.

And because any Green's function strongly depends on the model parameters, the slightest change in a coefficient, say on a stretch of only 0.1 m, will affect the *whole* function, the sensitivities of a model can be made visible by plotting the updated Green's function and so the user gets an impression of the difficulties involved in evaluating certain functionals. The machinery, so to speak, which the FE-program uses, can be laid bare. FE-analysis becomes transparent, becomes visual.

What is nice about Green's functions and finite elements is that the calculation of Green's functions is easy to implement in any existing code. Detailed instructions on how to calculate influence functions with FE-programs close the chapter.

F. Hartmann, *Green's Functions and Finite Elements*,
DOI: 10.1007/978-3-642-29523-2_1, © Springer-Verlag Berlin Heidelberg 2013

1.1 What are Green's Functions?

A Green's function $G(\boldsymbol{y}, \boldsymbol{x})$ is a function that allows to solve boundary value problems such as

$$-\Delta u = -\left(\frac{\partial^2 u}{\partial x_1^2} + \frac{\partial^2 u}{\partial x_2^2}\right) = p \quad \text{on } \Omega \quad u = 0 \quad \text{on } \Gamma \text{ (the edge)} \tag{1.1}$$

by an integral

$$u(\boldsymbol{x}) = \int_\Omega G(\boldsymbol{y}, \boldsymbol{x})\, p(\boldsymbol{y})\, d\Omega_{\boldsymbol{y}}\,, \tag{1.2}$$

that is the Green's function is the *kernel* of the *inverse operator* which here is an integral operator.

In physical terms is the Green's function the response of the medium if a unit point load, a Dirac delta, is applied at the source point $\boldsymbol{x}$ (differentiation in the next equation is done with respect to the variable $\boldsymbol{y} = (y_1, y_2)$)

$$-\Delta G(\boldsymbol{y}, \boldsymbol{x}) = \delta(\boldsymbol{y} - \boldsymbol{x}) \qquad G(\boldsymbol{y}, \boldsymbol{x}) = 0\,, \qquad \boldsymbol{y} \in \Gamma. \tag{1.3}$$

In linear algebra the point loads are unit vectors $\boldsymbol{e}_i$ or—the most simple scenario—the number 1. To solve the equation

$$3 \cdot u = 12 \tag{1.4}$$

we multiply the right-hand side with the number $g = 1/3$

$$u = g \cdot 12 = \frac{1}{3} \cdot 12 \tag{1.5}$$

which is the "Green's function", the solution of the equation

$$3 \cdot u = 1 \qquad \leftarrow \qquad 1 = \text{"point load"}. \tag{1.6}$$

The columns of the inverse of a matrix as for example

$$\boldsymbol{K} = \begin{bmatrix} 2 & -1 & 0 & 0 \\ -1 & 2 & -1 & 0 \\ 0 & -1 & 2 & -1 \\ 0 & 0 & -1 & 2 \end{bmatrix}, \quad \Rightarrow \quad \boldsymbol{K}^{-1} = \frac{1}{5} \begin{bmatrix} 4 & 3 & 2 & 1 \\ 3 & 6 & 4 & 2 \\ 2 & 4 & 6 & 3 \\ 1 & 2 & 3 & 4 \end{bmatrix}, \tag{1.7}$$

are the Green's functions of the nodal values u_i. They map the right-hand side of the equation $\boldsymbol{K}\,\boldsymbol{u} = \boldsymbol{f}$ onto the solution vector $\boldsymbol{u}$

$$u = K^{-1} f = \frac{1}{5} \cdot \left(f_1 \cdot \begin{bmatrix} 4 \\ 3 \\ 2 \\ 1 \end{bmatrix} + f_2 \cdot \begin{bmatrix} 3 \\ 6 \\ 4 \\ 2 \end{bmatrix} + f_3 \cdot \begin{bmatrix} 2 \\ 4 \\ 6 \\ 3 \end{bmatrix} + f_4 \cdot \begin{bmatrix} 1 \\ 2 \\ 3 \\ 4 \end{bmatrix} \right) . \tag{1.8}$$

So Green's function are a frequently used tool in applied mathematics and in this introductory chapter we will show that Green's functions are also the "machinery" behind the FE-method.

1.2 The Importance of Green's Functions for FE-Analysis

In the language of mechanics is the integral

$$u(\boldsymbol{x}) = \int_{\Omega} G(\boldsymbol{y}, \boldsymbol{x})\, p(\boldsymbol{y})\, d\Omega_{\boldsymbol{y}} \tag{1.9}$$

an influence function. Each *observable* of a linear system, that is each displacement, each stress value, each gradient can be represented by such an influence function. A linear system is even defined by this *superposition principle.*

And when an FE-program solves a linear problem it applies this principle, it writes the FE-solution in integral form as in (1.9)

$$u_h(\boldsymbol{x}) = \int_{\Omega} G_h(\boldsymbol{y}, \boldsymbol{x})\, p(\boldsymbol{y})\, d\Omega_{\boldsymbol{y}} . \tag{1.10}$$

It is only that it replaces the exact Green's function by its *projection* $G_h(\boldsymbol{y}, \boldsymbol{x})$ onto the trial and test space $\mathcal{V}_h$ which means that $G_h(\boldsymbol{y}, \boldsymbol{x})$ is the FE-solution $\boldsymbol{K}\, \boldsymbol{u} = \boldsymbol{f}$ of the boundary value problem (1.3), with a Dirac delta on the right-hand side, or—as an engineer would say—when a unit load $P = 1$ is applied at the point $\boldsymbol{x}$.

But the exact response of a medium to a point load, a Dirac delta, cannot be modeled—in general—with finite element shape functions, so that the kernel $G_h(\boldsymbol{y}, \boldsymbol{x})$ is not exact and this gap $G(\boldsymbol{y}, \boldsymbol{x}) - G_h(\boldsymbol{y}, \boldsymbol{x})$ is responsible for the error in FE-solutions[1]

$$u(\boldsymbol{x}) - u_h(\boldsymbol{x}) = \int_{\Omega} [G(\boldsymbol{y}, \boldsymbol{x}) - G_h(\boldsymbol{y}, \boldsymbol{x})]\, p(\boldsymbol{y})\, d\Omega_{\boldsymbol{y}} . \tag{1.11}$$

[1] The response of an FE-mesh to a single nodal force $f_i = 1$ is only a rough approximation of the true Green's function, see for example Fig. 3.3 p. 114.

To minimize the error of the FE-solution at a particular point $\boldsymbol{x}$ we therefore refine the mesh in such a way that the Green's function for this point can be approximated exceptionally well on the mesh. This is the idea of *goal-oriented refinement.*

This strategy works best when the focus is on specific values, say the high stresses σ_{yy} at the tip of a crack. The loading that causes these high stresses constitutes the so-called *primal problem.* In standard adaptive refinement we would solve the primal problem first on an initial mesh and then refine the mesh where necessary. In goal-oriented refinement we solve an additional, second problem, a *dual problem,* in which a unit dislocation in y-direction is applied at the crack tip. Such a dislocation generates the influence function for σ_{yy} and if the mesh is well adapted to model this particular dislocation then *automatically* the stress σ_{yy} in the first load case (*the primal problem*) will be very accurate.

Or imagine that the temperature in a motor bloc at a specific point $\boldsymbol{x}$ is to be predicted very precisely. In this case we would place a single heat source at the point $\boldsymbol{x}$ and ensure that the temperature field which is generated by this point source can be modeled on the mesh very precisely. If this is guaranteed then the temperature at the point $\boldsymbol{x}$ can be predicted very accurately—in virtually any load case.

1.3 A 1-D Problem

To illustrate this close relationship between Green's functions and finite elements in more details we study a simple 1-D example, a rope suspended between two walls, prestressed by a horizontal force H and carrying a lateral load p, see Fig. 1.1.

1.3.1 The Analytical Solution

Finding the deflection $u(x)$ of the rope means to solve the boundary value problem

$$-\,H\,u''(x) = p(x) \qquad 0 < x < l \qquad u(0) = u(l) = 0\,. \tag{1.12}$$

When the load p is constant we choose the *ansatz* $u(x) = -p\,x^2/2H + c_1 x + c_0$ and so we easily find the solution

$$u(x) = \frac{p}{2\,H}\,(l\,x - x^2)\,. \tag{1.13}$$

1.3.2 Green's Function

We could have approached the task also by observing that a force $P = 1$ applied at a point y will give the rope a triangular shape, see Fig. 1.2,

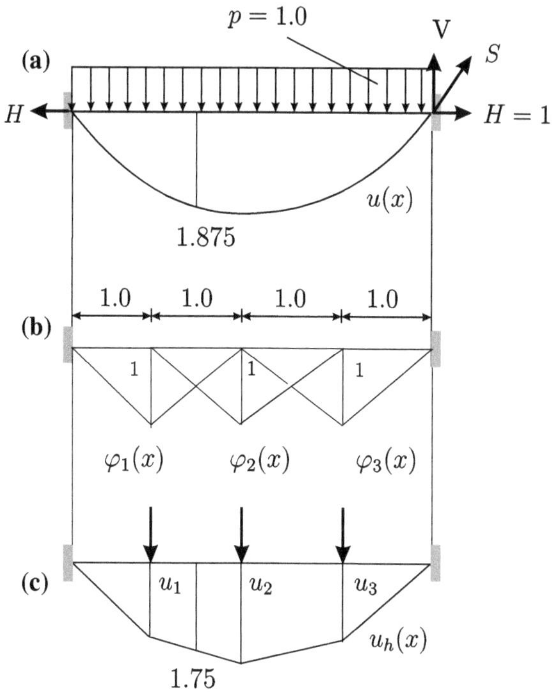

Fig. 1.1 FE-analysis of a taut rope: **a** exact solution, **b** hat functions, **c** FE-solution

$$G(x, y) = \frac{1}{H\,l} \begin{cases} y\,(l - x) & y \leq x \\ x\,(l - y) & x \leq y \end{cases} \tag{1.14}$$

and because the distributed load p resembles a series of infinitely many infinitely small point loads $p(y)\,dy$, each of which contributes the deflection

$$du(x) = G(x, y)\,p(y)\,dy\,, \tag{1.15}$$

the sum must be the deflection of the rope

$$u(x) = \int_0^l du(x) = \int_0^l G(x, y)\,p(y)\,dy = \frac{p}{2\,H}\,(l\,x - x^2) \tag{1.16}$$

which it is: the integral has at each point x the same value as the solution in (1.13).

1.3.2.1 A Switch

Before proceeding further we pause to do a switch. The Green's function is symmetric, $G(y, x) = G(x, y)$, so that in (1.16) the places of y and x can be swapped

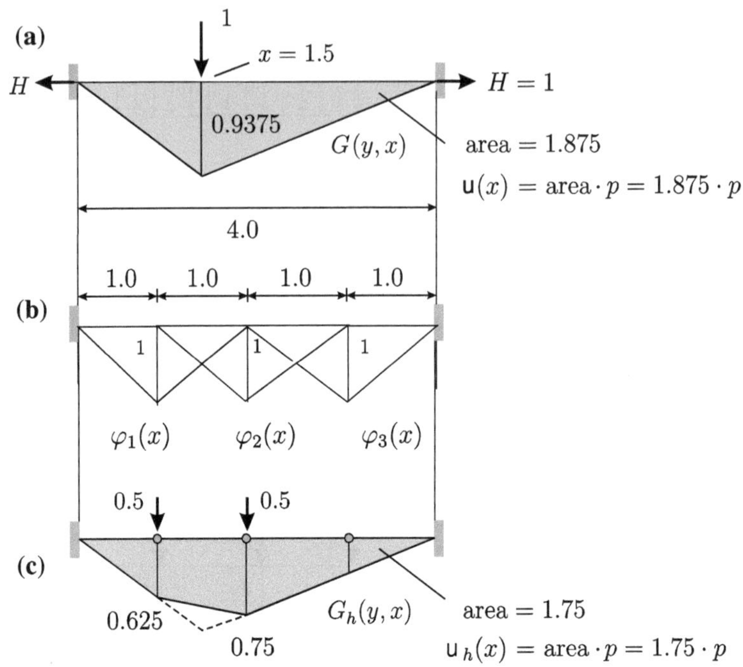

Fig. 1.2 Green's function, **a** exact Green's function, **b** hat functions, **c** approximate Green's function

$$u(x) = \int_0^l G(y,x)\,p(y)\,dy = \int_0^l$$ 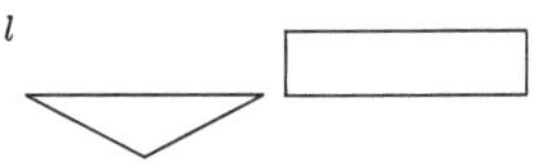 dy (1.17)

which makes that the tip of the triangle now stays fixed at the point x whereas in (1.16) the tip moved with the integration point y.

1.3.3 Finite Elements

To determine the shape of the rope with finite elements a subdivision of the rope into four linear finite elements may suffice for a start, see Fig. 1.1,

$$u_h(x) = \sum_{i=1}^{3} u_i\,\varphi_i(x)\,. \tag{1.18}$$

The formulation of the system $\boldsymbol{K}\,\boldsymbol{u} = \boldsymbol{f}$

$$\begin{bmatrix} 2 & -1 & 0 \\ -1 & 2 & -1 \\ 0 & -1 & 2 \end{bmatrix} \begin{bmatrix} u_1 \\ u_2 \\ u_3 \end{bmatrix} = \begin{bmatrix} 1 \\ 1 \\ 1 \end{bmatrix}. \tag{1.19}$$

and finding its solution

$$u_1 = u_3 = 1.5 \qquad u_2 = 2.0 \tag{1.20}$$

is standard and so we easily determine that

$$u_h(x) = 1.5 \cdot \varphi_1(x) + 2.0 \cdot \varphi_2(x) + 1.5 \cdot \varphi_3(x) \tag{1.21}$$

is the best approximation to $u(x)$ with four elements in terms of the strain energy metric.

1.3.4 Finite Elements and Green's Function

Now let us try a novel approach, a combination of finite elements and Green's function.

The Green's function $G(y, x)$ is the shape of the rope when a force $P = 1$ is applied at x that is the Green's function is the solution of the boundary value problem

$$-H \frac{d^2}{dy^2} G(y, x) = \delta(y - x) \qquad G(0, x) = G(l, x) = 0\,. \tag{1.22}$$

What happens if this boundary value problem is solved with finite elements and the FE-solution $G_h(y, x)$ instead of the exact kernel $G(y, x)$ is substituted into the influence function (1.16)? Which effect does this have on the influence function?

For a first try we choose the point x which lies half-way between node 1 and 2, see Fig. 1.2. How well can we approximate the Green's function of this point with finite elements? The exact Green's function for $u(x)$ is the response of the rope to a point load $P = 1$ applied at x. This triangular shape with its apex at x cannot be modeled with the shape functions because in between nodes they run straight. What the FE-program does is that it replaces the force $P = 1$ by two equivalent nodal forces, $f_i = 0.5$ each, at the two neighboring nodes

$$\begin{bmatrix} 2 & -1 & 0 \\ -1 & 2 & -1 \\ 0 & -1 & 2 \end{bmatrix} \begin{bmatrix} u_1 \\ u_2 \\ u_3 \end{bmatrix} = \begin{bmatrix} 0.5 \\ 0.5 \\ 0 \end{bmatrix} \tag{1.23}$$

and it solves this system for the nodal values of the Green's function,

$$u_1 = 0.625 \qquad u_2 = 0.750 \qquad u_3 = 0.375 \tag{1.24}$$

which give the Green's function a shape as in Fig. 1.2c.

This is of course not the correct shape, the peak under the point load is missing, and so the substitute influence function (1.16) the integral

$$1.75 = \int_0^l G_h(y,x)\, p(y)\, dy \tag{1.25}$$

gives a wrong value—the true value is 1.875. But when we look closer then we find—to our surprise—that the value 1.75 is *exactly* the value of the FE-solution at this point, see Fig. 1.1c.

And if we repeat this experiment at other points of the rope we always find that the integral of the approximate Green's function $G_h(y,x)$ and the distributed load p is exactly the value of the FE-solution at this point

$$u_h(x) = \sum_{i=1}^{3} u_i\, \varphi_i(x) = \int_0^l G_h(y,x)\, p(y)\, dy\,. \tag{1.26}$$

So it must be true that an FE-program uses the approximate Green's functions to calculate the deflection $u_h(x)$ of the rope.

But does'nt an FE-program calculate the nodal values $\boldsymbol{u}$—and therewith also indirectly the values in between—by solving the equation $\boldsymbol{K}\,\boldsymbol{u} = \boldsymbol{f}$? This is true as well but it does not contradict the previous statement. Both statements are true.[2] The nodal values u_i the FE-program outputs—and any other value in between as well—are as large *as if* the FE-program had calculated them with the approximate Green's function which can be generated on the mesh.

And this principle is not restricted to 1-D problems. To calculate the vertical displacement of the center node of the plate in Fig. 1.3 the FE-program applies a vertical unit point load at the node and it watches by how much the upper edge as a reaction to this nodal load gets displaced. The work done by the edge load p on acting through this edge displacement $\eta(x)$ is the nodal displacement

$$u_y = \int_0^l p(x)\, \eta(x)\, dx\,. \tag{1.27}$$

It is the same value $u_i = u_y$ you get by solving the system $\boldsymbol{K}\,\boldsymbol{u} = \boldsymbol{f}$.

Or watch how an FE-program calculates stresses, see Fig. 1.4. The program applies a unit dislocation in x or y direction, here of magnitude 1 m = 1,000 mm, and it measures the resulting vertical displacement at the foot of the point load. The ratio of this displacement versus the triggering dislocation times the magnitude of the point load is the stress σ_{xx} or σ_{yy} respectively at the source point

---

[2] Call this the "particle-wave" duality of FE-analysis.

Fig. 1.3 How an FE-program calculates the nodal displacement: it applies a point load at the node and watches by how much the edge as a reaction to this displaces. The work done by the edge load on acting through this displacement $\eta(x)$ equals the nodal displacement

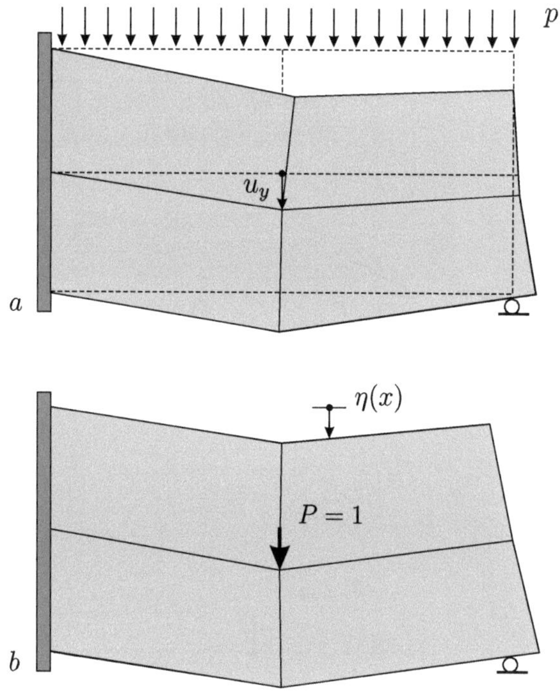

$$u_y = \int_0^l \eta(x)\, p\, dx$$

$$\sigma = \frac{\textit{perceived displ.}}{\textit{triggering disl.}} \cdot P \tag{1.28}$$

The better a mesh can react to these point loads and dislocations the better the results. So the kinematics of a mesh—influence functions are displacements (!)—decides how good the FE-results are, see Figs. 1.5, 1.6 and 1.7.

1.4 Entanglement

On second thoughts this close connection between the accuracy and the kinematics of a mesh seems logical, correct measurements require a precise meter (= Green's function). What is a surprise though is the "entanglement" (to borrow a term from quantum mechanics) between the FE-solution and the yardsticks, the meters. How does the FE-solution know in advance what the FE-Green's functions will measure, which values $u_h(x)$ or $\sigma_h(x)$ or $V_h(x)$, etc., they will extract from p? Because only

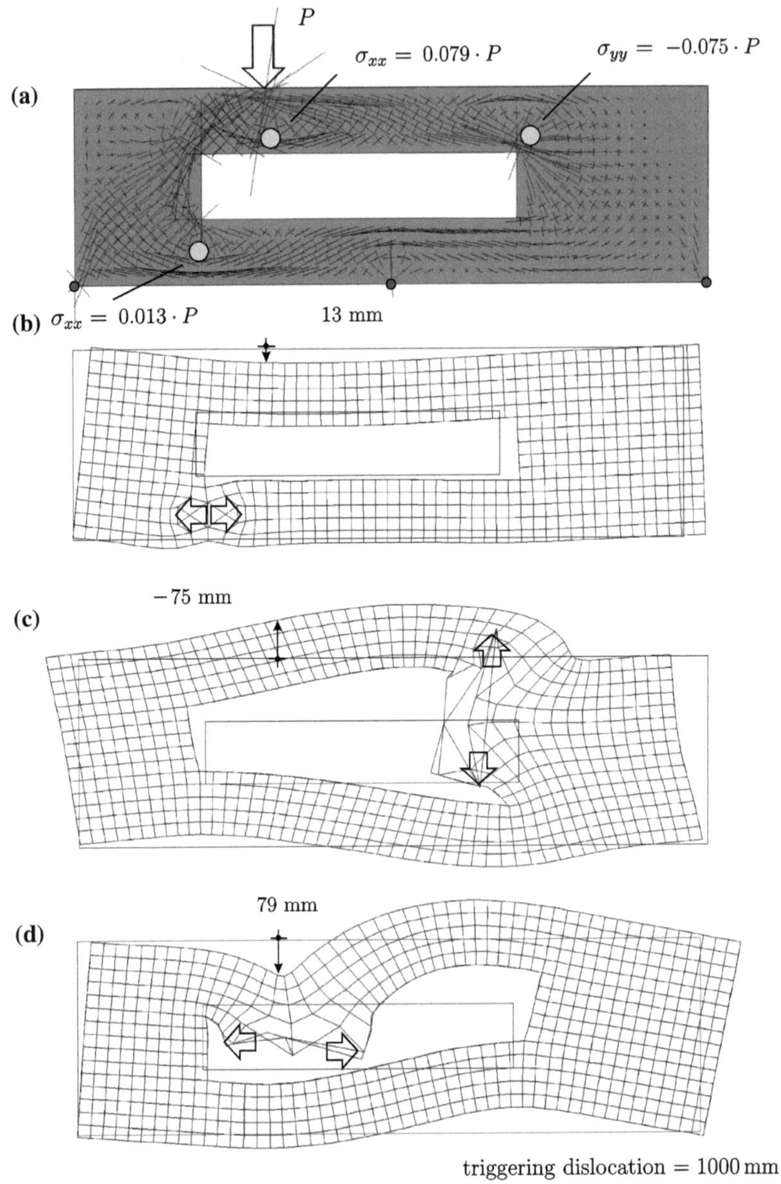

Fig. 1.4 How an FE-program calculates the stresses in a plate. Each stress value corresponds to the work $P \cdot u_y$ done by the applied load P on acting through the displacement u_y at its foot as generated by the corresponding influence function which is the displacement field due to a unit dislocation, here of 1,000 mm, in the corresponding direction. Note that in 2-D a dislocation of 1,000 mm is not simply a spread of 1,000 mm. It is an integral measure, you circle the source point once and you experience a shift of 1,000 mm, see Sect. 2.5

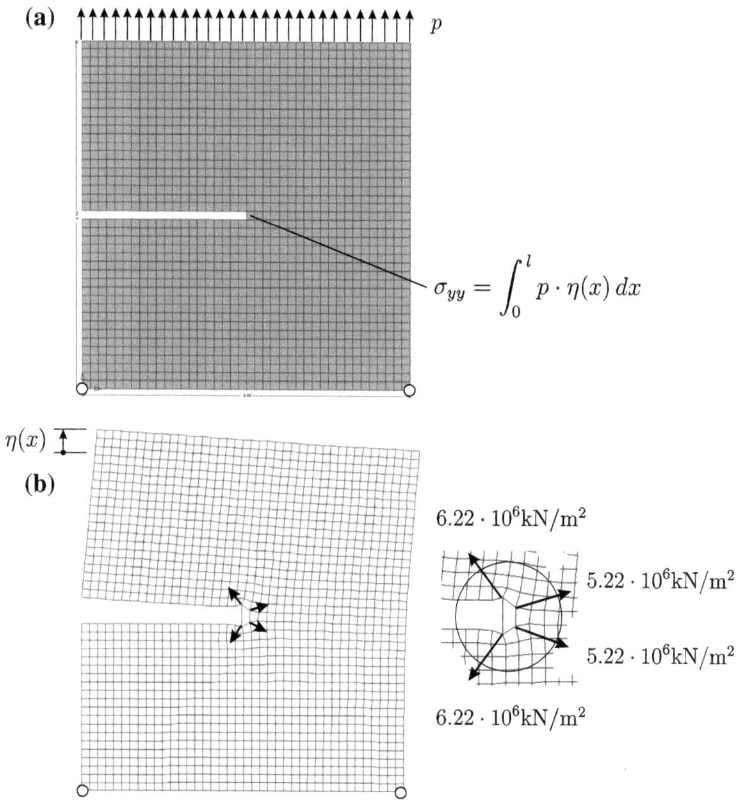

Fig. 1.5 How an FE-program calculates the stress σ_{yy} at the bottom of the slit: it applies nodal forces $f_i = \sigma_{yy}(\varphi_i)$ at the four corner of the bilinear element and the work done by the edge load on acting through the edge displacement $\eta(x)$ is the stress σ_{yy}

with clever foresight can it choose the nodal values u_i of the FE-solution

$$u_h(x) = \sum_i^n u_i\, \varphi_i(x) \tag{1.29}$$

in such a way that the FE-solution agrees with the results produced by the approximate FE-influence functions

$$\sum_i^n u_i\, \varphi_i(x) = u_h(x) = \int_0^l G_h(y, x)\, p(y)\, dy\,. \tag{1.30}$$

This is the surprise. Why do the two techniques give the same value $u_h(x)$?

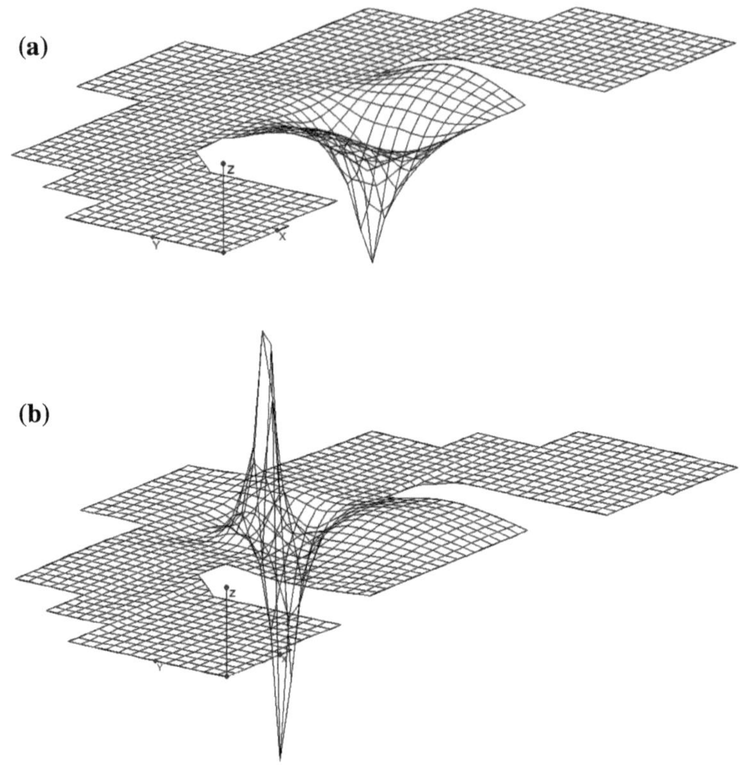

Fig. 1.6 How an FE-program calculates the bending moment m_{xx} and the shear force q_x respectively of a slab: it calculates the work done by the surface load on acting through the influence functions of **a** the bending moment and **b** of the shear force

1.4.1 Functionals

To understand this we must understand the concept of a functional. A functional $J(u)$ is a function of functions as

$$J(u) = u(0)\,. \tag{1.31}$$

The value of the functional is the value of the function $u(x)$ at the point $x = 0$, so that

$$J(\sin x) = \sin 0 = 0 \qquad J(\cos x) = \cos 0 = 1.0 \tag{1.32}$$

The area under a curve also is a functional

$$J(u) = \int_0^{\pi} u(x)\,dx \qquad J(\sin x) = \int_0^{\pi} \sin x\,dx = 2\,. \tag{1.33}$$

Assume you are learning a new programming language and your first assignment is to write a small program to calculate the deflection of a rope subjected to various types of loads p. For a test the instructor asks you for the deflection at a specific point x. In an abstract sense the teacher is asking for the value of the functional

$$J(u) = u(x)\,. \tag{1.34}$$

How does the program calculate $u(x)$? You—as the author of the program—will say that the program calls the subroutine where $u(x)$ is stored and that it simply looks up the value of u.

But a point value such as $u(x)$ can also be thought to come from the application of a Dirac delta to the function u

$$1 \cdot u(x) = \int_0^{l} \delta(y - x)\,u(y)\,dy \tag{1.35}$$

that is when a unit point load acts through $u(x)$. On first view this seems to complicate matters unnecessarily: "What sense is there in interpreting a function call—the simple evaluation of a function—an application of a Dirac delta?"

But Betti would argue that when a Dirac delta is applied to the solution u then this is the same as if the product of the Green's function and the right-hand side p, the applied load, is integrated

$$1 \cdot u(x) = \int_0^{l} \delta(y - x)\,u(y)\,dy = \int_0^{l} G(y, x)\,p(y)\,dy\,. \tag{1.36}$$

And Betti is right, indeed, the work done by the point load $\delta(y - x)$ on acting through $u(x)$ is equal to the work done by the distributed load p on acting through the displacement $G(y, x)$ caused by the point load. This is the important step.

Saying a solution has at a point x the value $u(x)$ is saying the functional $J(u) = u(x)$ has the value $u(x)$. *Each observable is identical with a functional.* A support reaction R_A, the shear force $V(x)$, the second derivative $u''(x)$, a nodal value u_i, all these are functionals.

How can Betti's logic be extended to handle these various functionals? To do this we first generalize the idea of the Dirac delta by claiming that *any* linear functional, as the following three functionals

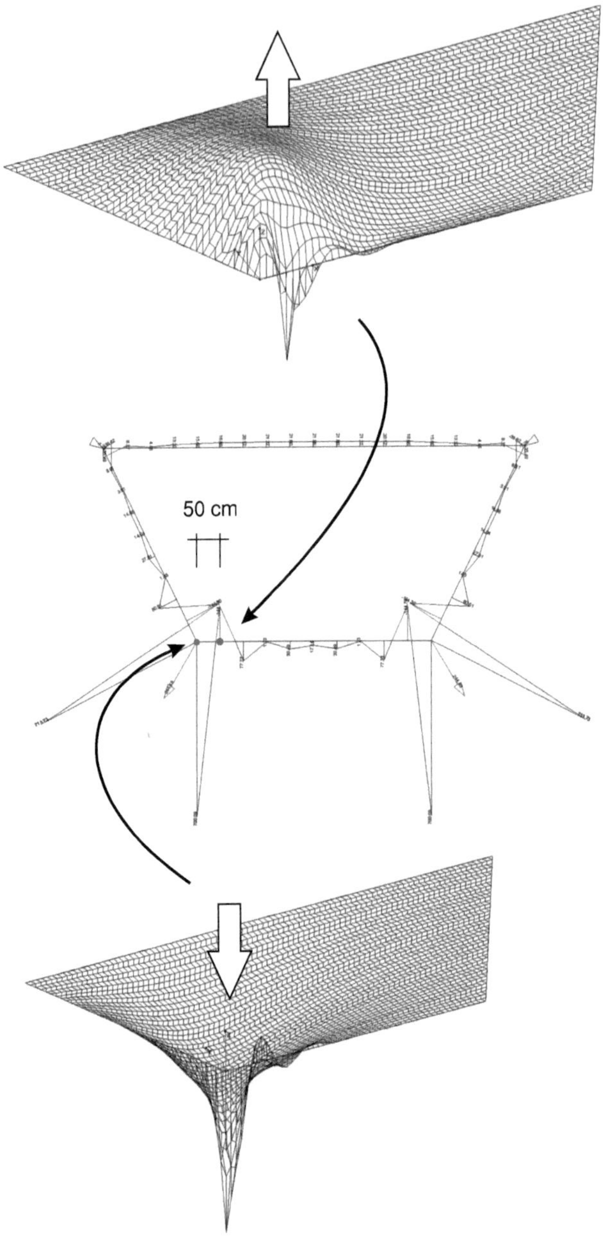

Fig. 1.7 Oscillations in the support reactions of a trapezoidal slab find their expression in the erratic behavior of the influence functions

$$J_a(u) = u'(0) \qquad = \int_0^l \delta_a(y-x)\,u(y)\,dy \tag{1.37}$$

$$J_b(u) = \int_0^l u(x)\,dx \qquad = \int_0^l \delta_b(y-x)\,u(y)\,dy \tag{1.38}$$

$$J_c(u) = u'(x) + u(x) \qquad = \int_0^l \delta_c(y-x)\,u(y)\,dy \qquad \text{etc.}, \tag{1.39}$$

can be written as integrals between certain Dirac deltas and the solution u. What these Dirac deltas $\delta_a, \delta_b, \delta_c$ look like is not important. It suffices to postulate that they exist.

What is important is that each of these integrals can be interpreted as an expression of exterior work: a certain point load $\delta(y-x)$ is acting through u and contributes a certain amount of work $W_{1,2} = J(u)$. And according to Betti's theorem this work must be equal to the work $W_{2,1}$, (we switch the indices), done by the right-hand side p, belonging to u, on acting through the shape $G(y,x)$ produced by the Dirac delta $\delta(y-x)$

$$J(u) = \int_0^l \delta(y-x)\,u(y)\,dy = W_{1,2} = W_{2,1} = \int_0^l G(y,x)\,p(y)\,dy\,. \tag{1.40}$$

This is the basic logic. *This is duality.*

The function $G(y,x)$ (or kernel) is called the *Riesz element* of $J(u)$. In physical terms it is the response of the medium to the "point load" $\delta(y-x)$, the Dirac delta.

It is essential that (1.40) remains valid[3] if the two exact solutions G and u are replaced by their FE-approximations while the loads δ and p stay the same

$$W^h_{1,2} = \int_0^l \delta(y-x)\,\underset{\uparrow}{u_h}(y)\,dy = \int_0^l \underset{\uparrow}{G_h}(y,x)\,p(y)\,dy = W^h_{2,1} \tag{1.41}$$

because this invariance implies, see *Tottenham's equation* p. 116,

$$J(u_h) = \int_0^l G_h(y,x)\,p(y)\,dy \tag{1.42}$$

[3] see Betti's Theorem Extended, p. 114.

which is the key result. It means that in linear FE-analysis any *observable* is the L_2-scalar product (integral) between the right-hand side p and the projection G_h of the Green's function belonging to the observable onto the test and trial space $\mathcal{V}_h$.

In this sense FE-analysis is "consistent". While the exact solution u is projected onto the FE-solution u_h and—parallel to this—the exact kernels G onto FE-kernels G_h, the bond between u and the kernels G is never lost. It is inherited by u_h and the kernels G_h. While the kernel G maps p onto $J(u)$, the projected kernel G_h maps p onto $J(u_h)$, the functional value of the projection u_h

$$J(u) = \int_0^l G(y,x)\, p(y)\, dy \tag{1.43}$$

$$J(u_h) = \int_0^l G_h(y,x)\, p(y)\, dy\,. \tag{1.44}$$

It is only that in FE-analysis the mapping is distorted. The "lens" $G_h(y,x)$ should send the "beam" p onto the point $J(u)$ but instead it sends it to a slightly different spot $J(u_h)$ on the real axis.

1.4.2 Proof

What we call *entanglement* is of course simply a consequence of the fact that the underlying differential equation is self-adjoint and therefore the FE-matrix $\boldsymbol{K}$ symmetric. This guarantees (1.42).

To exemplify this let the J be the point functional $J(u) = u(x)$. The Green's function $G(y,x)$ of this functional is the solution of the boundary value problem

$$-H\frac{d^2}{dy^2}G(y,x) = \delta(y-x) \qquad G(0,x) = G(l,x) = 0 \tag{1.45}$$

where the Dirac delta represents a point force in the weak sense. Like a real point force it is zero almost everywhere and the work done by $\delta(y-x)$ on acting through a virtual displacement v is $v(x)$

$$\delta(y-x) = 0 \qquad y \neq x \tag{1.46}$$

$$\int_0^l \delta(y-x)\, v(y)\, dy = v(x) \qquad x \in (0,l)\,. \tag{1.47}$$

This is what *weak* means, one studies something by observing its effect on a set of test functions, i.e. virtual displacements.

The FE-solution of the boundary value problem (1.45), the FE-Green's function, has the form

$$G_h(y, x) = \sum_i g_i(x)\,\varphi_i(y) \tag{1.48}$$

where the vector $\boldsymbol{g}$ of the nodal displacements ($g_i \equiv u_i$) is the solution of the system

$$\boldsymbol{K}\,\boldsymbol{g} = \boldsymbol{j} \tag{1.49}$$

and where $\boldsymbol{j}$ is the vector of equivalent nodal forces for the Green's function,

$$j_i = J(\varphi_i) \tag{1.50}$$

or given that $J(u) = u(x)$

$$j_i = J(\varphi_i) = \varphi_i(x)\,. \tag{1.51}$$

So in contrast to the standard FE-notation we use g_i for the nodal values of the Green's function and the equivalent nodal forces which generate the Green's function we call j_i because the forces are just the values of the functional applied to the single shape functions φ_i.

The following theorem summarizes these results.

Theorem 1.1 (The central equation).

$$\begin{aligned} u_h(x) &= \int_0^l G_h(y, x)\,p(y)\,dy = \int_0^l \sum_i g_i(x)\,\varphi_i(y)\,p(y)\,dy = \sum_i g_i(x)\,f_i \\ &= \boldsymbol{g}^T\,\boldsymbol{f} = \boldsymbol{g}^T\,\boldsymbol{K}\,\boldsymbol{u} = \boldsymbol{g}^T\,\boldsymbol{K}^T\,\boldsymbol{u} = \boldsymbol{j}^T\,\boldsymbol{u} = \sum_i j_i\,u_i \\ &= \sum_i \varphi_i(x)\,u_i = \int_0^l \sum_i u_i\,\varphi_i(y)\,\delta(y - x)\,dy = \int_0^l u_h(y)\,\delta(y - x)\,dy\,. \end{aligned} \tag{1.52}$$

The displacement $u_h(x)$ is the scalar product, $\boldsymbol{g}^T\,\boldsymbol{f}$, between the nodal displacements of the Green's function and the equivalent nodal forces of the load p or—vice versa—the scalar product, $\boldsymbol{j}^T\,\boldsymbol{u}$, between the nodal displacements of the FE-solution u_h and the nodal forces j_i of the Green's function

$$u_h(x) = \begin{cases} \sum_i \varphi_i(x)\, u_i = \int_0^l u_h(y)\, \delta(y-x)\, dy = \boldsymbol{j}^T \boldsymbol{u} \\ \int_0^l G_h(y,x)\, p(y)\, dy = \boldsymbol{g}^T \boldsymbol{f}\,. \end{cases} \tag{1.53}$$

The important point is that *any* linear functional $J(u)$ can be written in this way

$$J(u_h) = \begin{cases} \sum_i J(\varphi_i)\, u_i = \int_0^l u_h(y)\, \delta(y-x)\, dy = \boldsymbol{j}^T \boldsymbol{u} \\ \int_0^l G_h(y,x)\, p(y)\, dy = \boldsymbol{g}^T \boldsymbol{f}\,. \end{cases} \tag{1.54}$$

The equivalent nodal force j_i is the value of the functional applied to the nodal shape function φ_i

$$j_i = J(\varphi_i) \tag{1.55}$$

and the vector $\boldsymbol{g}$ is the solution of the system

$$\boldsymbol{K}\, \boldsymbol{g} = \boldsymbol{j}\,. \tag{1.56}$$

The first equation is evident

$$J(u) = J\left(\sum_i u_i\, \varphi_i\right) = \sum_i J(\varphi_i)\, u_i = \boldsymbol{j}^T \boldsymbol{u} = \boldsymbol{g}^T \boldsymbol{f}\,, \tag{1.57}$$

the second is Betti.

Remark 1.1 The symbol $\delta(y-x)$ in (1.54) is used in a generic sense, it is that Dirac delta which extracts the value $J(u_h)$ from u_h whatever $J(u_h)$ is, a displacement, a stress or else.

1.5 Goal Oriented Refinement

Obviously does the accuracy the FE-Green's functions G_h achieves in mapping p onto the observables onto the values $J(u)$

$$J(u_h) = \int_0^l G_h(y,x)\, p(y)\, dy\,, \tag{1.58}$$

depend on the kinematics of a mesh—how accurately can a mesh react to the point sources, point loads or dislocations, see Fig. 1.8, which generate the Green's functions. The more refined a mesh is the smaller the error will be.

While in Fig. 1.8 the refinement was done in a uniform way the algorithm alternatively could concentrate on those zones where the error is the largest. This is known as adaptive refinement. When this technique is applied *simultaneously* to the Green's function, *the dual problem*, and the original problem, *the primal problem*, then this is called *goal-oriented refinement*. This strategy produces the mesh in Fig. 1.9.

The algorithm refines the mesh near the point load and near the source point, the regions were the errors are most prominent. In comparison with a standard adaptive refinement the goal-oriented adaptive refinement achieves more with less effort but of course the mesh can be truly optimal only with regard to *one* functional.

1.6 Model Adaptivity

One-sided traffic loads cause large bending moments in the hangers of the *Ponte della Musica*, see Figs. 1.10 and 1.11, because the tensile forces in the hangers try to pull the rods straight so that the curvature at the end points of the hangers is large and consequently also the bending moments M. This means that eventually after many cycles cracks will develop which reduce the stiffness of the load carrying members and lead to changes in the stress distribution.

In a mathematical sense are such cracks or other changes in an element stiffness equivalent to modifications of the coefficients of the governing differential equations or in the FE-context to modifications of the stiffness matrix

$$\boldsymbol{K} \quad \rightarrow \quad \boldsymbol{K} + \Delta\boldsymbol{K}\,. \tag{1.59}$$

What we are interested in is to assess which influence such modifications have on the solution $\boldsymbol{u}$. Of course the modified set of equations could simply be solved anew to find the new vector

$$\boldsymbol{u}_c = (\boldsymbol{K} + \Delta\boldsymbol{K})^{-1}\,\boldsymbol{f} \tag{1.60}$$

and this is what is most often done but qualitatively we would like to gain some insight into the response of a system: How sensitive is the system to changes in the stiffness matrix?

A simple problem, the task to determine the weight W of a suitcase, may provide the necessary clue. To find the weight we would lift the suitcase by, say, 1 m which

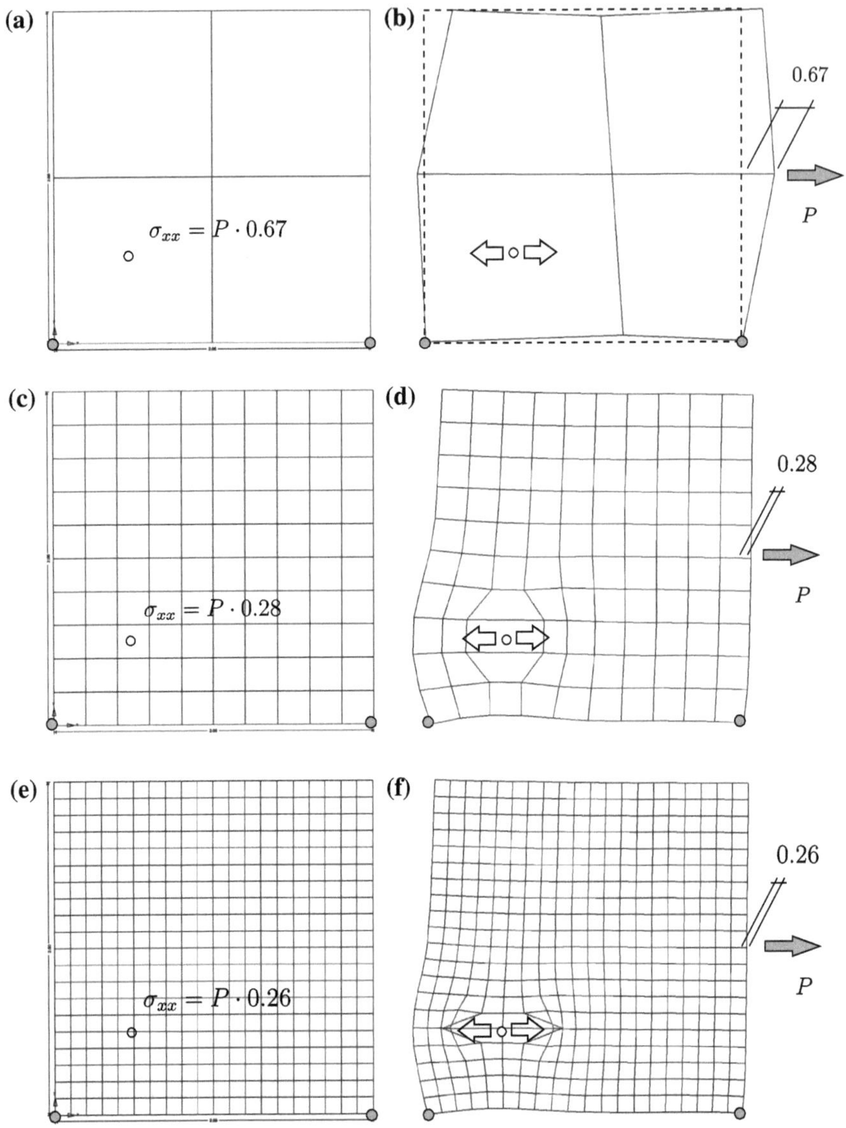

Fig. 1.8 Plate analysis **a** with four elements, **b** influence function for σ_{xx}, **c** refined mesh, **d** influence function for σ_{xx} on refined mesh and the same **e** and **f** after one further refinement

should be enough to gain a feeling for the weight of the suitcase and so, for our purposes, the lift can be considered the influence function for the weight.

Assume that the suitcase contains six gold bars and on the trip one bar (mysteriously) changes its weight by a (negative) amount ΔB, see Fig. 1.12,

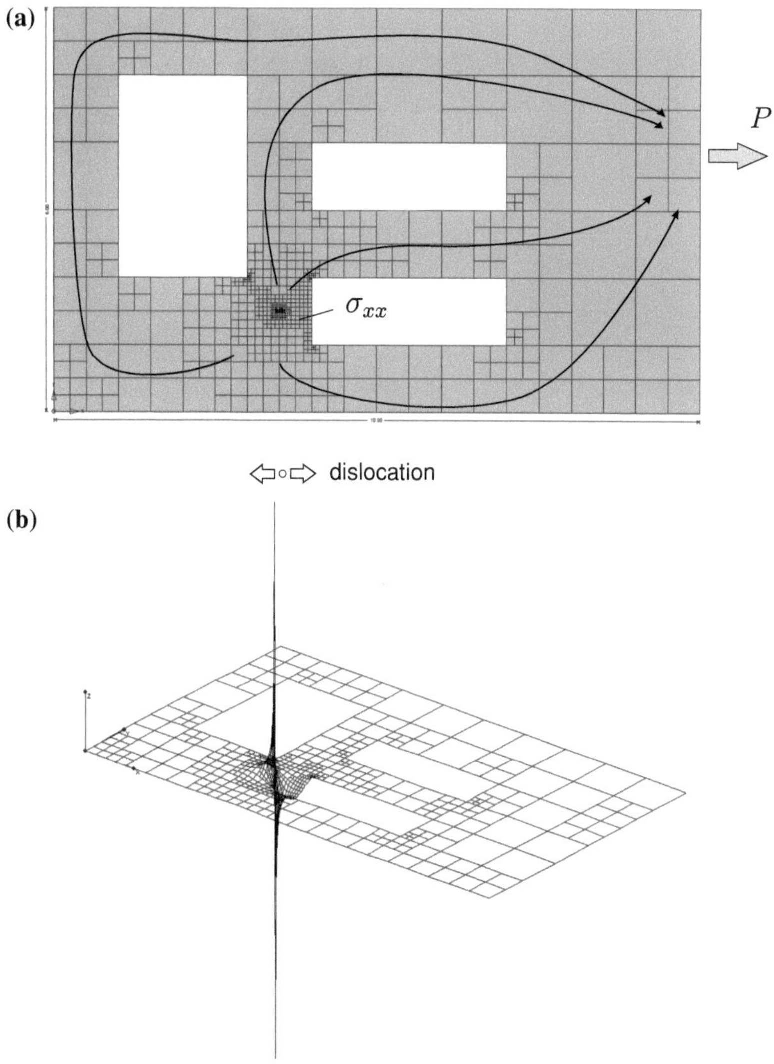

Fig. 1.9 Goal oriented adaptive refinement: the accuracy of the point value σ_{xx} depends on how well the effect of the dislocation is transmitted to the foot of the point load **a** adaptively refined mesh **b** 3-D representation of the (horizontal component of the) Green's function for σ_{xx}

$$B \to B + \Delta B\,. \tag{1.61}$$

How will that affect the weight of the suitcase, what is its new weight W_c, with c as in changed? To find this out we could lift the whole suitcase anew or we only lift the bar whose weight has changed

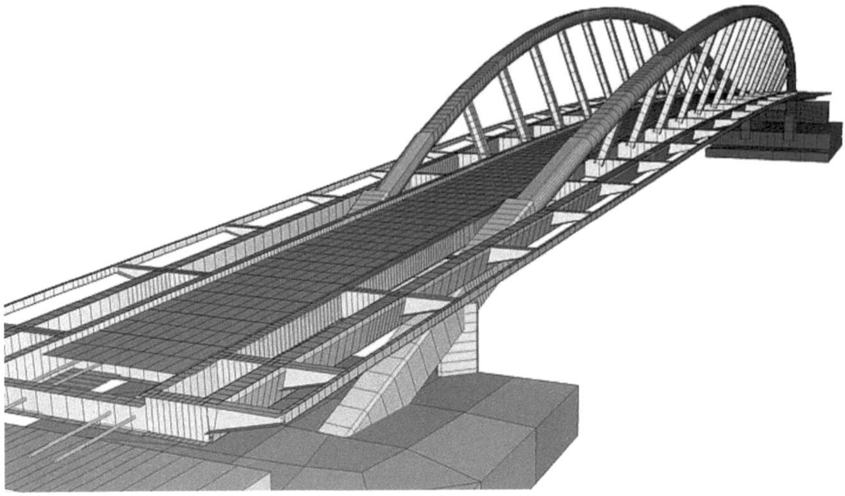

Fig. 1.10 Ponte della Musica, Roma, Italia, [2]

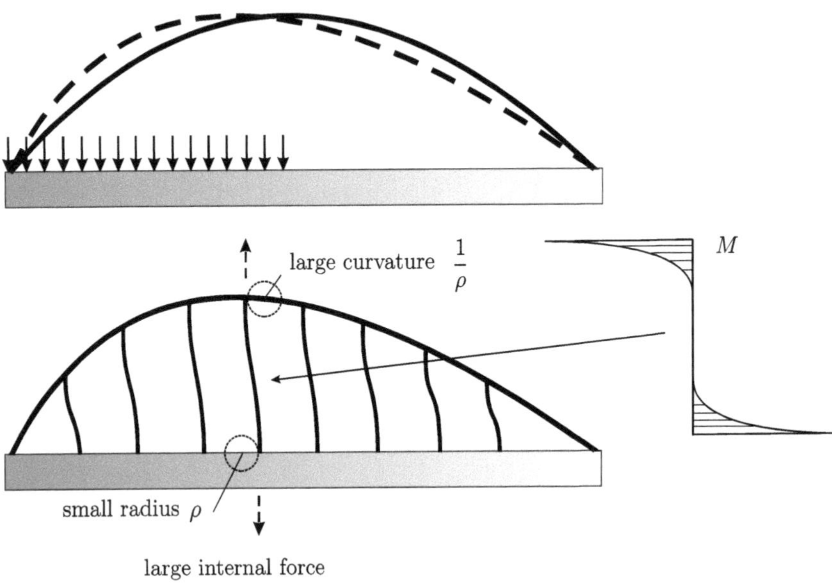

Fig. 1.11 One sided loads tilt the arc to one side causing large bending moments in the hangers

$$\Delta W = W_c - W = \Delta B \tag{1.62}$$

because this information evidently suffices to determine the new weight

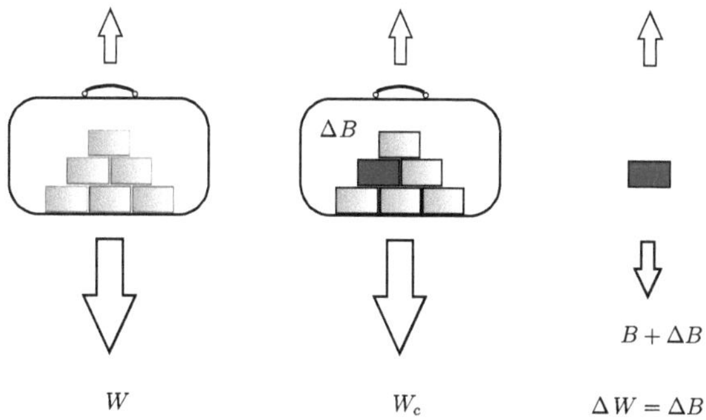

Fig. 1.12 One gold bar (mysteriously) changes its weight

$$W_c = W + \Delta W\,. \tag{1.63}$$

This is the technique which—figuratively speaking—we apply in model adaptivity.

For illustrative purposes consider a shear wall (plate) which is discretized with bilinear elements. Such elements have four nodes and $4 \cdot 2 = 8$ degrees of freedom. The plate is subjected to some loads and suddenly, as it happens, one plate element Ω_e cracks.

These cracks produce a change, $\boldsymbol{K}_e \to \boldsymbol{K}_e + \Delta\boldsymbol{K}_e$, in the element matrix $\boldsymbol{K}_e(8 \times 8)$ and consequently also in the full system

$$(\boldsymbol{K} + \Delta\boldsymbol{K})\,\boldsymbol{u}_c = \boldsymbol{f} \tag{1.64}$$

where $\Delta\boldsymbol{K}$ basically is the element matrix $\Delta\boldsymbol{K}_e$ but enlarged—with the help of many zeros—to the full size $(n \times n)$ of the system. To determine the new state of the plate this system would have to be solved for $\boldsymbol{u}_c$.

Assume the focus is on the horizontal displacement $u(\boldsymbol{x})$ at a particular point. Which change in the displacement, $u_c(\boldsymbol{x}) - u(\boldsymbol{x})$, will the cracks produce? This can be determined with the formula

$$u_c(\boldsymbol{x}) - u(\boldsymbol{x}) = \boldsymbol{g}^T\,\Delta\boldsymbol{K}\,\boldsymbol{u}_c \tag{1.65}$$

where $\boldsymbol{g}$ is the nodal vector of the Green's function for $u(\boldsymbol{x})$ and $\boldsymbol{u}_c$ is the nodal vector of the modified solution.

In some sense this formula corresponds to the lift of the single gold bar: no need—theoretically at least—to invert the full matrix $\boldsymbol{K} + \Delta\boldsymbol{K}$. Measuring the modification in the strain energy product of the afflicted element, $\boldsymbol{g}^T\,\Delta\boldsymbol{K}\,\boldsymbol{u}_c$, suffices (Fig. 1.12).

However this formula requires the nodal vector $\boldsymbol{u}_c$ of the cracked system (1.64) or at least that part of $\boldsymbol{u}_c$ which is in contact with Ω_e, so it seems nothing is gained in the end by following this local approach.

But there is a technique by which we can overcome this hurdle. Introducing the vector

$$\boldsymbol{f}^+ := \Delta\boldsymbol{K}\,\boldsymbol{u}_c \qquad (n \times 1) = (n \times n) \times (n \times 1) \tag{1.66}$$

the equation takes the form

$$u_c(\boldsymbol{x}) - u(\boldsymbol{x}) = \boldsymbol{g}^T\, \boldsymbol{f}^+ \qquad (1 \times n)\,(n \times 1)\,. \tag{1.67}$$

This vector

$$\boldsymbol{f}^+ = \{0, 0, \ldots, 0, \underbrace{*, *, *, *, *, *, *, *}_{\boldsymbol{f}_e^+}, 0, \ldots, 0, 0\}^T \tag{1.68}$$

has only eight non-zero entries because the matrix $\Delta\boldsymbol{K}$ is essentially the 8×8 element matrix $\Delta\boldsymbol{K}_e$ and not much else and so only those parts of the vectors $\boldsymbol{g}$ and $\boldsymbol{u}_c$ which are in contact with the defective element Ω_e enter the equation

$$u_c(\boldsymbol{x}) - u(\boldsymbol{x}) = \boldsymbol{g}_e^T\, \boldsymbol{f}_e^+ \qquad (1 \times 8)\,(8 \times 1)\,. \tag{1.69}$$

Here $\boldsymbol{g}_e$ and $\boldsymbol{f}_e^+$ designate the parts of $\boldsymbol{g}$ and $\boldsymbol{f}$ which have contact with Ω_e.

As will be shown later $\boldsymbol{f}_e^+$ can be determined by solving a small system of size 8×8

$$(\boldsymbol{F}_e + \Delta\boldsymbol{K}_e^{-1})\, \boldsymbol{f}_e^+ = \boldsymbol{u}_e \tag{1.70}$$

where $\boldsymbol{u}_e(8 \times 1)$ are the nodal displacements of the element Ω_e before the cracks developed. The matrix $\boldsymbol{F}_e$ is the local part, (8×8), of the flexibility matrix of the system, that is the inverse $\boldsymbol{K}^{-1}$ of the full system matrix; for details, how the (normally) singular element matrix $\Delta\boldsymbol{K}_e$ can be inverted and how the entries of $\boldsymbol{F}_e$ can be found without calculating the inverse $\boldsymbol{K}^{-1}$, see Chap. 5.

Adding the vector $\boldsymbol{f}^+$ to the right-hand side of the original equation

$$\boldsymbol{K}\,\boldsymbol{u}_c = \boldsymbol{f} + \boldsymbol{f}^+ \tag{1.71}$$

produces the vector $\boldsymbol{u}_c$, the displacement vector of the cracked system! This implies—and this is the important part—that the Green's function of the uncracked system (vector $\boldsymbol{g}$) allows to predict the displacements (and all other observables as well) of the cracked system

$$u_c(\boldsymbol{x}) = \boldsymbol{g}^T(\boldsymbol{f} + \boldsymbol{f}^+)\,. \tag{1.72}$$

Whether this approach has any advantage over the "brute force" approach where the modified system $(\boldsymbol{K} + \Delta\boldsymbol{K})\,\boldsymbol{u}_c = \boldsymbol{f}$ is simply solved for $\boldsymbol{u}_c$ depends on the circumstances.

But irrespective of the computational merits of the equation

$$u_c(\boldsymbol{x}) - u(\boldsymbol{x}) = \boldsymbol{g}^T\, \Delta\boldsymbol{K}\, \boldsymbol{u}_c \tag{1.73}$$

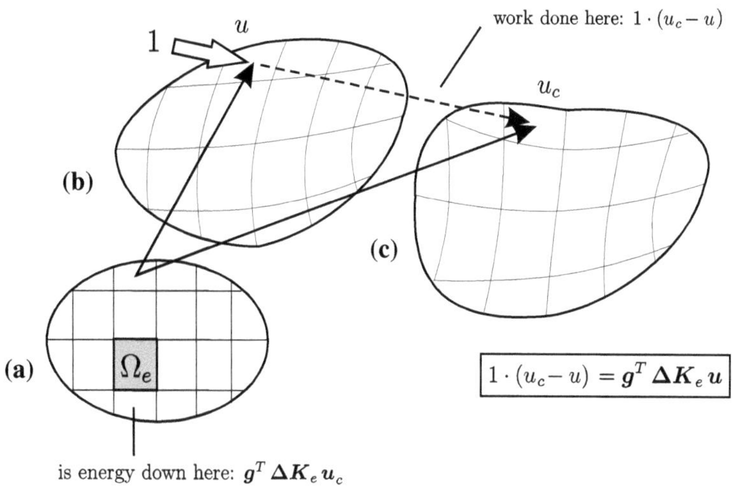

Fig. 1.13 The defective element Ω_e acts like a CCD (lens) it collects and processes all informations

it provides important information about the interplay between the Green's function (vector $\boldsymbol{g}$) and the modified solution (vector $\boldsymbol{u}_c$). The vector $\boldsymbol{g}$ represents the *sensitivity* of $u(\boldsymbol{x})$ with respect to the change $\Delta \boldsymbol{K}$ in the stiffness matrix.

Technically does (1.73) express an energy balance. The work done by a virtual force $F = 1$ applied at $\boldsymbol{x}$ in the direction of the displacement $u(\boldsymbol{x})$ acting through the increment $u_c(\boldsymbol{x}) - u(\boldsymbol{x})$ is equal to the strain energy product (virtual strain energy) in the element Ω_e between the Green's function and u_c, see Fig. 1.13,

$$1 \cdot (u_c(\boldsymbol{x}) - u(\boldsymbol{x})) = \boldsymbol{g}^T \, \Delta \boldsymbol{K} \, \boldsymbol{u}_c \, . \tag{1.74}$$

The important point is that the change $u_c(\boldsymbol{x}) - u(\boldsymbol{x})$ can be determined by integrating *only* over Ω_e, the defective element. And with regard to any other observable, say the stress σ at a point $\boldsymbol{x}$, the formula is the same

$$1 \cdot (\sigma_c(\boldsymbol{x}) - \sigma(\boldsymbol{x})) = \boldsymbol{g}^T \, \Delta \boldsymbol{K} \, \boldsymbol{u}_c \tag{1.75}$$

only the Green's function, the vector $\boldsymbol{g}$, changes. (The 1 is now a unit dislocation).

So the defective element Ω_e acts like a test bed which allows to determine *any* change in the system by monitoring the virtual strain energy in Ω_e triggered by the corresponding Dirac delta. If the Green's function of a functional $J(u)$ barely causes any ripples on Ω_e then the change in the functional, $J(u_c) - J(u)$ will be negligible.

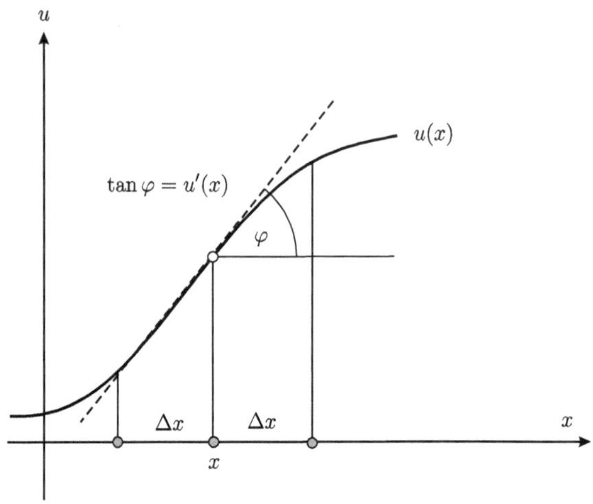

Fig. 1.14 In calculus we learn that the slope $u'(x)$ is a *local* quantity. In mechanics the slope is a *global* quantity because it is under the influence of sources far-off.

1.6.1 Local & Global

In calculus we learn that the slope of a function, see Fig. 1.14,

$$u'(x) = \lim_{\Delta x \to 0} \frac{u(x+\Delta x) - u(x-\Delta x)}{2\Delta x}, \tag{1.76}$$

expresses a *local* property of a function.

But if u is the solution of a boundary value problem,

$$-u''(x) = p(x) \qquad u(0) = u(l) = 0\,, \tag{1.77}$$

then the slope is determined by an influence function

$$u'(x) = \int_0^l G(y, x)\, p(y)\, dy \tag{1.78}$$

and so $u'(x)$ depends on *global* effects: modify p in any small subinterval $[x_a, x_b] \subset (0, l)$ and the slope at x will change!

The reason for this observation is that the inverse of a differential operator is an integral operator, a sum. The tri-diagonal stiffness matrix $\boldsymbol{K}$ of a rope is a difference matrix $(-u_{i-1} + 2\,u_i - u_{i+1})$ and its inverse is a fully populated summation matrix $\boldsymbol{K}^{-1}$ the columns of which are the Green's functions of the nodal values u_i. And quite naturally does the summation matrix, the influence function, register any small change in p or in the nodal forces $\boldsymbol{f}_i$ respectively [3].

1.7 How to Calculate Influence Functions with Finite Elements

Influence functions are displacements, they have nodal values u_i as ordinary functions do and these nodal values are the solution of the system $\boldsymbol{K}\,\boldsymbol{u} = \boldsymbol{f}$ just as in standard FE-analysis. The only question is: which nodal forces f_i (or j_i as we usually call them) do generate the influence functions?

Two examples may suffice to give the idea. (1) The equivalent nodal forces f_i which generate the influence function for the value $u(\boldsymbol{x})$ of the solution at a point $\boldsymbol{x}$ are the values of the shape functions at this point $\boldsymbol{x}$

$$f_i = \varphi_i(\boldsymbol{x})\,. \tag{1.79}$$

(2) The forces f_i which generate the influence function for the stress $\sigma(\boldsymbol{x})$ at a point $\boldsymbol{x}$ are the stresses

$$f_i = \sigma(\varphi_i)(\boldsymbol{x}) \tag{1.80}$$

of the shape functions at the point $\boldsymbol{x}$—and so on. So the rule is easy:

Theorem 1.2 (Nodal forces for influence functions) *The Green's function for a linear functional $J(u)$ is generated by applying the values $f_i = J(\varphi_i)$ of the shape functions φ_i as equivalent nodal forces*

Example 1.1 The Green's function for the deflection $u(x)$ of the rope in Fig. 1.15 at the point $x = 1.25$ is to be determined and this means the values of the shape functions φ_i at the point x are to be applied as nodal forces

$$f_1 = \varphi_1(x) = 0.75 \qquad f_2 = \varphi_2(x) = 0.25 \qquad f_3 = \varphi_3(x) = 0\,. \tag{1.81}$$

Solving the system

$$\begin{bmatrix} 2 & -1 & 0 \\ -1 & 2 & -1 \\ 0 & -1 & 2 \end{bmatrix} \begin{bmatrix} g_1 \\ g_2 \\ g_3 \end{bmatrix} = \begin{bmatrix} 0.75 \\ 0.25 \\ 0 \end{bmatrix} \tag{1.82}$$

for the nodal values g_i of the Green's function gives

$$g_1 = 0.6875 \qquad g_2 = 0.625 \qquad g_3 = 0.3125 \tag{1.83}$$

and so the Green's function is

$$G(y, x) = g_1 \cdot \varphi_2(y) + g_2 \cdot \varphi_2(y) + g_3 \cdot \varphi_3(y)\,. \tag{1.84}$$

If $u(x)$ were the deflection at the first node, $x = 1.0$, then the nodal forces would be

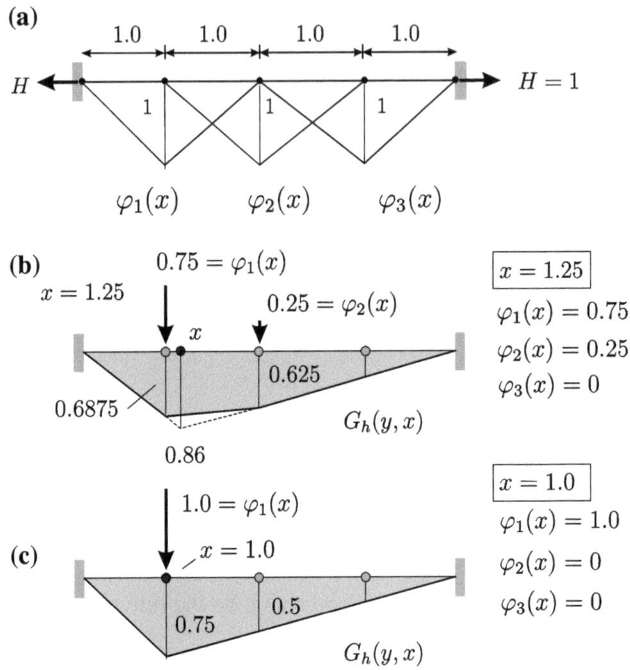

Fig. 1.15 FE-model of a rope **a** shape functions **b** FE-Green's function for u at the point $x = 1.25$ and exact value (0.86) **c** FE-Green's function for u at the first node, the function is exact, $G_h(y, x) = G(y, x)$

$$f_1 = \varphi_1(x) = 1.0 \qquad f_2 = \varphi_2(x) = 0.0 \qquad f_3 = \varphi_3(x) = 0 \tag{1.85}$$

and in this case the FE-Green's function for $u(x)$ would even be exact because the triangle with the nodal values

$$g_1 = 0.75 \qquad g_2 = 0.5 \qquad g_3 = 0.25 \tag{1.86}$$

has its apex at the first node and it has the correct value $g_1 = 0.75$, see Fig. 1.15c.

Remark 1.2 The nodal values g_i are normally called u_i but to better distinguish between Green's functions and standard functions we prefer the notation g_i. Unlike standard u_i the nodal values g_i are in general functions of the source point x which can change its position, so that

$$G(y, x) = g_1(x) \cdot \varphi_2(y) + g_2(x) \cdot \varphi_2(y) + g_3(x) \cdot \varphi_3(y) \tag{1.87}$$

is the general form of an FE-Green's function: the nodal values depend on the source point x, the position of the Dirac delta, and the shape functions vary with the position of the field point y.

In some books the source point is denoted by the Greek letter ξ, allowing the x to be the field variable, but that would make the solution u to be a function of ξ when the influence function is evaluated

$$u(\xi) = \int_0^l G(x,\xi)\, p(x)\, dx\,. \tag{1.88}$$

This is the reason why we prefer the combination x and y.

Example 1.2 The influence function for the bending moment $M = -EI\, u''$ of the continuous beam in Fig. 1.17 at the point x is generated by the equivalent nodal forces, see Fig. 1.16,

$$f_i = -EI\, \varphi_i''(x) \tag{1.89}$$

and the influence function for the shear force $V(x) = -EI\, u'''(x)$ by the forces

$$f_i = -EI\, \varphi_i'''(x) \tag{1.90}$$

where the φ_i are the shape functions of the nodes. Elementwise these shape functions are (length of the element $= l_e$)

$$\begin{aligned} \varphi_1^e(x) &= 1 - \frac{3x^2}{l_e^2} + \frac{2x^3}{l_e^3} & \varphi_3^e(x) &= \frac{3x^2}{l_e^2} - \frac{2x^3}{l_e^3} \\ \varphi_2^e(x) &= -x + \frac{2x^2}{l_e} - \frac{x^3}{l_e^2} & \varphi_4^e(x) &= \frac{x^2}{l_e} - \frac{x^3}{l_e^2}\,. \end{aligned} \tag{1.91}$$

Only the end nodes of the element that contains the point x carry nodal forces f_i because the shape functions φ_i of the other nodes, lying farther off, have zero moments and shear forces at x.

The two influence functions are exact outside the element which contains the source point x [4].

Example 1.3 In this example we calculate various Green's functions for a plate which is modeled with bilinear elements.

Such an element has 4 nodes and $2 \cdot 4$ degrees of freedom and the same number of vector valued shape functions $\boldsymbol{\varphi}_i(\boldsymbol{x})$. These eight shape functions are displacement fields which push the node in question either in horizontal or vertical direction

$$\boldsymbol{\varphi}_1(\boldsymbol{x}) = \begin{bmatrix} \psi_1(\boldsymbol{x}) \\ 0 \end{bmatrix} \quad \boldsymbol{\varphi}_2(\boldsymbol{x}) = \begin{bmatrix} 0 \\ \psi_1(\boldsymbol{x}) \end{bmatrix} \quad \boldsymbol{\varphi}_3(\boldsymbol{x}) = \begin{bmatrix} \psi_2(\boldsymbol{x}) \\ 0 \end{bmatrix} \quad \text{etc.} \tag{1.92}$$

The $\psi_i(\boldsymbol{x})$ are the four shape functions of the four corner points, see Fig. 1.18,

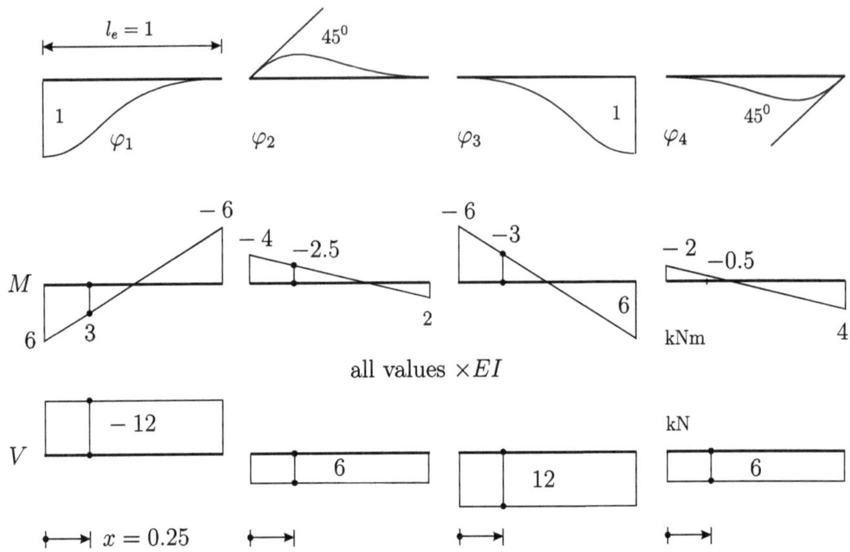

Fig. 1.16 Beam element **a** the four shape functions φ_i and **b** the corresponding bending moments M and **c** shear forces V. The values at the quarter point $x = 0.25\,l$ are the nodal forces f_i which generate the influence functions for M and V respectively in the next figure

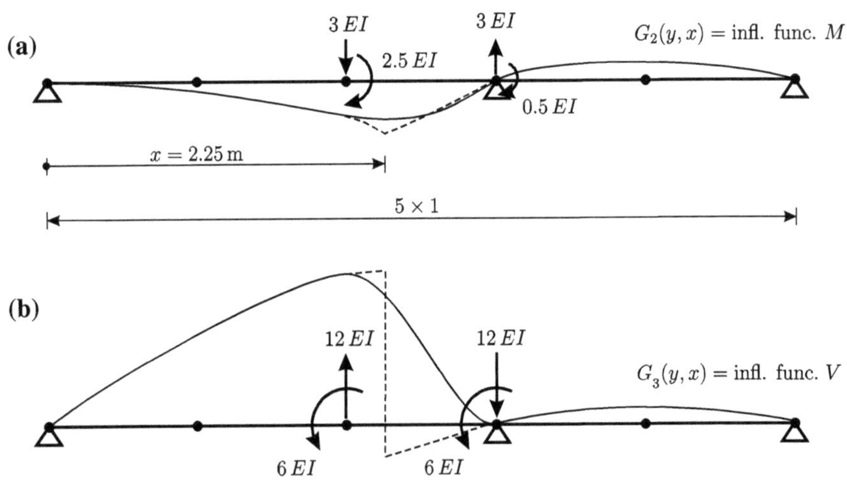

Fig. 1.17 FE-Green's function for **a** the bending moment M and **b** the shear force V at the point x

$$\psi_1(\boldsymbol{x}) = \frac{1}{4\,a\,b}\,(a - 2x)(b - 2y) \qquad \psi_2(\boldsymbol{x}) = \frac{1}{4\,a\,b}\,(a + 2x)(b - 2y) \tag{1.93}$$

$$\psi_3(\boldsymbol{x}) = \frac{1}{4\,a\,b}\,(a + 2x)(b + 2y) \qquad \psi_4(\boldsymbol{x}) = \frac{1}{4\,a\,b}\,(a - 2x)(b + 2y)\,. \tag{1.94}$$

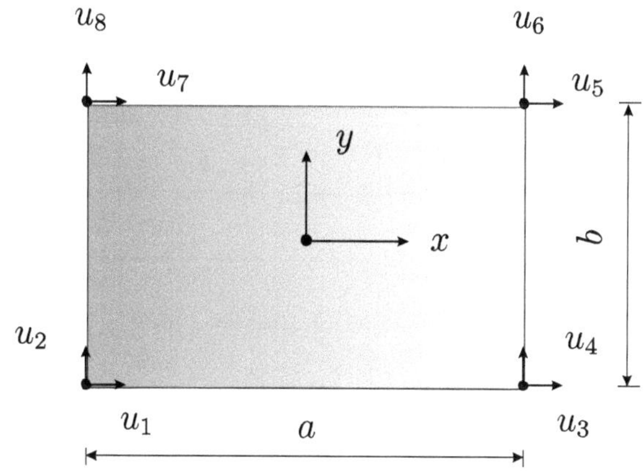

Fig. 1.18 Bilinear element

First we calculate the Green's function for the horizontal displacement at the quarter point of an element having the extensions $a = 2$ and $b = 1$. The four horizontal forces which are applied at the four corner points are the values of the four horizontal shape functions, indices 1, 3, 5, 7, at the point $\boldsymbol{x} = (-0.5, -0.25)$ (element coordinates)

$$f_1 = 0.5625 \qquad f_3 = 0.1875 \qquad f_5 = 0.0625 \qquad f_7 = 0.1875\,. \tag{1.95}$$

These forces generate the shape in Fig. 1.19. (The four vertical shape functions have of course zero horizontal displacement at $\boldsymbol{x}$ and so the f_i in vertical direction, f_2, f_4, f_6, f_8, are zero).

Next we calculate the Green's function for the stress σ_{yy} at the same point. To generate this function the stresses $\sigma_{yy}(\varphi_i)$ of the eight shape functions φ_i are to be applied as nodal forces f_i.

In a bilinear element of length a and width b as in Fig. 1.18, the stress distribution is

$$\sigma_{xx}(x, y) = \frac{E}{a\,b\,(-1+\nu^2)} \cdot \Big[b\,(u_1 - u_3) + a\,\nu\,(u_2 - u_8) + x\,\nu\,(-u_2 + u_4 - u_6 + u_8) + y\,(-u_1 + u_3 - u_5 + u_7) \Big] \tag{1.96}$$

$$\sigma_{yy}(x, y) = \frac{E}{a\,b\,(-1+\nu^2)} \cdot \Big[b\,\nu\,(u_1 - u_3) + a\,(u_2 - u_8) + x\,(-u_2 + u_4 - u_6 + u_8) + y\,\nu\,(-u_1 + u_3 - u_5 + u_7) \Big] \tag{1.97}$$

and

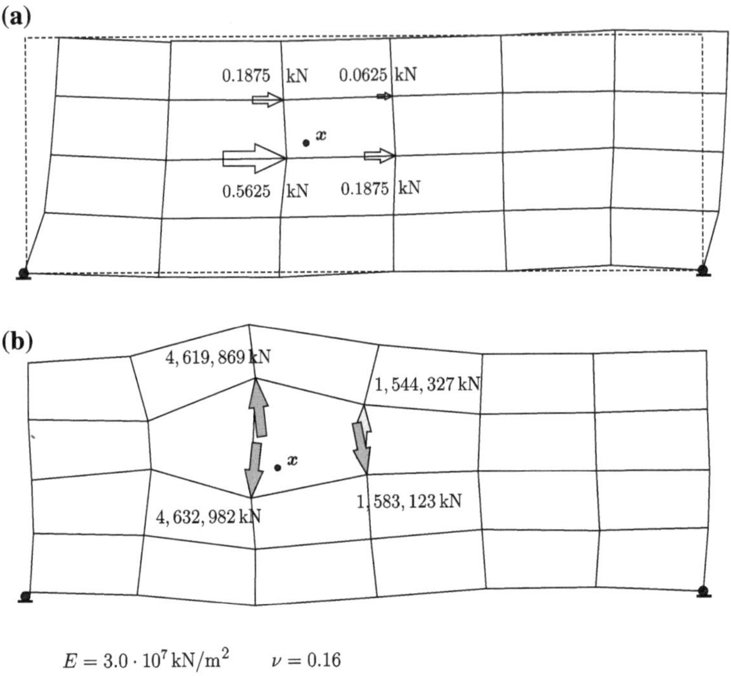

Fig. 1.19 Bilinear elements **a** Green's function for $u_x(\boldsymbol{x})$ and **b** for $\sigma_{yy}(\boldsymbol{x})$

$$\sigma_{xy}(x, y) = \frac{-E}{2\,a\,b\,(1+\nu)} \cdot \Big[b\,(u_2 - u_4) + a\,(u_1 - u_7) + x\,(-u_1 + u_3 - u_5 + u_7) + y\,(-u_2 + u_4 - u_6 + u_8) \Big]. \tag{1.98}$$

Setting $u_1 = 1$ and all other $u_i = 0$ gives the stresses which belong to the shape function $\varphi_1(\boldsymbol{x})$. So at the first node the equivalent nodal force in horizontal direction $(u_1 = 1)$ is

$$f_1 = \sigma_{yy}(x, y) = \frac{E}{a\,b\,(-1+\nu^2)} \cdot \Big[b\,\nu\,u_1 + y\,\nu\,(-u_1) \Big] = -3.07 \cdot 10^6\,\text{kN} \tag{1.99}$$

and in vertical direction $(u_2 = 1)$

$$f_2 = \sigma_{yy}(x, y) = \frac{E}{a\,b\,(-1+\nu^2)} \cdot \Big[a\,(u_2) + x\,(-u_2) \Big] = -3.85 \cdot 10^7\,\text{kN} \tag{1.100}$$

and the other f_i follow the same pattern.

The last example is the Green's function for the integral of the shear force

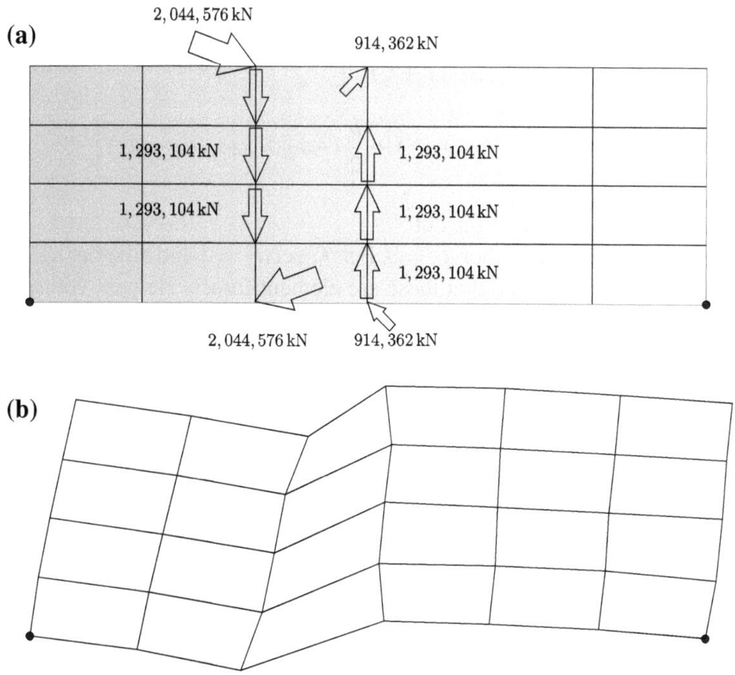

Fig. 1.20 Influence function for the integral of σ_{xy} in a vertical cross section **a** equivalent nodal forces **b** Green's function

$$N_{xy} = \int_0^l \sigma_{xy}\, dy \tag{1.101}$$

in a vertical cross-section that passes through a given point $\boldsymbol{x}$. In this case the equivalent nodal forces are integrals, see Fig. 1.20,

$$f_i = \int_0^l \sigma_{xy}(\varphi_i)\, dy\,, \tag{1.102}$$

which represent the resulting shear forces of the shape functions belonging to the four corners of an element. So if an element $a \times b$ happens to lie in the path of the cut the following equivalent nodal forces are applied at the four corners

$$f_i^e = \int_0^b \sigma_{xy}(\varphi_i)\, dy = \frac{-E}{2a\,(1+\nu)} \cdot \Big[b\,(u_2 - u_4) + a\,(u_1 - u_7) + x\,(-u_1 + u_3 - u_5 + u_7) + \frac{b}{2}\,(-u_2 + u_4 - u_6 + u_8) \Big] \tag{1.103}$$

where x is the x-coordinate of the cut.

For f_1^e set $u_1 = 1$ and all other $u_i = 0$. For f_2^e set $u_2 = 1$ and all other $u_i = 0$, etc. The notation f_i^e is to indicate that these are element forces, element contributions. The total nodal force f_i is the sum of the single element contributions to a node.

References

1. James DL (2001) Multiresolution green's function methods for interactive simulations of large-scale elastostatic objects and other physical systems in equilibrium, Ph.D. Thesis, Inst Appl Math, University of British Columbia
2. Ponte della Musica, Roma, Italia, Ing. Giorgio Rizzo, Software: SOFiSTiK
3. Strang G (2007) Computational science and engineering. Wellesley-Cambridge Press, Wellesley
4. Hartmann F, Katz C (2007) Structural analysis with finite elements. 2nd ed. Springer-Verlag, Berlin

Chapter 2
Basic Concepts

The Hilbert space theory of weak boundary value problems forms the mathematical background of the FE-method. This theory requires some (rudimentary) knowledge of functional analysis and so the chapter starts with a short recap of the principal concepts as vector spaces, scalar product, linear functionals, projection, equivalent norms, Galerkin's method and the definition of Sobolev spaces. The main result is that in the weak theory Green's functions are the Riesz elements of linear functionals.

The exposition is based on the first and second Green's identities of the governing differential operator as these underlie the variational and energy principles of mathematical physics and their algebra is essential for the formulation of weak boundary value problems.

Green's identities are then used to derive the classical influence functions for displacements and force terms. The observation that influence functions for force terms can only be formulated with Betti's theorem (Green's second identity) but not with the principle of virtual forces (Green's first identity) leads to the distinction between strong and weak influence functions.

Physically Green's functions represent monopoles, dipoles and higher-order poles. Monopoles are like integral operators, they sum, while dipoles differentiate, they are sensitive to imbalances. Most Green's function, being the solution to point forces or higher-order terms, have infinite energy which makes that they lie outside the theory of weak boundary value problems. When and why that happens can be explained with Sobolev's Embedding Theorem.

In the algebra of linear problems Green's functions are identical with Lagrange multipliers. This observation allows to extend—formally—the concept of a Green's function also to nonlinear problems; not in the true sense of forming the kernel of an inverse operator but to speak of a Green's function at the linearization point of a stiffness matrix.

F. Hartmann, *Green's Functions and Finite Elements*,
DOI: 10.1007/978-3-642-29523-2_2,

2.1 Elements of Functional Analysis

The weak form of the classical boundary value problems of mechanics are of "quadratic type". What is meant by this is best explained by a simple spring.

In a linear spring with a stiffness k the displacement u is proportional to the applied force f

$$k\,u = f \tag{2.1}$$

and consequently the displacement u minimizes the function

$$\Pi(u) = \frac{1}{2}\,k\,u^2 - f\,u \tag{2.2}$$

because if u satisfies (2.1) then it is also a zero of $\Pi'(u) = k\,u - f$.

If u is a solution of (2.1) then multiplying both sides with any number v leaves the solution unchanged and therefore u is also the solution of the variational problem

Find a number u such that

$$\delta W_i := v\,k\,u = f\,v =: \delta W_e \quad \forall\, v \in \mathbb{R}\,. \tag{2.3}$$

The symbol $\forall$ means "for all". For all numbers v this equation must be true. This is the principle of virtual displacements: at the equilibrium position u the virtual strain energy δW_i of the spring equals the virtual exterior energy δW_e—and this holds true for all virtual displacements v.

Quadratic refers to the fact that the internal energy of the spring

$$W_i = \frac{1}{2}\,k\,u^2 =: a(u, u) \tag{2.4}$$

is a quadratic function, see Fig. 2.1, and this means it is positive definite

$$a(u, u) = k\,u^2 > 0 \quad u \neq 0. \tag{2.5}$$

With regard to the applied load f we can state that if it is not just infinitely large then the external work $f \cdot u$ is a *continuous function* of u. These two properties which we can attribute to $a(u, u)$ and the function $f \cdot u$ guarantee that the weak problem (2.3) has always a solution $u = f/k$. Ultimately the parabola $1/2\,k\,u^2$ will rise faster than the linear function. The point where the path of the parabola crosses the straight line $1/2\,f u$ is the equilibrium point u, the internal energy and external work balance at this point

$$W_i = \frac{1}{2}\,k\,u^2 = \frac{1}{2}\,f u = W_e \quad \text{at } u = f/k\,. \tag{2.6}$$

But at the start, when the spring begins to move, it is the straight line—representing the external work $1/2\,f\,u$—which grows faster (and it *must* grow faster!) than the

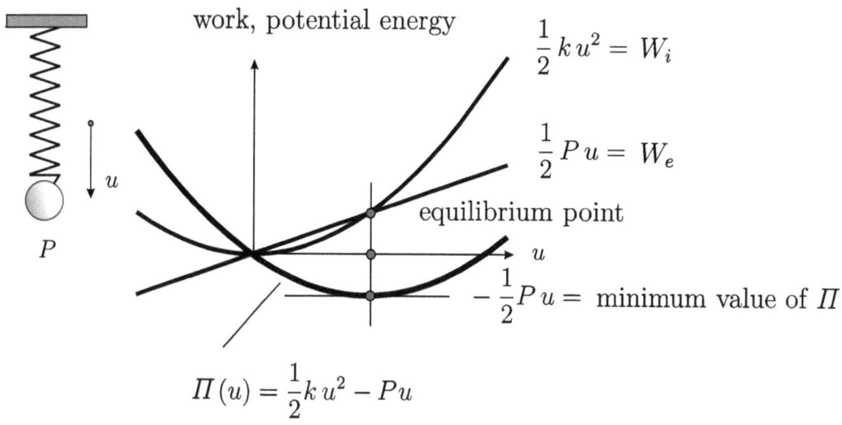

Fig. 2.1 Because the internal energy W_i grows quadratically with u, while the external work W_e grows only linearly, W_i always catches up with W_e, and there will always be an equilibrium point where $W_i = W_e$ [1]

parabola $1/2\, ku^2$ of the internal energy. If that were not the case then the spring would not move one *iota*! So in some sense—let $f = k = 1$—computational mechanics is rooted in the fact that $u > u^2$ in the interval (0, 1), and that beyond this point the opposite is true. *The transition point* $u = 1$ *is the equilibrium point.*

In the continuous case, when the equilibrium position u of the mechanical system is a function, then technically things are more complicated but the arguments and the results are very nearly the same because the modern theory of partial differential equations is formulated in terms of weak boundary value problems on Hilbert spaces and this theory and its algebraic structure bears a close resemblance to what the engineer calls the *principle of virtual displacements*. In the following we will summarize the main results and concepts of this theory as far as we need them for the study of Green's functions and finite elements.

2.1.1 Notation

Small bold letters $\boldsymbol{u} = \{u_1, u_2, \ldots, u_n\}^T$ denote column vectors and capital bold letters $\boldsymbol{K} = [k_{ij}]$ matrices. The scalar product between two vectors $\boldsymbol{u}$ and $\boldsymbol{f}$ is written in two ways

$$\boldsymbol{u} \cdot \boldsymbol{f} = \boldsymbol{u}^T \boldsymbol{f} = u_1 f_1 + u_2 f_2 \tag{2.7}$$

and

$$\boldsymbol{S} \cdot \boldsymbol{E} = \sigma_{11}\, \varepsilon_{11} + \sigma_{12}\, \varepsilon_{12} + \sigma_{21}\, \varepsilon_{21} + \sigma_{22}\, \varepsilon_{22} \tag{2.8}$$

is the scalar product of matrices. A small dot, $3 \cdot 4$, denotes a multiplication. Green's functions are written $G(\boldsymbol{y}, \boldsymbol{x})$ or if the emphasis is only on the source point $\boldsymbol{x}$ then $G[\boldsymbol{x}]$. Eventually the same is done with the Dirac delta $\delta(\boldsymbol{y} - \boldsymbol{x})$ which has the short form $\delta[\boldsymbol{x}]$.

2.1.2 Vector Spaces and Scalar Product

A real vector space $\mathcal{V}$ is a set of objects, called vectors, on which are defined two operations, addition and scalar multiplication (scalar product). Associated with such a space is a field, the real numbers $\alpha \in \mathbb{R}$, such that if u is a member of $\mathcal{V}$ then also $\alpha\, u$ is a member of $\mathcal{V}$.[1]

The space $\mathbb{R}^n$, the set of all vectors with n components, is such a vector space. The sum of two vectors $\boldsymbol{u}$ and $\boldsymbol{v}$ in $\mathcal{V}$ is the vector $\boldsymbol{u} + \boldsymbol{v}$ and the scalar multiplication between two vectors

$$\boldsymbol{u} \cdot \boldsymbol{v} = u_1\, v_1 + u_2\, v_2 + \ldots + u_n\, v_n = |\boldsymbol{u}|\, |\boldsymbol{v}|\, \cos\varphi \tag{2.9}$$

is of course the scalar product of two vectors where φ is the angle between the two vectors.

Occasionally we will write the scalar product of two such vectors as the product of the row vector $\boldsymbol{v}^T$ and the column vector $\boldsymbol{u}$

$$\boldsymbol{v} \cdot \boldsymbol{u} = \boldsymbol{v}^T\, \boldsymbol{u} = (\boldsymbol{v}, \boldsymbol{u}) \tag{2.10}$$

or we will use the notation $(\boldsymbol{v}, \boldsymbol{u})$ or (u, v) respectively as a generic expression for the scalar product of two quantities.

For an expression (u, v) to be a scalar product it must have the following properties $(\alpha, \beta \in \mathbb{R})$

Linearity:	$(\alpha\, u + \beta\, v, w) = \alpha\, (u, w) + \beta\, (v, w)$	(2.11a)
Symmetry:	$(u, v) = (v, u)$	(2.11b)
Definitness:	$(u, u) > 0 \quad \text{for } u \neq 0\,.$	(2.11c)

With these properties it defines also the norm of an element

$$\|u\| := \sqrt{(u, u)}\,. \tag{2.12}$$

The square root makes that the element $2\,u$ has double the norm of u, $\|2\,u\| = 2\,\|u\|$.

[1] For not to confuse the vector space with the shear force V we write $\mathcal{V}$ and $\mathcal{V}_h$ as well.

A norm has the properties

$$\begin{aligned} &\text{positive semi-definit:} && \|u\| = 0 \Leftrightarrow u = 0 \quad \text{or} \quad \|u\| \geq 0 \quad \forall u \in \mathcal{V} && \text{(2.13a)} \\ &\text{homogeneous:} && \|\lambda u\| = |\lambda| \|u\| && \text{(2.13b)} \\ &\text{triangular inequality:} && \|u + v\| \leq \|u\| + \|v\| \,. && \text{(2.13c)} \end{aligned}$$

In particular does the first property guarantee that the norm can *separate* the elements of $\mathcal{V}$

$$\|u - v\| = 0 \quad \Rightarrow \quad u = v \,. \tag{2.14}$$

This is what makes a norm a norm.

A central property of a normed space $\mathcal{V}$ is the Cauchy-Schwarz inequality

$$|(u, v)| \leq \|u\| \, \|v\| \quad \forall u, v \in \mathcal{V} \,. \tag{2.15}$$

A vector space $\mathcal{V}$ with a norm is called a *normed vector space*. If the space is also complete, if the limit of any *Cauchy sequence* (the distance between consecutive elements gets smaller and smaller) is an element of $\mathcal{V}$ then it is said that the space is *complete*. Such normed vector spaces are called *Hilbert spaces*.

2.1.3 Linear Functionals

A linear functional $J(u)$ on $\mathcal{V}$ is a real-valued linear function

$$J(\alpha\, u + \beta\, v) = \alpha\, J(u) + \beta\, J(v) \quad \alpha, \beta \in \mathbb{R} \,. \tag{2.16}$$

It is a continuous functional on $\mathcal{V}$ if there exists a constant c such that

$$J(u) \leq c\, \|u\| \quad \forall u \in \mathcal{V} \,. \tag{2.17}$$

The set of all continuous linear functionals defined on $\mathcal{V}$ forms itself a normed vector space which is called the *dual space* of $\mathcal{V}$ and is denoted by $\mathcal{V}^*$. The norm on $\mathcal{V}^*$ is defined as

$$\|J\|_{\mathcal{V}^*} = \sup \left(\frac{J(u)}{\|u\|} \right) \tag{2.18}$$

which means that the norm of a functional is the maximum value the functional $J(u)$ attains in the unit ball (all u with norm $\|u\| = 1$) because dividing $J(u)$ by the norm $\|u\|$ of u is the same as first dividing u by its norm and then applying the functional

$$\frac{J(u)}{\|u\|} = J\left(\frac{u}{\|u\|} \right) . \tag{2.19}$$

The following theorem asserts that we can associate with each linear and bounded functional $J(u)$ a so-called Riesz element v in the sense that $J(u) = (u, v)$.

Theorem 2.1 (Riesz representation theorem) *If $\mathcal{V}$ is a Hilbert space then its dual space $\mathcal{V}^*$ can be identified with $\mathcal{V}$:*

1. *For each $v \in \mathcal{V}$ the linear functional $J(u)$ defined by $J(u) = (u, v)$ belongs to $\mathcal{V}^*$ and*
$$\|J\|_{\mathcal{V}^*} = \|v\|_{\mathcal{V}} . \tag{2.20}$$
2. *For each $J \in \mathcal{V}^*$ there exists a unique $v \in \mathcal{V}$ such that*
$$\|J\|_{\mathcal{V}^*} = \|v\|_{\mathcal{V}} \tag{2.21}$$
and it is
$$J(u) = (u, v) \quad \forall u \in \mathcal{V} . \tag{2.22}$$

The Riesz element v in (2.22) is identical with what is later called the generalized Green's function of the functional $J(u)$.

Example 2.1 Let $\mathcal{V}$ be the Euclidean space $\mathbb{R}^3$. Let $\boldsymbol{v} = \{v_1, v_2, v_3\}^T$ be a vector anchored at the origin and pointing in a certain direction. Then the functional

$$J(\boldsymbol{u}) = \boldsymbol{v}^T \boldsymbol{u} \tag{2.23}$$

is simply the projection of a vector $\boldsymbol{u}$ onto the direction of the vector $\boldsymbol{v}$. Clearly the maximum value of $J(\boldsymbol{u})$, when it is restricted to the unit ball ($\|\boldsymbol{u}\| = 1$), is obtained for $\boldsymbol{u} = \boldsymbol{v}/\|\boldsymbol{v}\|$ and so indeed

$$\|J\|_{\mathcal{V}^*} = \|\boldsymbol{v}\|_{\mathcal{V}} . \tag{2.24}$$

Vice versa, let there be a linear functional $J(\boldsymbol{u})$ on $\mathbb{R}^3$. Let

$$j_i = J(\boldsymbol{e}_i) \quad i = 1, 2, 3 , \tag{2.25}$$

($\boldsymbol{e}_i$ = unit vector) then

$$J(\boldsymbol{u}) = u_1\, j_1 + u_2\, j_2 + u_3\, j_3 \tag{2.26}$$

and so the vector $\boldsymbol{j} = \{j_1, j_2, j_3\}^T$ represents the functional in the sense of (2.22)

$$J(\boldsymbol{u}) = \boldsymbol{j}^T \boldsymbol{u} . \tag{2.27}$$

Example 2.2 Let Ω be the unit disk and Γ the edge of the disk. On Ω we consider the following boundary value problem

$$-\Delta u = p \quad \text{on } \Omega \quad u = 0 \quad \text{on } \Gamma . \tag{2.28}$$

The space

$$\mathcal{V} = \{u \in H^1(\Omega) | u = 0 \text{ on } \Gamma\} \tag{2.29}$$

with the scalar product $a(u, v) := (\nabla u, \nabla v)$ and the norm $\|u\|_E := a(u, u)^{1/2}$ is a Hilbert space and so given any linear and bounded functional $J(u)$ there exists a Riesz element $G \in \mathcal{V}$ such that

Weak influence function

$$J(u) = a(G, u) \quad \forall u \in \mathcal{V} \tag{2.30}$$

where

$$a(G, u) = \int_\Omega \nabla G \cdot \nabla u \, d\Omega_y = \int_\Omega (G_{,1} \, u_{,1} + G_{,2} \, u_{,2}) \, d\Omega_y \, . \tag{2.31}$$

Because of Green's first identity, see (2.86), this formula is equivalent to

$$J(u) = \int_\Omega G \, (-\Delta u) \, d\Omega_y \quad \forall u \in \mathcal{V} \tag{2.32}$$

or if u is the solution of the boundary value problem (1.30) then holds

Strong influence function

$$J(u) = \int_\Omega G \, p \, d\Omega_y \, . \tag{2.33}$$

That is the Riesz element is the Green's function of the functional. We call these two influence functions, (2.30) and (2.33), *weak* and *strong* influence functions.

2.1.4 Projection

When a vector $\boldsymbol{u} = \{u_x, u_y, u_z\}^T \in \mathbb{R}^3$ is projected onto the $x-y$-plane (or $\mathbb{R}^2$ for short)

$$\boldsymbol{u} \rightarrow \bar{\boldsymbol{u}} = \{u_x, u_y, 0\}^T \tag{2.34}$$

then the error vector $\boldsymbol{e} = \boldsymbol{u} - \bar{\boldsymbol{u}}$ is perpendicular to the plane

$$\boldsymbol{e} \cdot \boldsymbol{v} = 0 \quad \forall \boldsymbol{v} \in \mathbb{R}^2 \tag{2.35}$$

and so the (length)^2 of $\boldsymbol{u}$ is the (length)^2 of $\bar{\boldsymbol{u}}$ plus the (length)^2 of $\boldsymbol{e}$

$$\|\boldsymbol{u}\|^2 = \|\bar{\boldsymbol{u}}\|^2 + \|\boldsymbol{e}\|^2 . \tag{2.36}$$

This is the *Pythagorean theorem* which holds true in any inner product space for two orthogonal elements of the space, $(\boldsymbol{e}, \bar{\boldsymbol{u}}) = 0$, and so the idea of a projection can be extended to any inner product space.

Theorem 2.2 (Projection) *Suppose $\mathcal{V}$ is an inner product space and $\mathcal{V}_h$ is a finite-dimensional subspace of $\mathcal{V}$ and u is an element of $\mathcal{V}$. Then*

1. *there exists a unique element $u_h \in \mathcal{V}_h$ satisfying*

$$\|u - u_h\| < \|u - v_h\| \quad \forall\, v_h \in \mathcal{V}_h,\, v_h \neq u_h \tag{2.37}$$

   that is u_h wins the competition, it has the shortest distance to u and
2. *this statement is equivalent to*

$$(u - u_h, v_h) = 0 \quad \forall\, v_h \in \mathcal{V}_h . \tag{2.38}$$

   that is the error is orthogonal to $\mathcal{V}_h$.

2.1.5 Variational Problems

In the framework of the FE-method the classical boundary value problems of physics are formulated as variational problems on appropriate Hilbert spaces $\mathcal{V}$

$$u \in \mathcal{V}: \quad a(u, v) = J(v) \quad \forall\, v \in \mathcal{V} \tag{2.39}$$

where $a(u, v)$ is a (not necessarily symmetric) bilinear form and $J(v)$ is a linear functional [2]. The next theorem lists sufficient conditions for this problem to have a unique solution.

Theorem 2.3 (Lax-Milgram) *Suppose $\mathcal{V}$ is a Hilbert space and $a(u, v)$ is a bilinear form on $\mathcal{V}$ which is bounded and coercive (or $\mathcal{V}$-elliptic), that is there exist two constants c_E and c_B such that*

$$|a(u, v)| \leq c_B\, \|u\|\, \|v\| \quad \forall\, u, v \in \mathcal{V} \quad \textit{(bounded)} \tag{2.40}$$

$$a(u, u) \geq c_E\, \|u\|^2 \quad \forall\, u \in \mathcal{V} \quad \textit{($\mathcal{V}$-elliptic)} \tag{2.41}$$

Then given any $J \in \mathcal{V}^*$ there exists a unique $u \in \mathcal{V}$ such that

$$a(u, v) = J(v) \quad \forall\, v \in \mathcal{V} \tag{2.42}$$

and the solution u depends continuously on J that is

$$\|u\|_{\mathcal{V}} \leq \frac{1}{c_E} \|J\|_{V*} . \qquad (2.43)$$

The difficult part in applying this theorem is to demonstrate that the bilinear form $a(u, v)$ in question is $\mathcal{V}$-elliptic.

Example 2.3 The energy of a beam, $EI\,u^{IV} = p$, is the integral

$$a(u, u) = \int_0^l EI\,(u'')^2\,dx \qquad (2.44)$$

and the "energy space" is $H^2(0, l)$ with the norm

$$\|u\|_2^2 = \int_0^l (u^2 + (u')^2 + (u'')^2)\,dx . \qquad (2.45)$$

Let

$$\mathcal{V} = \{u \,|u \in H^2(0, l), u(0) = u(l) = 0\} \qquad (2.46)$$

be the solution space of a hinged beam. It can be shown that the energy is $\mathcal{V}$-elliptic

$$a(u, u) > c\,\|u\|_2^2 \quad \forall u \in \mathcal{V} \qquad (2.47)$$

that is if $a(u, u)$ tends to zero then also the norm $\|u\|_2$ tends to zero and this means that

$$\int_0^l EI\,(u'')^2\,dx \to 0 \quad \Rightarrow \quad \int_0^l u^2\,dx \to 0\,, \int_0^l (u')^2\,dx \to 0\,. \qquad (2.48)$$

This is a remarkable property because it shows that the bending moments of a properly supported beam control the beam. Without supports the beam could perform rigid body motions, $u(x) = a + b\,x$, and because the bending moment is then zero, $M = 0$, (2.47) would be violated.

2.1.6 Equivalent Norms

If the bilinear form $a(u, v)$ is $\mathcal{V}$-elliptic then it forms an alternative inner product on $\mathcal{V}$ and the expression

$$\|u\|_E := \sqrt{a(u, u)} \qquad (2.49)$$

the so-called *energy norm*, is an equivalent norm[2] on $\mathcal{V}$, that is there exist two global constants c_E and c_B (not depending on the particular u) such that

$$\sqrt{c_E}\,\|u\|_{\mathcal{V}} \leq \|u\|_E \leq \sqrt{c_B}\,\|u\|_{\mathcal{V}} \tag{2.50}$$

which essentially means that the two norms generate the same topology on $\mathcal{V}$:

- if a sequence of elements converges in one norm to an element u then it converges also in the other norm to the same element,
- if $\mathcal{V}$ is complete then it is also complete with respect to the energy norm,
- if a linear functional $J(u)$ is bounded then it is also bounded in the energy norm,

In the FE-context $\mathcal{V}$ is the trial and solution space on which the variational problem

$$u \in \mathcal{V}: \quad a(u, v) = J(v) \quad \forall\, v \in \mathcal{V} \tag{2.51}$$

is posed. Because the Riesz representation theorem also holds true in the energy norm, the existence and uniqueness of the variational problem is settled.

In the same sense as $\|u\|$ and $\|u\|_E$ are equivalent norms on $\mathcal{V}$ the expression

$$||J||_{E^*} := \sup_{\substack{v \in \mathcal{V}\\ v \neq 0}} \frac{|J(v)|}{||v||_E} = \sup_{\substack{v \in \mathcal{V}\\ ||v||_E = 1}} |J(v)| \tag{2.52}$$

is an equivalent norm on the dual $\mathcal{V}^*$, equivalent to $\|J\|_{\mathcal{V}^*}$. If u is the solution of (2.51) then

$$\frac{|J(v)|}{\|v\|_E} = \frac{|a(u, v)|}{\|v\|_E} \leq \frac{\|u\|_E\, \|v\|_E}{\|v\|_E} = \|u\|_E \tag{2.53}$$

and so $\|u\|_E$ is an upper bound and because of

$$\frac{|J(u)|}{\|u\|_E} = \frac{|a(u, u)|}{\|u\|_E} = \|u\|_E \tag{2.54}$$

it is also the lowest upper bound. Hence the norm $\|J\|_{E^*}$ of the functional J is just the energy norm of the solution

$$\|J\|_{E^*} = \|u\|_E = \sqrt{a(u, u)}\,. \tag{2.55}$$

This implies that the exact solution u is that function in $\mathcal{V}$ which gets "the most mileage" out of J in the sense of (2.52), see Fig. 2.2.

Most milage means: the load p to which a beam is subjected constitutes a functional

[2] The so called taxi norm—distance by grid lines (blocks)—and the Euclidean norm—as the raven flies—is also a pair of equivalent norms.

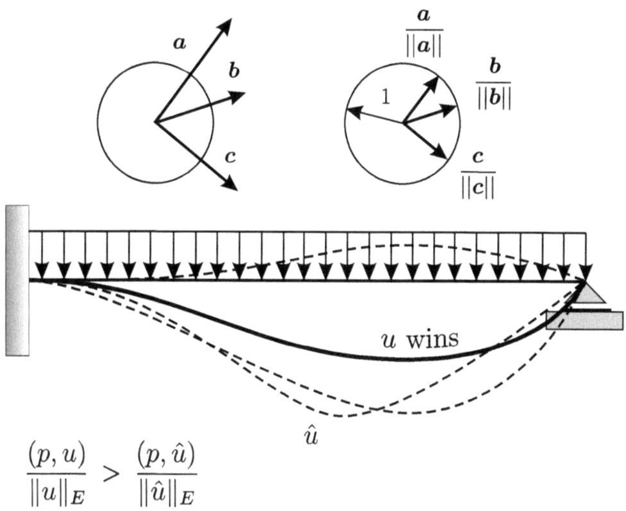

Fig. 2.2 In the unit ball $B_1 = \{u \mid ||u||_E = 1\}$ the normalized exact solution $u/||u||_E$ gets "the most mileage" out of p, i.e. the virtual work exceeds that of any other normalized deflection $\hat{u}/||\hat{u}||_E$ [1]

$$J(u) := \int_0^l p\, u\, dx \tag{2.56}$$

and the maximum value of $J(u/\|u\|)$ on the solution space

$$\mathcal{V} := \{u \in H^2(0,l) | u(0) = u'(0) = u(l) = 0\} \tag{2.57}$$

is attained if the function u is the deflection of the beam

$$EI\, u^{IV} = p \quad u(0) = u'(0) = u(l) = M(l) = 0\,. \tag{2.58}$$

This observation confirms that the principle of minimum potential energy is a maximum principle. The beam deflection u does not minimize the potential energy in the sense of "as small as possible" but it is the one function which *maximizes* the potential energy

$$\Pi(u) = \frac{1}{2}\int_0^l EI\,(u'')^2\, dx - \int_0^l p\, u\, dx$$
$$= -\frac{1}{2}\int_0^l p\, u\, dx\, |_{u=solution} = -\frac{1}{2}J(u) \tag{2.59}$$

in the sense of having the largest possible distance from zero, $|\Pi(u)| = \max$, and it is the one function in $\mathcal{V}$ which gets the most mileage out of p.

2.1.7 Galerkin Method

The idea of the Galerkin method in solving the variational problem

$$u \in \mathcal{V}: \quad a(u, v) = J(v) \quad v \in \mathcal{V} \tag{2.60}$$

approximately is to construct a finite-dimensional subspace $\mathcal{V}_h$ which is spanned by a set of n shape functions $\varphi_i, i = 1, 2, \ldots n$, and to choose that function

$$u_h = \sum_{i=1}^{n} u_i \, \varphi_i(\boldsymbol{x}) \tag{2.61}$$

which satisfies (2.60) with regard to all shape functions φ_i. This leads to the $(n \times n)$ system

$$\boldsymbol{K}\,\boldsymbol{u} = \boldsymbol{f} \tag{2.62}$$

for the vector $\boldsymbol{u} = \{u_i\}$ where

$$k_{ij} = a(\varphi_i, \varphi_j) \quad f_i = J(\varphi_i)\,. \tag{2.63}$$

If $J(v)$ is a linear bounded functional and if the energy norm $\|u\|_E$ is an equivalent norm on $\mathcal{V}$ then the solution u is also the Riesz element of the functional $J(v)$.

2.1.8 Sobolev Spaces

The concept of a scalar product can be extended to functions defined over a domain Ω

$$(u, v)_{L_2} := \int_\Omega u \, v \, d\Omega \tag{2.64}$$

and this introduces the space $L_2(\Omega)$ as the set of all functions for which the integral

$$\int_\Omega u^2 \, d\Omega < \infty \tag{2.65}$$

is bounded. Given two such functions u and v in $L_2(\Omega)$ their scalar product is finite, $(u, v) < \infty$, and also (u, u) is finite and the expression

$$\|u\| := \sqrt{(u, u)} = \left(\int_\Omega u^2 \, d\Omega \right)^{1/2} \tag{2.66}$$

defines a norm on $L_2(\Omega)$.

The Sobolev-spaces are the extension of $L_2(\Omega)$ by incorporating also the weak partial derivatives of the functions up to a certain order m.

Before we can explain what a weak partial derivative is we must first explain what a multi-index is. A multi-index α is an array of non-negative integers which serves as short-hand for derivatives as in

$$\frac{\partial^3 u}{\partial x^2 \partial y} = \partial_\alpha u \quad \alpha = (2, 1) \quad |\alpha| = 2 + 1 = 3 \tag{2.67}$$

$$\frac{\partial^4 u}{\partial x^2 \partial y^2} = \partial_\alpha u \quad \alpha = (2, 2) \quad |\alpha| = 2 + 2 = 4 \,. \tag{2.68}$$

Definition 2.1 (*weak derivative*) Let α a multi-index and $u \in L_2(\Omega)$. The function $u_\alpha = \partial_\alpha u \in L_2(\Omega)$ is called the weak derivative of u if

$$\int_\Omega u \, \partial_\alpha \varphi \, d\Omega = (-1)^{|\alpha|} \int_\Omega \partial_\alpha u \, \varphi \, d\Omega \quad \forall \varphi \in C_0^\infty(\Omega) \,. \tag{2.69}$$

Basically this is the integration by parts formula without boundary integrals because $\varphi \in C_0^\infty$ implies that φ and all its derivatives have zero boundary values and that φ is infinitely often differentiable on Ω. For "normal" functions such as $\sin x, e^x, x^2$ etc. weak derivatives and regular derivatives coincide.

This equation is of Betti-type: the work done by the "load" $\partial_\alpha \varphi$ on acting through u is the same as the work done by the "load" $\partial_\alpha u$ on acting through φ. For odd orders, $|\alpha| = 1, 3, 5, \ldots$, a (-1) factors in, symmetric and skew-symmetric operators.

Given a natural number m the inner product $(\cdot, , \cdot)_m$ of two functions u and v is the following sum

$$(u, v)_m := \sum_{|\alpha| \leq m} (\partial_\alpha u, \partial_\alpha v)_{L_2(\Omega)} \tag{2.70}$$

and correspondingly is

$$\|u\|_m = \sqrt{\sum_{|\alpha| \leq m} \|\partial_\alpha u\|^2_{L_2(\Omega)}} \tag{2.71}$$

the associated norm.

The *Sobolev-space* $H^m(\Omega)$ is the set of all m-times weakly differentiable functions $u \in L_2(\Omega)$, that is u and all its derivatives $\partial_\alpha u$ up to the order $|\alpha| = m$ are in $L_2(\Omega)$.

Example 2.4 Let $u(x) = \sin(x)$, $u'(x) = \cos(x)$, on the interval $(0, 2\pi)$. Then

$$\|u\|_0 = \sqrt{\int_0^{2\pi} \sin^2(x)\, dx} = \sqrt{\pi} \tag{2.72}$$

$$\|u\|_1 = \sqrt{\int_0^{2\pi} [\sin^2(x) + \cos^2(x)]\, dx} = \sqrt{\pi + \pi} \tag{2.73}$$

and evidently because of the cyclic nature of the derivatives

$$\|u\|_n = \sqrt{\int_0^{2\pi} [u^2(x) + (u'(x))^2 + \ldots (u^n(x))^2]\, dx} = \sqrt{n\,\pi}\,. \tag{2.74}$$

The space

$$\mathcal{V} = \{u \in H^1(\Omega) |\, u = 0 \text{ on } \Gamma\} \tag{2.75}$$

with the scalar product $a(u, v) := (\nabla u, \nabla v)$ and the norm $\|u\|_E := a(u, u)^{1/2}$ is a Hilbert space and therefore there exists for any linear and bounded functional $J(u)$ a Riesz element G such that

$$\textit{(weak infl. func.)} \quad J(u) = a(G, u) \quad \forall\, u \in \mathcal{V}\,. \tag{2.76}$$

Because of Green's first identity, see (2.86), this is equivalent to

$$J(u) = \int_\Omega G\,(-\Delta u)\, d\Omega_y \quad \forall\, u \in \mathcal{V} \tag{2.77}$$

and if u is the solution of the boundary value problem (2.28) then follows

$$\textit{(strong infl. func.)} \quad J(u) = \int_\Omega G\, p\, d\Omega\,. \tag{2.78}$$

2.2 Green's Identities

Green's first and second identities—which are essentially applications of Gauss' Theorem—form two important corner stones of mathematical physics and of continuum mechanics. The classical energy and variational principles and conservation laws are based on these identities.

2.2.1 *Gauss' Theorem*

Let u and $\hat{u}$ be two arbitrary functions which have continuous first order derivatives on the interval $(0, l)$ then holds

$$\int_0^l u'\,\hat{u}\,dx = [u\,\hat{u}]_0^l - \int_0^l u\,\hat{u}'\,dx \tag{2.79}$$

and in higher dimensions, given a domain Ω and two functions u and $\hat{u}$ in $C^1(\Omega)$ (continuous first derivatives), as well

$$\int_\Omega u_{,x_i}\,\hat{u}\,d\Omega = \int_\Gamma u\,n_i\,\hat{u}\,ds - \int_\Omega u\,\hat{u}_{,x_i}\,d\Omega\,, \tag{2.80}$$

where n_i is the i-th component of the normal vector $\boldsymbol{n}$ on the edge Γ of the domain Ω.

The *divergence theorem* that connects the operators div and ∇ when applied to a vector field $\boldsymbol{u}$ and a function $\hat{u}$

$$\int_\Omega \operatorname{div}\boldsymbol{u}\,\hat{u}\,d\Omega = \int_\Gamma \boldsymbol{u}\cdot\boldsymbol{n}\,\hat{u}\,ds - \int_\Omega \boldsymbol{u}\cdot\nabla\hat{u}\,d\Omega \tag{2.81}$$

is an application of Gauss' theorem.

2.2.2 *The Laplace Operator*

Consider a thin membrane, stretched by a force H (uniformly in all directions), that seals an opening Ω and which is subjected to a pressure p, see Fig. 2.3. The deflection $u(\boldsymbol{x})$ of the membrane is the solution of the boundary value problem

$$-\,H\,\Delta u := -H\,(u_{,xx} + u_{,yy}) = p \quad \text{on } \Omega \quad u = 0 \quad \text{on } \Gamma\,. \tag{2.82}$$

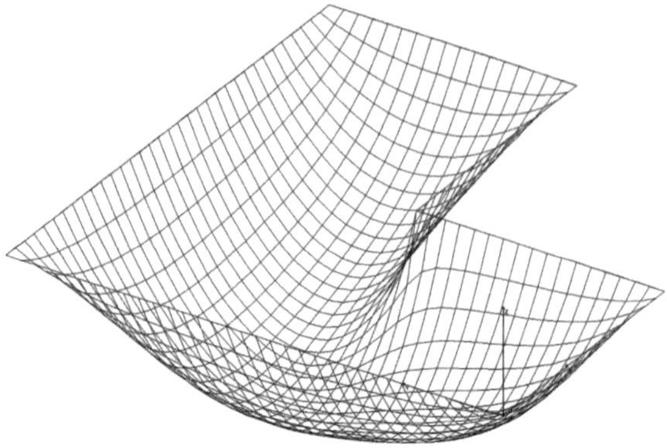

Fig. 2.3 Membrane under pressure load

Because H is a constant we can set $H = 1$ in the following.

Let $u \in C^2(\Omega)$ be a smooth (two times continuously differentiable) function on Ω. To this function we apply the Laplace operator

$$-\Delta u = -(u,_{xx} + u,_{yy}) \tag{2.83}$$

multiply the result with a second function $v \in C^1(\Omega)$, form the integral

$$\int_\Omega -\Delta u \, v \, d\Omega \tag{2.84}$$

and apply integration by parts to this integral

$$\int_\Omega -\Delta u \, v \, d\Omega = -\int_\Gamma \frac{\partial u}{\partial n} v \, ds + \int_\Omega \nabla u \cdot \nabla v \, d\Omega \,. \tag{2.85}$$

The result is Green's first identity

$$\mathcal{G}(u, v) = \underbrace{\int_\Omega -\Delta u \, v \, d\Omega + \int_\Gamma \frac{\partial u}{\partial n} v \, ds}_{\delta W_e} - \underbrace{\int_\Omega \nabla u \cdot \nabla v \, d\Omega}_{\delta W_i} = 0 \,. \tag{2.86}$$

Any pair of functions $\{u, v\}$ in $C^2(\Omega) \times C^1(\Omega)$ is a zero of this identity. If u is the solution to the boundary value problem (3.67) and v a virtual displacement of the membrane then the first two integrals represent the virtual exterior work δW_e and the integral

$$\int_{\Omega} \nabla u \cdot \nabla v \, d\Omega = \int_{\Omega} (u_{,x}\, v_{,x} + u_{,y}\, v_{,y})\, d\Omega =: a(u, v) \tag{2.87}$$

is called the virtual internal energy δW_i or the *strain energy product* between u and v. It is often abbreviated as $a(u, v)$.

Choosing for v the translation $v = 1$ gives

$$\mathscr{G}(u, 1) = \int_{\Omega} p \cdot 1 \, d\Omega + \int_{\Gamma} \frac{\partial u}{\partial n} \cdot 1 \, ds = 0 \tag{2.88}$$

which is the equilibrium condition: the tractions $\partial u / \partial n$ on the edge balance the pressure distribution p; they are the support reactions of a membrane.

On the "diagonal", when $v = u$,

$$\mathscr{G}(u, u) = \underbrace{\int_{\Omega} -\Delta u \, u \, d\Omega}_{2\, W_e} - \underbrace{\int_{\Omega} \nabla u \cdot \nabla v \, d\Omega}_{2\, W_i} = 0 \tag{2.89}$$

Green's first identity formulates the *principle of conservation of energy* and on the "off-diagonals" when $v = \delta u$ is a virtual displacement which we assume to be compatible with the boundary conditions, $\delta u = 0$ on Γ,

$$\mathscr{G}(u, \delta u) = \underbrace{\int_{\Omega} -\Delta u \, \delta u \, d\Omega}_{\delta W_e} - \underbrace{\int_{\Omega} \nabla u \cdot \nabla \delta u \, d\Omega}_{\delta W_i} = 0 \tag{2.90}$$

it formulates the *principle of virtual displacements*.

The *principle of virtual forces* is the identity in "reverse order", that is with the places of u and δu interchanged

$$\begin{aligned} \mathscr{G}(\delta u, u) &= \delta W_e^* - \delta W_i^* \\ &= \underbrace{\int_{\Omega} -\Delta \delta u \, u \, d\Omega + \int_{\Gamma} \frac{\partial \delta u}{\partial n} \, u \, ds}_{\delta W_e^*} - \underbrace{a(u, \delta u)}_{\delta W_i^*} = 0\,. \end{aligned} \tag{2.91}$$

Green's second identity is the expression

$$\mathscr{B}(u, v) = \mathscr{G}(u, v) - \mathscr{G}(v, u) = 0 - 0 = 0 \tag{2.92}$$

or

$$\mathscr{B}(u, v) = \int_\Omega -\Delta u \, v \, d\Omega + \int_\Gamma \frac{\partial u}{\partial n} v \, ds - \int_\Gamma \frac{\partial v}{\partial n} u \, ds - \int_\Omega u \, (-\Delta v) \, d\Omega = 0 \,. \tag{2.93}$$

This is Betti's theorem which states that if a membrane is subjected to two different pressure distributions p_1 and p_2

$$-\Delta u_1 = p_1 \quad \text{on } \Omega \quad u_1 = 0 \quad \text{on } \Gamma \tag{2.94}$$

$$-\Delta u_2 = p_2 \quad \text{on } \Omega \quad u_2 = 0 \quad \text{on } \Gamma \tag{2.95}$$

then the reciprocal exterior work of the two solutions u_1 and u_2 is the same

$$\mathscr{B}(u_1, u_2) = \int_\Omega p_1 \, u_2 \, d\Omega - \int_\Omega u_1 \, p_2 \, d\Omega = 0 \,. \tag{2.96}$$

FE-analysis operates with piecewise polynomials u_h and v_h which are smooth inside each element Ω_e, but the derivatives often jump at interelement boundaries and so the identities can only be formulated element for element

$$\mathscr{G}(u_h, v_h)_{\Omega_e} = \int_{\Omega_e} -\Delta \, u_h \; v_h \, d\Omega + \int_{\Gamma_e} \frac{\partial u_h}{\partial n} v_h \, ds - a(u_h, v_h)_{\Omega_e} = 0 \,. \tag{2.97}$$

Adding this string of zeros

$$\mathscr{G}(u_h, v_h) := \sum_e \mathscr{G}(u_h, v_h)_{\Omega_e} = 0 + 0 + \ldots + 0 = 0 \tag{2.98}$$

results in

$$\mathscr{G}(u_h, v_h) := \sum_e \int_{\Omega_e} -\Delta \, u_h \, v_h \, d\Omega + \sum_k \int_{\Gamma_k} l_k \, v_h \, ds - a(u_h, v_h) = 0 \tag{2.99}$$

where

$$a(u_h, v_h) := \sum_e a(u_h, v_h)_{\Omega_e} \,. \tag{2.100}$$

The functions l_k on the mesh lines Γ_k represent the jumps of the normal derivative in between the elements

$$l_k := \frac{\partial u_h}{\partial n}_+ - \frac{\partial u_h}{\partial n}_- \,. \tag{2.101}$$

If a side Γ_k of an element borders on the outer edge then l_k is simply the normal derivative on the edge.

2.2.3 Linear Self-Adjoint Operators

The Laplace operator is a linear self-adjoint differential operator of even order. Many operators in mathematical physics share this property and have the same mathematical structure so that in the following we can let L represent a generic operator of this type.

By applying integration by parts to the domain integral $(L\,u,\hat{u})$ the integral can be transformed into a series of boundary integrals $[\cdot\,,\cdot]$ and a symmetric bilinear form $a(u,\hat{u})$, the so called *strain energy product* or virtual strain energy, which is a domain integral. Collecting all terms on one side results in Green's first identity

$$\mathcal{G}(u,\hat{u}) = (L\,u,\hat{u}) + \sum_{i=0}^{m-1}(-1)^i[\partial_i u, \partial_{2m-1-i}\hat{u}] - a(u,\hat{u}) = 0 \tag{2.102}$$

and after a simple maneuver

$$\mathcal{B}(u,\hat{u}) := \mathcal{G}(u,\hat{u}) - \mathcal{G}(\hat{u},u) = 0 - 0 = 0 \tag{2.103}$$

also in Green's second identity

$$\mathcal{B}(u,\hat{u}) = (L\,u,\hat{u}) + \sum_{i=0}^{2m-1}(-1)^i[\partial_i u, \partial_{2m-1-i}\hat{u}] - (u, L\hat{u}) = 0\,. \tag{2.104}$$

The $2\,m$ boundary functions

$$\partial_i u \quad i = 0, 1, \ldots 2\,m-1 \tag{2.105}$$

in the boundary integrals are the *canonical boundary values*; they are also known as *Cauchy data*. They play a crucial role in the conservation laws, they register what happens on the boundary, what flows in and out of a region. Given these boundary functions and the right-hand side, $L\,u = p$, the function u inside Ω can be reconstructed from these data.

If you know the height of the end points, $u(0)$ and $u(l)$, of a rope and the curvature, $-u'' = p$, of the rope in between then you can draw the shape of the rope; this is the idea.

To the Laplace operator, $-\Delta u$, belong two boundary functions

$$\partial_0 u = u \quad \partial_1 u = \frac{\partial u}{\partial n} \tag{2.106}$$

and to the fourth-order beam equation $EI\,u^{IV} = p$ four boundary values

$$\partial_0 u = u \quad \partial_1 u = u' \quad \partial_2 u = -EI\,u'' = M \quad \partial_3 u = -EI\,u''' = V\,. \tag{2.107}$$

In the mechanical context the lower order terms, $\partial_i u, i < m$, are displacement terms while the higher order terms, $\partial_i u, m \leq i < 2m - 1$, are force terms and two such terms are called adjoint or conjugate, as for example the terms

$$\partial_0 u\, \partial_3 u = u\, V \qquad 0 + 3 = 3 \tag{2.108}$$
$$\partial_1 u\, \partial_2 u = u'\, M \qquad 1 + 2 = 3, \tag{2.109}$$

if the sum of the indices in the pair $\partial_i u\, \partial_j u$ equals $i + j = 2m - 1$ (= 3 in the case of a fourth order equation).

Obviously can these identities be extended to the Sobolev spaces as $H^2(\Omega)$ and $H^1(\Omega)$ by allowing the functions to have weak derivatives. But on two- and three-dimensional domains Ω this requires additional assumptions about the boundary values of such functions (*trace theorem*) which are very technical and which we do not want to discuss here in details.

For our purposes it suffices to assume that the identities are applicable to the standard FE-shape functions inside each element and that the extension to the whole domain, the whole mesh, can simply be done by adding zeros

$$0 + 0 + \ldots + 0 + 0 = 0\,. \tag{2.110}$$

2.3 Duality

Imagine two opposite forces, P_l and P_r, that pull on both sides of a rod, so that equilibrium is maintained

$$-\, P_l + P_r = 0 \tag{2.111}$$

but then the equation can be multiplied with any number δu

$$\delta u\, (-P_l + P_r) = \delta u \cdot 0 = 0 \tag{2.112}$$

without changing the mathematics. That is one can slide the rod effortlessly across the table: the virtual exterior work is zero—for each displacement δu.

What looks like a child's play is in truth one of the most important steps in applied mathematics, it is the earliest manifestation of *duality*. A whole plethora of concepts and ideas such as *scalar product, adjoint quantities, virtual displacement, virtual work, Green's function, spectral theory, etc.* comes to light by this simple maneuver.

The common theme of all these concepts is perhaps best explained in terms of linear algebra.

2.3.1 Linear Algebra

Let

$$\boldsymbol{K}\,\boldsymbol{u} = \boldsymbol{f} \tag{2.113}$$

a linear system of equations where $\boldsymbol{K}$ is an $n \times n$ matrix and $\boldsymbol{u}$ and $\boldsymbol{f}$ are vectors in $\mathbb{R}^n$.

The transpose of a real number is the same number as the original number, $\pi^T = \pi$, and so given any two vectors $\boldsymbol{u}$ and $\hat{\boldsymbol{u}}$ the following expression is an identity

$$\hat{\boldsymbol{u}}^T\,\boldsymbol{K}\,\boldsymbol{u} = (\hat{\boldsymbol{u}}^T\,\boldsymbol{K}\,\boldsymbol{u})^T = \boldsymbol{u}^T\,\boldsymbol{K}^T\,\hat{\boldsymbol{u}} \tag{2.114}$$

or in a slightly different arrangement

$$\mathscr{B}(\boldsymbol{u}, \hat{\boldsymbol{u}}) = \hat{\boldsymbol{u}}^T\,\boldsymbol{K}\,\boldsymbol{u} - \boldsymbol{u}^T\,\boldsymbol{K}^T\,\hat{\boldsymbol{u}} = 0\,. \tag{2.115}$$

This identity implies that the system (2.113) can only have a solution if the right hand side $\boldsymbol{f}$ is orthogonal to all solutions $\hat{\boldsymbol{u}}$ of the homogeneous adjoint system $\boldsymbol{K}^T\,\hat{\boldsymbol{u}} = \boldsymbol{0}$ because

$$\mathscr{B}(\boldsymbol{u}, \hat{\boldsymbol{u}}) = \hat{\boldsymbol{u}}^T\,\boldsymbol{K}\,\boldsymbol{u} - \boldsymbol{u}^T\,\boldsymbol{K}^T\,\hat{\boldsymbol{u}} = \hat{\boldsymbol{u}}^T\,\boldsymbol{f} - 0 = \hat{\boldsymbol{u}}^T\,\boldsymbol{f} = 0\,. \tag{2.116}$$

And if (2.113) has a solution $\boldsymbol{u}$ then $\boldsymbol{u}$ is also a solution of the variational problem

$$\hat{\boldsymbol{u}}^T\boldsymbol{K}\,\boldsymbol{u} = \hat{\boldsymbol{u}}^T\,\boldsymbol{f} \quad \forall\,\hat{\boldsymbol{u}} \in \mathbb{R}^n \tag{2.117}$$

and vice versa which establishes that in the discrete case the *strong formulation*, (2.113), and the *weak formulation*, (2.117), are equivalent.

Next let the vector $\boldsymbol{g}_i$ be the solution of the adjoint equation if the right hand side is the i-th unit vector, $\boldsymbol{e}_i = \{0, 0, \ldots 1, \ldots 0, 0\}^T$

$$\boldsymbol{K}^T\,\boldsymbol{g}_i = \boldsymbol{e}_i \tag{2.118}$$

and let $\boldsymbol{u}$ the solution of (2.113) then follows

$$\mathscr{B}(\boldsymbol{u}, \boldsymbol{g}_i) = \boldsymbol{g}_i^T\,\boldsymbol{K}\,\boldsymbol{u} - \boldsymbol{u}^T\,\boldsymbol{K}^T\,\boldsymbol{g}_i = \boldsymbol{g}_i^T\,\boldsymbol{f} - \boldsymbol{u}^T\,\boldsymbol{e}_i = 0 \tag{2.119}$$

or

$$u_i = \boldsymbol{g}_i^T\,\boldsymbol{f}\,. \tag{2.120}$$

So the "point solutions" of the adjoint system and the solution $\boldsymbol{u}$ of the original problem are intrinsically intertwined. Projecting $\boldsymbol{f}$ on all the axes $\boldsymbol{g}_i, i = 1, 2 \ldots n$ allows to construct the solution $\boldsymbol{u} = u_1\,\boldsymbol{e}_1 + \ldots + u_n\,\boldsymbol{e}_n$ with respect to the basis $\boldsymbol{e}_i$ because projecting $\boldsymbol{f}$ onto the $\boldsymbol{g}_i$ is the same as projecting $\boldsymbol{u}$ onto the $\boldsymbol{e}_i$

$$u_i = \boldsymbol{g}_i^T \boldsymbol{f} = \boldsymbol{u}^T \boldsymbol{e}_i \,. \tag{2.121}$$

This is what is done when $\boldsymbol{f}$ is multiplied with the inverse matrix $\boldsymbol{K}^{-1}$

$$\boldsymbol{u} = \boldsymbol{K}^{-1} \boldsymbol{f} \tag{2.122}$$

because the rows of $\boldsymbol{K}^{-1}$ are just the vectors $\boldsymbol{g}_i$ and so

$$\boldsymbol{u} = (\boldsymbol{g}_1^T \boldsymbol{f})\, \boldsymbol{e}_1 + (\boldsymbol{g}_2^T \boldsymbol{f})\, \boldsymbol{e}_2 + \ldots + (\boldsymbol{g}_n^T \boldsymbol{f})\, \boldsymbol{e}_n \,. \tag{2.123}$$

In the continuous case, when the focus is on differential equations, as for example

$$-\, u''(x) = p(x) \quad u(0) = u(l) = 0\,, \tag{2.124}$$

the matrix $\boldsymbol{K}$ has infinitely many columns and the unit vectors become Dirac deltas

$$-\frac{d^2}{dy^2}\, G(y, x) = \delta(y - x) \tag{2.125}$$

but the formalism is the same. By projecting the right-hand p side onto the solutions $G(y, x)$, that is by forming the L_2-scalar product (integral) of the two functions, the value of the solution can be recovered at any point x

$$u(x) = \underbrace{\int_0^l G(y, x)\, p(y)\, dy}_{\boldsymbol{g}_i^T \boldsymbol{f}} = \underbrace{\int_0^l \delta(y - x)\, u(y)\, dy}_{\boldsymbol{u}^T \boldsymbol{e}_i} \tag{2.126}$$

and the first integral is, as the equation shows, the same as the L_2-scalar product between the function u itself and the Dirac delta, so that these two equations are indeed equivalent to (2.121).

To the symmetry of the strain energy product

$$a(u, \hat{u}) = a(\hat{u}, u) \tag{2.127}$$

corresponds the symmetry of $\boldsymbol{K}$ and it so follows that if u and G are the variational solutions of the *primal* and the *dual* problem respectively

$$u \in \mathcal{V} \quad a(u, v) = (p, v) \quad \forall\, v \in \mathcal{V} \quad \textit{primal problem} \tag{2.128}$$
$$G \in \mathcal{V} \quad a(G, v) = (\delta, v) \quad \forall\, v \in \mathcal{V} \quad \textit{dual problem} \tag{2.129}$$

then this implies

$$u(\boldsymbol{x}) = (p, G) = (\delta, u)\,. \tag{2.130}$$

Equation (2.126) is the same result in a "longhand" notation.

Much more could be said at this point about linear self-adjoint operators with constant coefficients, Green's functions and matrices, about eigenvalues, eigenvectors, convolution and the construction of Green's function from eigenfunctions. Linear, time-invariant systems (signal processing) provide the most elementary application and they are also a good example for the transition from the discrete case to the continuous case, $\boldsymbol{K}_{(n\times n)} \rightarrow \boldsymbol{K}_{(\infty\times\infty)}$. Such systems are completely characterized by their response to an impulse which is either modeled as a Dirac delta function for continuous-time systems or as the Kronecker delta for discrete-time systems [3].

Remark 2.1 In signal processing the analysis is not done by integrating the impulse response function (the Green's function) and the input function but by multiplying the integral transforms of the two functions because a convolution in the time domain is equivalent to a multiplication in the frequency domain.

2.3.2 Vectors and Linear Functionals

In linear algebra, $\mathcal{V} = \mathbb{R}^n$, all linear functionals have the form of a scalar product

$$J(\boldsymbol{u}) = \boldsymbol{j}^T \boldsymbol{u} \tag{2.131}$$

between $\boldsymbol{u}$ and a certain vector $\boldsymbol{j}$ whose components are simply the values of the functional applied to the base vectors $\boldsymbol{e}_i, i = 1, 2, \ldots n$ of $\mathbb{R}^n$

$$j_i = J(\boldsymbol{e}_i)\,. \tag{2.132}$$

In the case of the functional $J(\boldsymbol{u}) = u_1$ which returns the first component of the input, the vector $\boldsymbol{j}$ is just the unit vector $\boldsymbol{e}_1$

$$J(\boldsymbol{u}) = \boldsymbol{e}_1^T \boldsymbol{u} = u_1\,. \tag{2.133}$$

To understand the notion of a Green's function or rather Green's vector $\boldsymbol{g}$ imagine a situation where the vector $\boldsymbol{u}$ is unknown and so $J(\boldsymbol{u})$ cannot be calculated directly. It is only known that the matrix $\boldsymbol{K}$ has mapped the vector $\boldsymbol{u}$ onto a vector $\boldsymbol{f} = \boldsymbol{K}\,\boldsymbol{u}$. Can the value $J(\boldsymbol{u})$, the first component u_1 of $\boldsymbol{u}$, be recovered from $\boldsymbol{f}$? Yes, this is possible. Solving the system

$$\boldsymbol{K}^T \boldsymbol{g} = \boldsymbol{j} = \boldsymbol{e}_1 \tag{2.134}$$

for the vector $\boldsymbol{g}$ and formulating the identity

$$\mathscr{B}(\boldsymbol{u}, \boldsymbol{g}) = \boldsymbol{g}^T \boldsymbol{K}\,\boldsymbol{u} - \boldsymbol{u}^T \boldsymbol{K}^T \boldsymbol{g} = \boldsymbol{g}^T \boldsymbol{f} - \boldsymbol{u}^T \boldsymbol{e}_1 = 0 \tag{2.135}$$

it is found that

$$J(\boldsymbol{u}) = u_1 = \boldsymbol{g}^T \boldsymbol{f} . \tag{2.136}$$

Note that the vector $\boldsymbol{g}$ depends on the matrix $\boldsymbol{K}$, that is to find the way back from $\boldsymbol{f}$ to $\boldsymbol{u}$, or here u_1, we must know which matrix $\boldsymbol{K}$ mapped $\boldsymbol{u}$ onto $\boldsymbol{f}$.

If a different matrix $\bar{\boldsymbol{K}}$ maps $\boldsymbol{u}$ onto a different vector $\bar{\boldsymbol{f}} = \bar{\boldsymbol{K}}\,\boldsymbol{u}$ then—even though the functional $J(\boldsymbol{u}) = u_1$ is the same—a different Green's vector is required

$$\bar{\boldsymbol{K}}^T \bar{g} = \boldsymbol{j} = \boldsymbol{e}_1 . \tag{2.137}$$

The same logic applies in the continuous case. Take the function $u(x) = \sin(\pi\, x/l)$ and differentiate it two times or four times respectively

$$-u'' = \left(\frac{\pi}{l}\right)^2 \sin(\pi\, x/l) \quad = f(x) \tag{2.138}$$

$$EI\, u^{IV} = \left(\frac{\pi}{l}\right)^4 \sin(\pi\, x/l) \quad = \bar{f}(x) . \tag{2.139}$$

In the first case $u(x)$ is the deflection of a taut rope subjected to a lateral load $f(x)$ and in the second case it is the deflection of a beam subjected to a lateral load $\bar{f}(x)$.

To recover the deflection at mid-span, $u(l/2)$, from the right-hand sides $f(x)$ and $\bar{f}(x)$, different Green's functions are required though we are asking for the same value, $u(l/2)$. So the mapping is important. Which operator mapped u onto the right-hand side? Where do the data come from?

2.4 Influence Functions

Linear algebra is easy. Let us now turn to functions and let us replace the matrix $\boldsymbol{K}$ by the second-order differential operator in the equation for the rope

$$-\, u''(x) = p(x) \quad u(0) = u(l) = 0 . \tag{2.140}$$

The aim is to formulate two influence functions, one for the deflection $u(x)$ of the rope and and one for the transverse force $V(x) = H\, u'(x) = u'(x)$, by using Green's first identity (principle of virtual forces) and, alternatively, by using Green's second identity (Betti's theorem).

Green's first identity for the operator $-d^2/dx^2$ is the expression

$$If\ \ \hat{u} \in C^2(0,l), u \in C^1(0,l)\ then$$

$$\mathscr{G}(\hat{u}, u) = \int_0^l -\hat{u}''\, u\, dx + [\hat{u}'\, u]_0^l - \int_0^l \hat{u}'\, u'\, dx = 0 \tag{2.141}$$

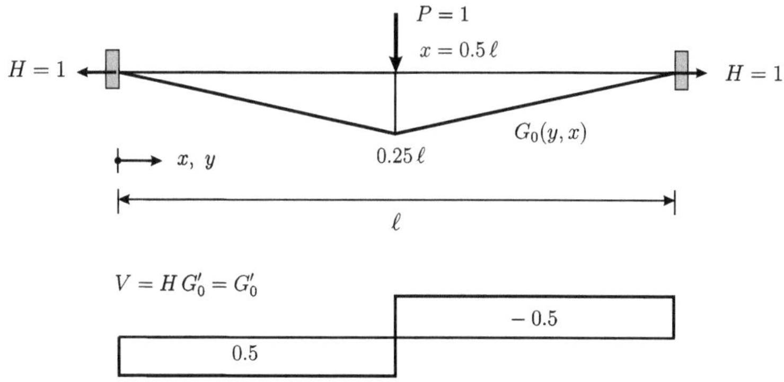

Fig. 2.4 Influence function G_0

and the second identity, $\mathscr{B}(\hat{u}, u) = \mathscr{G}(\hat{u}, u) - \mathscr{G}(u, \hat{u})$, is the expression

$$\textit{If } \hat{u} \in C^2(0, l), u \in C^2(0, l) \textit{ then}$$

$$\mathscr{B}(\hat{u}, u) = \int_0^l -\hat{u}''\, u\, dx + [\hat{u}'\, u] - [\hat{u}\, u']_0^l - \int_0^l \hat{u}\,(-u'')\, dx = 0\,. \tag{2.142}$$

Because these identities are based on integration by parts the functions u and $\hat{u}$ must meet certain regularity requirements. In the following, when we formulate these or other identities, we will tacitly assume that the functions satisfy these conditions or that we have partitioned the domain appropriately into subregions—typically the elements—to circumvent restrictions which apply because the shape functions are only piecewise C^2 and C^1 respectively.

2.4.1 Influence Function for $u(x)$

Let $G_0(y, x)$ be the deflection of the rope when a single force $P = 1$ acts at a point x of the rope. That is G_0 is the solution of the boundary value problem

$$-\frac{d^2}{dy^2}\, G_0(y, x) = \delta_0(y - x) \quad G_0(0, x) = G_0(l, x) = 0\,. \tag{2.143}$$

The index 0 on the Dirac delta is to indicate that it represents a point load. Higher-order Dirac deltas, $\delta_1, \delta_2, \ldots$, will represent moments, dislocations, etc. The Green's functions carry the same index as the Dirac delta to which they belong.

The Green's function has the form of a string plugged at x, see Fig. 2.4,

$$G_0(y,x) = \frac{1}{l} \begin{cases} y\,(l-x) & y \leq x \\ x\,(l-y) & x \leq y \end{cases} . \tag{2.144}$$

The first derivative jumps at x and the second derivative is the Delta function, so clearly $G_0(y,x)$ is not a C^2 function. To derive Betti's theorem Green's second identity is formulated on the punctured domain or—here—interval Ω_ε that is a small ε-neighborhood of the source point x is excluded from $[0, l]$

$$\Omega_\varepsilon := [0, x-\varepsilon] \cup [x+\varepsilon, l] \tag{2.145}$$

and we then let ε tend to zero.

The functions u and G_0 are C^2-functions on Ω_ε. The second derivative of $G_0(y,x)$ is zero on Ω_ε because G_0 is a linear function and so

$$\begin{aligned} \mathscr{B}(G_0, u)_{\Omega_\varepsilon} = & \underbrace{G_0'(x, y-\varepsilon)\,u(y-\varepsilon) - G_0'(x, y+\varepsilon)\,u(y+\varepsilon)}_{\lim_{\varepsilon\to 0} = 1\cdot u(x)} \\ & \underbrace{- G_0(x, y-\varepsilon)\,u'(y-\varepsilon) + G_0(x, y+\varepsilon)\,u'(y+\varepsilon)}_{\lim_{\varepsilon\to 0} = 0} \\ & - \int_{\Omega_\varepsilon} G_0(y,x)\,p(y)\,dy = 0 . \end{aligned} \tag{2.146}$$

Because the sum is zero for all $\varepsilon > 0$ the limit must be zero as well and so we conclude that

$$\lim_{\varepsilon \to 0} \mathscr{B}(G_0, u)_{\Omega_\varepsilon} = u(x) - \int_0^l G_0(y,x)\,p(y)\,dy = 0 . \tag{2.147}$$

The term $u(x)$—or to be more precise $1 \cdot u(x)$—is the work done by the unit-jump in the derivative G_0' at the source point x on acting through $u(x)$; the derivative G_0' corresponds to the transverse force and to accommodate the point load the derivative must jump at x.

The expression on the second line in (2.146) is zero in the limit because both G_0 and u' are continuous at x.

Reordering the terms in (2.147) gives the influence function for $u(x)$

$$u(x) = \int_0^l G_0(y,x)\,p(y)\,dy . \tag{2.148}$$

We can now also explain what it means that $G_0(y,x)$ solves the boundary value problem (2.143). This is an open question: a direct verification is not possible because the second derivative of $G_0(y, x)$ is not defined at the source point x and also $\delta_0(y, x)$ is not a proper function.

The answer is that the differential equation (2.143) must be interpreted in the *distributional sense*. The function $G_0(y, x)$ is a solution of (2.143) if and only if

$$\int_0^l -\frac{d^2}{dy^2} G_0(y, x)\, u(y)\, dy = \int_0^l \delta_0(y, x)\, u(y)\, dy = u(x) \quad \forall u \in C_0^\infty(0, l)\,. \tag{2.149}$$

That is the two sides are weighted with a test function $u \in C_0^\infty(0, l)$ and the results on both sides should match. The test functions u are infinitely smooth functions and all their boundary values, $u, u', u'', \ldots$, are zero—at both ends.

The trick is to apply integration by parts to the left side, neglecting all warnings that this is not allowed because $-G_0''$ is not a proper function, and to shift the differential operator from G_0 onto the infinitely patient test function

$$\int_0^l -\frac{d^2}{dy^2} G_0(y, x)\, u(y)\, dy = \int_0^l G_0(y, x)\, (-u''(y))\, dy \tag{2.150}$$

and so it must be true that (2.149) is equivalent to

$$\int_0^l G_0(y, x)\, (-u''(y))\, dy = u(x) \quad \forall u \in C_0^\infty(0, l)\,. \tag{2.151}$$

And this was demonstrated above (substitute for $-u''$ the right hand side p).

2.4.2 *Influence Function for $u'(x)$*

Instead of a point load now a unit dislocation $\delta_1(y, x)$ is applied at the source point x so that the influence function $G_1(y, x)$ for the first derivative $u'(x)$ is the solution to the boundary value problem

$$-\frac{d^2}{dy^2} G_1(y, x) = \delta_1(y, x) \quad G_1(0, x) = G_1(l, x) = 0 \tag{2.152}$$

and consequently

$$
\begin{aligned}
\mathscr{B}(G_1, u)_{\Omega_\varepsilon} = &\underbrace{G_1'(x, y-\varepsilon)\, u(y-\varepsilon) - G_1'(x, y+\varepsilon)\, u(y+\varepsilon)}_{\lim_{\varepsilon\to 0}=0} \\
&- \underbrace{G_1(x, y-\varepsilon)\, u'(y-\varepsilon) + G_1(x, y+\varepsilon)\, u'(y+\varepsilon)}_{\lim_{\varepsilon\to 0}=1\cdot u'(x)} \\
&- \int_{\Omega_\varepsilon} G_1(y, x)\, p(y)\, dy = 0
\end{aligned}
\tag{2.153}
$$

and therefore as well

$$
\lim_{\varepsilon\to 0} \mathscr{B}(G_1, u)_{\Omega_\varepsilon} = u'(x) - \int_0^l G_1(y, x)\, p(y)\, dy = 0 \tag{2.154}
$$

which is the influence function for $u'(x)$

$$
u'(x) = \int_0^l G_1(y, x)\, p(y)\, dy\,. \tag{2.155}
$$

Remark 2.2 The Dirac delta δ_1

$$
\delta_1(y, x) = 0 \quad x \neq y \tag{2.156}
$$

$$
\int_0^l \delta_1(y, x)\, u(y)\, dy = u'(x) \quad x \in (0, l) \tag{2.157}
$$

can be considered the first derivative of δ_0 because integration by parts allows to write (note that $u(0) = u(l) = 0$)

$$
\int_0^l \delta_0'(y, x)\, u(y)\, dy = \int_0^l \delta_0(y, x)\, u'(y)\, dy = u'(x) \tag{2.158}
$$

so that δ_0' performs the same action as δ_1. For a very detailed discussion of these shifts and other transformation rules applicable to Dirac deltas see [4].

2.4.3 Weak Influence Function for $u(x)$

Next Green's first identity (or the principle of virtual forces) will be used to formulate an alternative influence function for $u(x)$. On the punctured domain the starting point is the expression

$$\begin{aligned}\mathscr{G}(G_0, u)_{\Omega_\varepsilon} = &\underbrace{G_0'(x, y-\varepsilon)\,u(y-\varepsilon) - G_0'(x, y+\varepsilon)\,u(y+\varepsilon)}_{\lim_{\varepsilon\to 0} = u(x)} \\ &- a(G_0(y, x), u)_{\Omega_\varepsilon} = 0 \end{aligned} \tag{2.159}$$

and taking the limit

$$\lim_{\varepsilon\to 0} \mathscr{G}(G_0, u)_{\Omega_\varepsilon} = u(x) - a(G_0, u) = 0 \tag{2.160}$$

gives

$$u(x) = a(G_0, u) := \int_0^l G_0'\,u'\,dy\,. \tag{2.161}$$

We call this influence function for $u(x)$ a *weak influence function* in contrast to (2.148) which we call a *strong influence function*.

A weak influence function is as good as a strong influence function. Weak means that it is based on an evaluation of the stresses and strains, which are of the order m if $2\,m$ is the order of the differential equation, while a strong influence function extracts $u(x)$ (or possibly other values) from the right-hand side p which—in terms of the solution u—is of order $2\,m$.

Weak influence functions can safely be formulated for displacement terms, the lower order terms (derivatives less than m if the differential equation is of order $2\,m$), but not for force terms (the higher derivatives), the result is nil, is zero.

To verify this we try to derive a weak influence function for the first derivative u' of a rope. The starting point is as before a formulation on the punctured interval

$$\begin{aligned}\mathscr{G}(G_1, u)_{\Omega_\varepsilon} = &- G_1'(x, y-\varepsilon)\,u(y-\varepsilon) + G_1'(x, y+\varepsilon)\,u(y+\varepsilon) \\ &- a(G_1(y, x), u)_{\Omega_\varepsilon} = 0\,. \end{aligned} \tag{2.162}$$

The limit of the first two terms is zero

$$\lim_{\varepsilon\to 0} \left[-G_1'(x, y-\varepsilon)\,u(y-\varepsilon) + G_1'(x, y+\varepsilon)\,u(y+\varepsilon)\right] = 0 \tag{2.163}$$

because both functions, the slope G_1' of the Green's function G_1 and the solution u, are continuous at x. But if the limit of the first two terms is zero (the exterior work) then the limit of the strain energy product between G_1 and u (the interior work) must

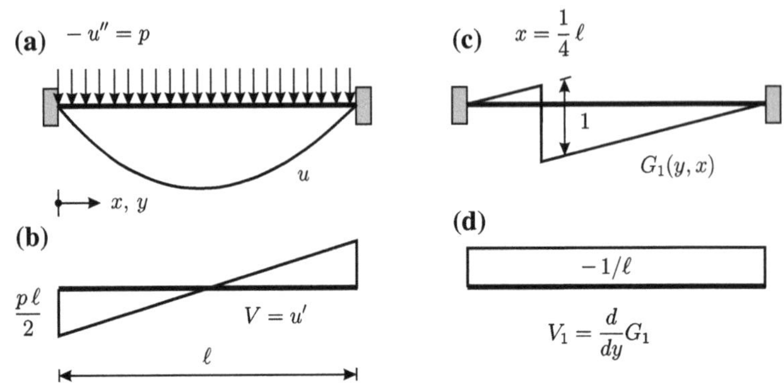

Fig. 2.5 Rope **a** deflection u, **b** shear force $V = u'$, **c** influence function G_1 for $u'(x)$, **d** derivative of G_1

be zero as well

$$\lim_{\varepsilon \to 0} \mathscr{G}(G_1, u)_{\Omega_\varepsilon} = 0 - \lim_{\varepsilon \to 0} a(G_1, u)_{\Omega_\varepsilon} = 0\,, \tag{2.164}$$

which is a correct but useless result. Stated separately the limit is

$$a(G_1, u) := \lim_{\varepsilon \to 0} a(G_1, u)_{\Omega_\varepsilon} = 0 \tag{2.165}$$

and not, as one might have guessed, $u'(x)$.

An example may prove the point. We try to formulate a weak influence function for the first derivative $V = u'$ of the rope, see Fig. 2.5, at the point $x = 0.25\,\ell$. But the strain energy product between the slope $V = u'$ of the rope and the slope of the Green's function G_1 evidently is zero

$$a(G_1, u) = \int_0^l \frac{d}{dy} G_1(y, x)\, u'(y)\, dy = 0 \tag{2.166}$$

while the true value of V at the quarter point is $pl/4$.

2.4.3.1 Background

Does this result not contradict the claim that any linear bounded functional $J(u)$ allows such a "weak representation" by a Riesz element (G_1),

$$J(u) = u'(x) = a(G_1, u) \quad ? \tag{2.167}$$

No it does not. Taking the derivative of a function is a bounded functional only if the three indices

- $i =$ the order of the derivative
- $m =$ the order of the energy space $H^m(\Omega)$
- $n =$ dimension of the continuum $\Omega \subset \mathbb{R}^n$

satisfy the inequality

$$m - i > \frac{n}{2} \,. \tag{2.168}$$

The differential operators, typically, have the order $2\,m = 2$ or $2\,m = 4$ and the associated energy space is $H^1(\Omega)$ and $H^2(\Omega)$ respectively, that is the order m of the energy space is half the value of the order $2\,m$ of the differential equation.

The solution space of the rope, $-u'' = p$, is $H^1(0,l)$ and on this space the zero-order derivative is a bounded functional but the first-order derivative is not

$$\underbrace{J(u) = u(x)}_{bounded} \quad \underbrace{J(u) = u'(x)}_{unbounded} \tag{2.169}$$

while on $H^2(0,l)$ (beam problems, $EI\,u^{IV} = p$, $2m = 4$) the derivatives are classified as follows

$$\underbrace{J(u) = u(x) \quad J(u) = u'(x)}_{bounded} \quad \underbrace{J(u) = u''(x) \quad J(u) = u'''(x)}_{unbounded} \,. \tag{2.170}$$

There is nothing wrong with higher order derivatives. It all depends on the definition of the underlying solution space. If the index m of the Sobolev space H^m is high enough, if it satisfies the inequality

$$m > i + \frac{n}{2} \tag{2.171}$$

then also higher order derivatives are bounded functionals.

To be precise, in 1-D problems the ith order derivative is a bounded functional on $H^m(0,l)$

$$|J(u)| = \left| \frac{d^i u}{dx^i}(x) \right| < c\, \|u\|_m \tag{2.172}$$

if $C^i(0,l) \subset H^m(0,l)$ ("continuous embedding") and according to Sobolev's Embedding Theorem this is the case if $m > i + 1/2$. So for the third derivative, $J(u) = u'''(x)$, to be bounded the solution space H^m must have an index $m > 3+1/2$, that is on $H^4(0,l)$ the third derivative is a continuous functional. In 2-D the necessary condition would be $m > i + 2/2$ and in 3-D $m > i + 3/2$, see p. xx.

Note that it would not do to choose very smooth shape functions, $\varphi_i \in H^4(0,l)$, and then to claim that the problem

$$a(G_h, \varphi_i) = J(\varphi_i) = \varphi_i'''(x) \tag{2.173}$$

is well-posed, that it is in agreement with the Riesz-representation theorem. For that the strain energy

$$a(u, u) = \int_0^l (u')^2 \, dx \tag{2.174}$$

must be equivalent on $\mathcal{V}$ to the norm on $H^4(0, l)$

$$\|u\|_4 := \left\{ \int_0^l (u^{IV})^2 + (u''')^2 + (u'')^2 + (u')^2 + u^2] \, dx \right\}^{1/2} \tag{2.175}$$

but because the strain energy disregards the higher-order derivatives it cannot really separate the elements of $H^4(0, l)$, that is it cannot be a norm on $H^4(0, l)$. The correct weak form for the Green's function would be

$$(G_h, \varphi_i)_4 = \varphi_i'''(x) \quad \forall \varphi_i \in \mathcal{V}_h \tag{2.176}$$

where $(\cdot, \cdot)_4$ is the scalar-product on $H^4(0, l)$ and in $u_h'''(x) = (G_h, p)$ the load p would be the eighth-order derivative of u or something similar to that.

Remark 2.3 An engineer will not wonder about this negative outcome, that the strain energy $a(G_1, u)$ is zero, because for him a weak influence functions such as

$$u(x) = \int_0^l G'(y, x) \, u'(y) \, dy \equiv a(G, u) \quad (-u'' = p) \tag{2.177}$$

is based on the *principle of virtual forces* and which virtual force do you apply, so he would ask, to calculate a force term?

To clarify this issue: What the engineer calls the principle of virtual forces is Green's first identity with δu and u in reverse order

$$\mathcal{G}(\delta u, u) = 0 \, . \tag{2.178}$$

The virtual displacement comes first and the solution is second while in the *principle of virtual displacements* the sequence is: first u and then δu.

The reverse order makes that ("virtual") forces, $\delta u''$ and δV $(= \delta u')$, act on u

$$\mathcal{G}(\delta u, u) = \int_0^l -\delta u''\, u\, dx + [\delta V\, u]_0^l - a(\delta u, v) = 0 \quad \text{(rope)}\,. \tag{2.179}$$

The single terms $u(x) = \ldots$ or $V(x) = \ldots$ in the influence functions are the limits of certain "boundary integrals"[3]

$$\lim_{\varepsilon \to 0} \left\{ [V(G_0)\, u]_0^{x-\varepsilon} + [V(G_0)\, u]_{x-\varepsilon}^{l} \right\} = 1 \cdot u(x) \tag{2.180}$$

$$\lim_{\varepsilon \to 0} \left\{ [G_1\, V]_0^{x-\varepsilon} + [G_1\, V]_{x-\varepsilon}^{l} \right\} = 1 \cdot V(x)\,. \tag{2.181}$$

But the force term $V(x)$ could only jump out of the boundary integral $[\delta u\, V]$ and this is not a part of (2.179). It is contained in Green's second identity and this is why Betti can do it but the principle of virtual forces can not.

2.4.4 A Sequence that Converges to G_1

The jump in the Green's function G_1 at x forced us to exclude an ε-neighborhood of the source point from the interval $[0, l]$.

Alternatively we could study what happens if a sequence of continuous and piecewise polynomial functions $G_1^{\varepsilon}(y, x)$ tends to $G_1(y, x)$. Such a sequence is easily constructed: we place two point forces $P = \pm 1/\Delta x$ at each side of the source point x, an unit Δx apart, and we then let the distance Δx shrink to zero so that the sequence of shapes $G_1^{\Delta x}$ converges to the exact Green's function $G_1(y, x)$, see Fig. 2.6.

Let us see how we go about the formulation of the identities in this case. For a particular value of Δx Green's first identity reads

$$\mathcal{G}(G_1^{\Delta x}, u) = \frac{1}{\Delta x}\, u(x + 0.5\, \Delta x) - \frac{1}{\Delta x}\, u(x - 0.5\, \Delta x) - a(G_1^{\Delta x}, u) = 0\,. \tag{2.182}$$

The limit of the first term is $u'(x)$

$$\lim_{\Delta x \to 0} \left[\frac{1}{\Delta x}\, u(x + 0.5\, \Delta x) - \frac{1}{\Delta x}\, u(x - 0.5\, \Delta x) \right] = u'(x) \tag{2.183}$$

and because of

$$\lim_{\Delta x \to 0} \mathcal{G}(G_1^{\Delta x}, u) = u'(x) - \lim_{\Delta x \to 0} a(G_1^{\Delta x}, u) = 0 \tag{2.184}$$

the second term must have the same limit

[3] In higher dimensions these terms would be genuine boundary integrals.

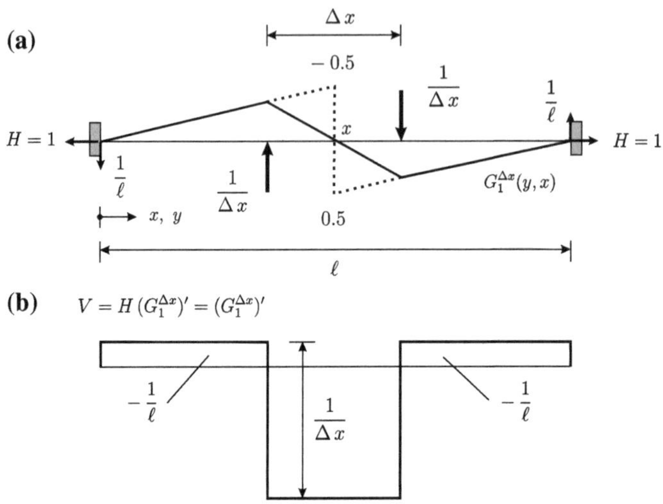

Fig. 2.6 Influence function G_1 (*dashed line*) for the first derivative $V = H\,u' = u'$ **a** FE-approximation $G_1^{\Delta x}$ and **b** its derivative

$$\lim_{\Delta x \to 0} a(G_1^{\Delta x}, u) = u'(x)\,. \tag{2.185}$$

This is a remarkable result because we know that

$$a(G_1, u) := \lim_{\varepsilon \to 0} a(G_1, u)_{\Omega_\varepsilon} = 0\,. \tag{2.186}$$

So the sequence, $\Delta x \to 0$, of the integrals $a(G_1^{\Delta x}, u)$ converges to $u'(x)$ but when we substitute for $G_1^{\Delta x}$ the limit function G_1 then the integral is zero.

Numerically this is easily verified: the slope of G_1 is $1/\ell$ and so

$$a(G_1, u) = \int_0^\ell \frac{1}{\ell}\, u'(x)\, dx = u(\ell) - u(0) = 0 - 0 = 0\,. \tag{2.187}$$

The Green's function G_1 of the rope seemingly has a bounded energy, but all the (infinite) energy is concentrated in one point, the point where the deflection jumps, $[[G]] = 1$. But because the energy integral $a(G_1, u)$, when we follow the definition (2.186), is insensitive to issues at points (1-D) the infinity is non-existing in the energy metric.

The same happens in 2-D and 3-D problems: a stair welded together from flat steel panels has zero energy but when a flat sheet of metal is folded into the same shape then the energy of the stair is infinite though the energy hiding in the folds (= lines) of the plate is "lost" because the strain energy is an 2-D integral which is insensitive to sudden jumps of the integrand along curves (= the folds) because curves have measure zero.

Remark 2.4 The situation in Fig. 2.6 is the following: the two forces $\pm 1/\Delta x$ are statically equivalent to a unit couple $1/\Delta x \cdot \Delta x = 1$ and so the support reactions, $\pm 1/\ell$, are independent of Δx. The transverse force V is equal to the support reaction, $V = -1/\ell$, and on passing the foot of the force $-1/\Delta x$ the transverse force V jumps by $1/\Delta x$.

Basically what happens when $\Delta x \to 0$ is that the two infinite forces tear the rope apart but surprisingly the discontinuity, the dislocation, is finite; the gap has unit length. If G_1 were a finite element shape function we would say that it is a non-conforming shape function. Such functions have an infinite strain energy.

To see this let us assume, for simplicity, that the length ℓ of the rope is very large then $1/\ell \simeq 0$ and so the strain energy in the rope is approximately

$$\begin{aligned} a(G_1^{\Delta x}, G_1^{\Delta x}) &= \int_0^l H\,[(G_1^{\Delta x})']^2\,dy \simeq \int_{x-0.5\,\Delta x}^{x+0.5\,\Delta x} H\left(\frac{1}{\Delta x}\right)^2 dy \\ &= H\left(\frac{1}{\Delta x}\right)^2 \Delta x = \frac{H}{\Delta x} \end{aligned} \tag{2.188}$$

and this term becomes infinite when $\Delta x \to 0$. If $1/\ell$ is not negligible then the transverse force in the middle is $1/\Delta x - 1/\ell$ but the leading term in the square is the same as before

$$\int_{x-0.5\,\Delta x}^{x+0.5\,\Delta x} H\left(\frac{1}{\Delta x} - \frac{1}{\ell}\right)^2 dy = H\left(\frac{1}{\Delta x} - \frac{1}{\ell}\right)^2 \Delta x = \frac{H}{\Delta x} + \dots \tag{2.189}$$

and so the energy becomes infinite as well.

2.4.5 Elevators and Escalators

Weak influence functions are like escalators while strong influence functions resemble elevators. Depict the Green's function as a mountain range where the height above sea level indicates the influence a point load $P = 1$ on top of a mountain will have on $u(x)$. The lift $G(y, x)$ takes us directly to the mountain top while the escalator $a(G, u)$ does one step dx at a time, calculates the increment $du = u'\,G'\,dx$—the derivative G' plays the role of a weight—and accumulates all the single increments du to reach the value $u(x)$.

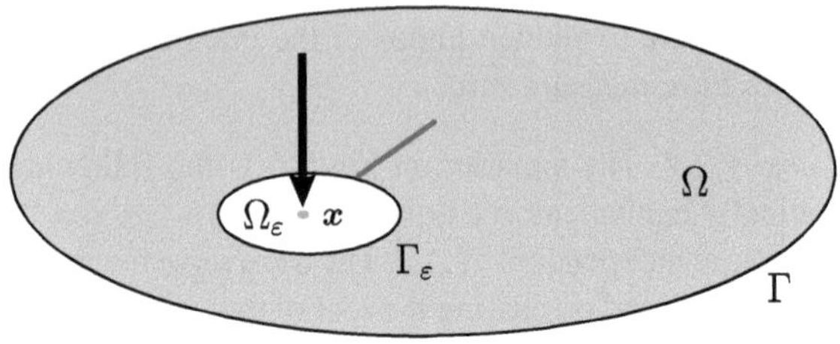

Fig. 2.7 Punctured domain with source point

The classical example of an equation of this incremental type is the formula

$$\int_{\Omega} \operatorname{div} \boldsymbol{u}\, d\Omega = \int_{\Gamma} \boldsymbol{u} \cdot \boldsymbol{n}\, ds = J(\boldsymbol{u}) \tag{2.190}$$

which is the influence function for the net flow, $J(\boldsymbol{u})$, in and out of a planar region Ω bounded by a curve Γ. This influence function counts all the positive and negative increments of the velocity field $\boldsymbol{u}$ inside Ω and if the average value of the divergence of the field $\boldsymbol{u}$

$$\operatorname{div} \boldsymbol{u} = u_{x,x} + u_{y,y} \tag{2.191}$$

is zero then the net flow $J(\boldsymbol{u})$ is also zero (Fig. 2.7).

Remark 2.5 Later in Chap. 3 it will be seen that the FE-method does not distinguish between weak and strong influence functions. In FE-analysis the approximate Green's function is the solution of the variational problem

$$G_h \in \mathcal{V}_h : \quad a(G_h, \varphi_i) = J(\varphi_i) \quad \varphi_i \in \mathcal{V}_h \tag{2.192}$$

and G_h gives the same result with both formulas

$$J(u_h) = \underbrace{\int_0^l G_h(y, x)\, p(y)\, dy}_{strong} = \underbrace{a(G_h, u_h)}_{weak}, \tag{2.193}$$

because on $\mathcal{V}_h$ the two formulas—strong and weak—coincide

$$J(u_h) = \int_0^l G_h(y, x)\, p(y)\, dy = \boldsymbol{g}^T \boldsymbol{f} = \boldsymbol{g}^T \boldsymbol{K}\, \boldsymbol{u} = a(G_h, u_h). \tag{2.194}$$

In a matrix formulation the difference between strong and weak influence functions only depends on how one reads the equations

$$J(u_h) = \underbrace{g^T f}_{strong} = \underbrace{g^T K u}_{weak} . \tag{2.195}$$

In the weak formulation $g^T \, K \, u$ we sum over all entries k_{ij} which procedure is like the domain integral of the divergence. While the strong formulation $g^T \, f$ in contrast weights the nodal displacements with the vector f. If f has only one non-zero component f_i then this formulation can be very economically.

Note that in (2.190) the divergence is weighted with $g = 1$ while here the gradient ∇u_h is weighted with the gradient ∇G_h.

Remark 2.6 In beam analysis engineers use the zero result of the weak influence function for the force term $M(x) = -EI\, u''(x)$

$$a(G_2, u) = \int_0^l \frac{M_2\, M}{EI}\, dy = 0 \tag{2.196}$$

as a check that the derivative u' of the beam deflection u is continuous at a point x (orthogonality of M_i belonging to the redundant X_i, and M in the force method). If the result is not zero then the deflection curve has a kink at x and so M cannot be the correct bending moment distribution in the beam.

2.4.6 Influence Functions in Higher Dimensions

In higher dimensions the technique to formulate influence functions is basically the same. To exemplify this we consider the boundary value problem

$$-\,\Delta u = p \quad \text{on } \Omega \quad u = 0 \quad \text{on } \Gamma\, . \tag{2.197}$$

The Green's function for $u(\boldsymbol{x})$ is the solution of the boundary value problem

$$-\,\Delta G_0 = \Delta_0(\boldsymbol{y} - \boldsymbol{x}) \quad G_0(\boldsymbol{y}, \boldsymbol{x}) = 0 \quad \boldsymbol{y} \in \Gamma \tag{2.198}$$

and Green's second identity for the Laplace operator reads

$$\mathscr{B}(u, v) = \int_\Omega -\Delta u\, v\, d\Omega + \int_\Gamma \frac{\partial u}{\partial n}\, v\, ds - \int_\Gamma \frac{\partial v}{\partial n}\, u\, ds - \int_\Omega u\,(-\Delta v)\, d\Omega = 0\, . \tag{2.199}$$

It is well known that the Green's function has a logarithmic singularity at the source point $\boldsymbol{x}$

$$G(\boldsymbol{y}, \boldsymbol{x}) = -\frac{1}{2\pi} \ln r + \text{regular terms} \tag{2.200}$$

and so the integral identity must be formulated on the punctured domain, see Fig. 2.7,

$$\Omega_\varepsilon = \Omega - N_\varepsilon(\boldsymbol{x}) \tag{2.201}$$

that is a small circular neighborhood N_ε of the source point $\boldsymbol{x}$ is spared from the domain and then the argument is the following: because Green's second identity is zero for all $\varepsilon > 0$

$$\mathscr{B}(G_0, u)_{\Omega_\varepsilon} = \underbrace{\int_{\Gamma_{N_\varepsilon}} \frac{\partial G_0}{\partial n} u \, ds_{\boldsymbol{y}}}_{\lim_{\varepsilon\to 0} = u(\boldsymbol{x})} - \underbrace{\int_{\Gamma_{N_\varepsilon}} \frac{\partial u}{\partial n} G_0 \, ds_{\boldsymbol{y}}}_{\lim_{\varepsilon\to 0} = 0} - \int_{\Omega_\varepsilon} G_0 \, p \, d\Omega_{\boldsymbol{y}} = 0 \tag{2.202}$$

it must be also zero in the limit, $\varepsilon \to 0$,

$$\lim_{\varepsilon\to 0} \mathscr{B}(u, G_0[\boldsymbol{x}])_{\Omega_\varepsilon} = u(\boldsymbol{x}) - \int_{\Omega} G_0 \, p \, d\Omega_{\boldsymbol{y}} = 0 \tag{2.203}$$

or

$$u(\boldsymbol{x}) = \int_{\Omega} G_0 \, p \, d\Omega_{\boldsymbol{y}} \tag{2.204}$$

which is the influence function for the solution $u(\boldsymbol{x})$.

The derivation of the influence function for the slope in the direction of an arbitrary unit vector $\boldsymbol{m}$

$$\frac{\partial u}{\partial m}(\boldsymbol{x}) = \nabla u(\boldsymbol{x}) \cdot \boldsymbol{m} \tag{2.205}$$

follows the same lines. In this case the Green's function G_1 is the solution of the boundary value problem

$$-\Delta G_1 = \delta_1(\boldsymbol{y} - \boldsymbol{x}) \quad G_1(\boldsymbol{y}, \boldsymbol{x}) = 0 \quad \boldsymbol{y} \in \Gamma \tag{2.206}$$

where the Dirac function has the properties

$$\delta_1(\boldsymbol{y} - \boldsymbol{x}) = 0 \qquad \boldsymbol{y} \neq \boldsymbol{x} \tag{2.207}$$

$$\int_{\Omega} \delta_1(\boldsymbol{y} - \boldsymbol{x}) \, u(\boldsymbol{y}) \, d\Omega_{\boldsymbol{y}} = \frac{\partial u}{\partial m}(\boldsymbol{x}) \qquad \boldsymbol{x} \in \Omega \,. \tag{2.208}$$

Evidently the Green's function is

$$G_1(\boldsymbol{y}, \boldsymbol{x}) = \nabla_{\boldsymbol{x}} G_0(\boldsymbol{y}, \boldsymbol{x}) \cdot \boldsymbol{m} = G_{0,x_1} \, m_1 + G_{0,x_2} \, m_2 \,. \tag{2.209}$$

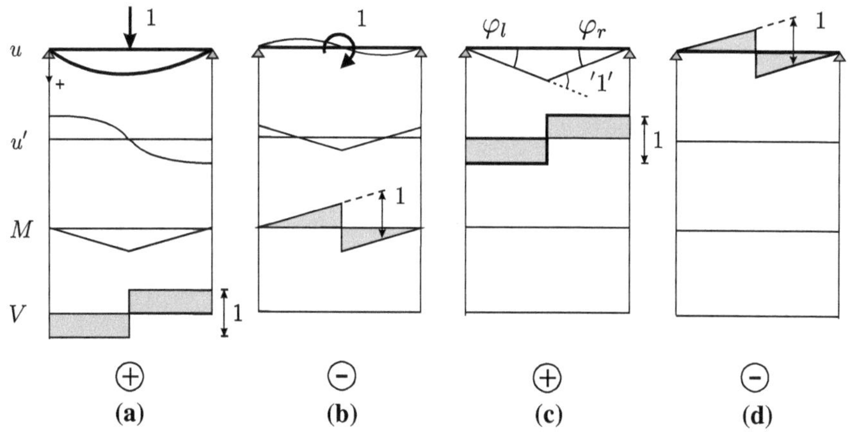

Fig. 2.8 The four singularities of a beam, '1' $\equiv$ tan φ_l + tan φ_r = 1, generate the four influence functions (uppermost row) for **a** u, **b** u', **c** $M = -EI\,u''$ and **d** $V = -EIu'''$ at the center of the beam; + = inf. func integrates, − = inf. func. differentiates

As before the starting point is

$$\mathscr{B}(G_1, u)_{\Omega_\varepsilon} = \underbrace{\int_{\Gamma_{N_\varepsilon}} \frac{\partial G_1}{\partial n}\, u\, ds_y}_{\lim_{\varepsilon\to 0} = \frac{1}{2}\frac{\partial u}{\partial m}(\boldsymbol{x})} - \underbrace{\int_{\Gamma_{N_\varepsilon}} \frac{\partial u}{\partial n}\, G_1\, ds_y}_{\lim_{\varepsilon\to 0} = -\frac{1}{2}\frac{\partial u}{\partial m}(\boldsymbol{x})} - \int_{\Omega_\varepsilon} G_1\, p\, d\Omega_y = 0 \tag{2.210}$$

and after taking the limit

$$\lim_{\varepsilon\to 0} \mathscr{B}(u, G_1[\boldsymbol{x}])_{\Omega_\varepsilon} = \frac{\partial u}{\partial m}(\boldsymbol{x}) - \int_{\Omega} G_1\, p\, d\Omega_y = 0\,, \tag{2.211}$$

we recover

$$\frac{\partial u}{\partial m}(\boldsymbol{x}) = \int_{\Omega} G_1(\boldsymbol{y}, \boldsymbol{x})\, p(\boldsymbol{y})\, d\Omega_y \tag{2.212}$$

which is the influence function for the slope. Basically it is the same logic as in the 1-D case, see Fig. 2.8.

2.4.7 Weak Influence Functions

As in the 1-D case we can derive a weak influence function for $u(\boldsymbol{x})$

$$\lim_{\varepsilon\to 0} \mathscr{G}(G_0, u)_{\Omega_\varepsilon} = u(\boldsymbol{x}) - a(G_0, u) = 0 \tag{2.213}$$

but not for the slope (Fig. 2.8).

2.4.8 Non-Zero Boundary Values

To be complete a word must be said about influence functions for problems where the displacements on the boundary are not zero as in

$$-\Delta u = 0 \quad u = \bar{u} \quad \text{on } \Gamma\,. \tag{2.214}$$

As before Green's second identity is formulated with the solution u and a test function $v \in \mathcal{V}$ ($v = 0$ on Γ)

$$\mathscr{B}(u, v) = \int_\Omega 0 \cdot v\, d\Omega + \int_\Gamma \frac{\partial u}{\partial n} \cdot 0\, ds - \int_\Gamma \bar{u} \frac{\partial v}{\partial n}\, ds - \int_\Omega u(-\Delta v)\, d\Omega = 0 \tag{2.215}$$

and so it follows that if the Green's function solves the boundary value problem

$$-\Delta G(\boldsymbol{y}, \boldsymbol{x}) = \delta(\boldsymbol{y} - \boldsymbol{x}) \quad \text{on } \Omega \quad G(\boldsymbol{y}, \boldsymbol{x}) = 0 \quad \boldsymbol{y} \in \Gamma \tag{2.216}$$

then $\mathscr{B}(u, G) = 0$ is equivalent to

$$u(\boldsymbol{x}) = \int_\Gamma \frac{\partial G(\boldsymbol{y}, \boldsymbol{x})}{\partial n} \bar{u}(\boldsymbol{y})\, ds_{\boldsymbol{y}}\,. \tag{2.217}$$

By weighting the boundary displacements $\bar{u}$ with the slope of the Green's function—which is the term conjugated to the displacement $u = \bar{u}$—the value of $u(\boldsymbol{x})$ can be calculated.

In a boundary value problem such as

$$-\Delta u = p \quad u = \bar{u} \quad \text{on } \Gamma_D \quad \frac{\partial u}{\partial n} = \bar{t} \quad \text{on } \Gamma_N \tag{2.218}$$

the test functions $v \in \mathcal{V}$ must vanish on the part Γ_D (D as in Dirichlet) of $\Gamma = \Gamma_D \cup \Gamma_N$ (N as in Neumann) and so if the Green's function solves the boundary value problem

$$-\Delta G(\boldsymbol{y}, \boldsymbol{x}) = \delta(\boldsymbol{y} - \boldsymbol{x}) \quad G(\boldsymbol{y}, \boldsymbol{x}) = 0 \quad \boldsymbol{y} \in \Gamma_D \tag{2.219}$$

then

$$u(\boldsymbol{x}) = \int_\Omega G(\boldsymbol{y}, \boldsymbol{x})\, p(\boldsymbol{y})\, d\Omega_{\boldsymbol{y}} + \int_{\Gamma_D} \frac{\partial G(\boldsymbol{y}, \boldsymbol{x})}{\partial n} \bar{u}(\boldsymbol{y})\, ds_{\boldsymbol{y}}$$
$$+ \int_{\Gamma_N} G(\boldsymbol{y}, \boldsymbol{x})\, \bar{t}(\boldsymbol{y})\, ds_{\boldsymbol{y}} \tag{2.220}$$

is the influence function. All the input $\{p, \bar{t}, \bar{u}\}$ contributes to $u(\boldsymbol{x})$.

Example 2.5 If a rigid punch indents the surface of the half-space then this is such a boundary value problem with a prescribed inhomogeneous displacement $\bar{u}$ along the Dirichlet part Γ_D of the soil surface, see Fig. 2.9, and so all influence functions must have zero displacements on Γ_D.

By analogy with (2.220) it follows that the influence function for the stress σ_{yy} at a point $\boldsymbol{x}$ inside the half-space must be

$$\sigma_{yy}(\boldsymbol{x}) = \bar{u} \int_{\Gamma_D} t(\boldsymbol{y}, \boldsymbol{x})\, ds_{\boldsymbol{y}} \,. \tag{2.221}$$

What in the Laplace equation is the slope $\partial G/\partial n$ are here the tractions $t(\boldsymbol{y}, \boldsymbol{x})$ (vertical stress) directly under the rigid punch due to a unit dislocation in vertical direction at the source point $\boldsymbol{x}$ (= influence function for σ_{yy}). We know from experience that the stress σ_{yy} at the ends of the punch is singular and so we conclude that the integral of the tractions $t(\boldsymbol{y}, \boldsymbol{x})$ must be unbounded and not measurable when the dislocation is applied directly under the edge of the punch.

Example 2.6 In 1-D problems the boundary consists of points. Supports are point supports. If a support settles by $\bar{u}$ units the influence $J(u)$ on any observable quantity is

$$J(u)(x) = R(x)\, \bar{u} \tag{2.222}$$

where $R(x)$ is the support reaction in the direction of the displacement $\bar{u}$ due to the action of the Dirac delta associated with the functional $J(u)(x)$.

2.4.9 Average Values of Stresses

The integration by parts formula

$$\int_0^l u'\, v\, dx = [u\, v]_0^l - \int_0^l u\, v'\, dx \tag{2.223}$$

is also a statement about the average value of a derivative u'

$$\frac{1}{l} \int_0^l u'\, dx = \frac{1}{l}[u \cdot 1]_0^l = \frac{1}{l}\,(u(l) - u(0)) \tag{2.224}$$

which is just the slope between the end points.

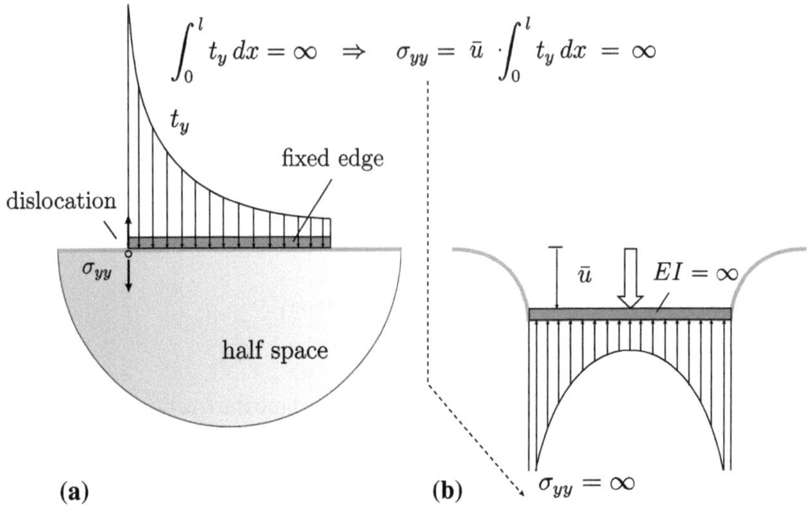

Fig. 2.9 Why the stress under a corner point of a rigid punch becomes infinite, **a** the influence function for σ_{yy} is generated by a unit dislocation while the punch is kept fixed and **b** the sum of the forces needed to hold the punch fixed times the indentation $\bar{u}$ of the punch is the stress σ_{yy} at the edge of the punch

In the same sense the average value of the stress

$$\sigma_{xx} = E\,(\varepsilon_{xx} + \nu\,\varepsilon_{yy}) = E\,(u_{x,x} + \nu\,u_{y,y}) \tag{2.225}$$

in a plate element Ω_e of size $|\Omega_e|$ can be expressed by a boundary integral

$$\frac{1}{|\Omega_e|}\int_{\Omega_e} E\,(u_{x,x} + \nu\,u_{y,y})\,d\Omega = \frac{E}{|\Omega_e|}\int_{\Gamma_e}(u_x\,n_x + \nu\,u_y\,n_y)\,ds\,. \tag{2.226}$$

The influence function for the boundary integral is the displacement field when line forces $E/|\Omega_e|\,n_x$ and $E\,\nu/|\Omega_e|\,n_y$ respectively pull on the edge Γ_e of the element, see Fig. 2.10 [1].

If the element Ω_e is embedded into a rather soft matrix, $E_M \ll 1$, then the forces can stretch the element unhindered, the influence function will extend quiet far, and so much of the load will flow through the element. But when the matrix is very stiff, $E_M \gg 1$, and the element in contrast is very soft, $E \ll 1$, then the influence function will almost be zero and so hardly any stresses will flow through Ω_e.

There exists a close connection between these influence functions and the hydrostatic pressure distribution in a plate. The edge forces which generate the influence function for the average value of $\sigma_{xx} + \sigma_{yy}$

$$t_x = \frac{E}{|\Omega_e|\,(1+\nu)}\,n_x \quad t_y = \frac{E}{|\Omega_e|\,(1+\nu)}\,n_y \tag{2.227}$$

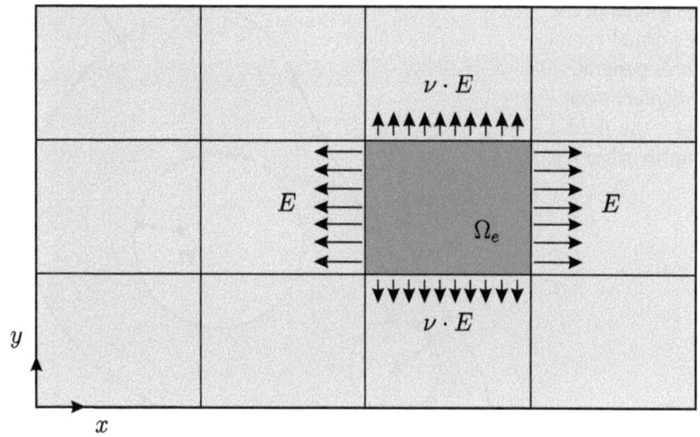

Fig. 2.10 These forces generate the influence function for the average value of σ_{xx} in the element Ω_e

are essentially the same forces a plate which is submerged in a fluid experiences along its edge

$$t_x = p\, n_x \quad t_y = p\, n_y \quad p = \text{water pressure}\,. \tag{2.228}$$

The stress tensor in the plate, $\boldsymbol{S} = p\,\boldsymbol{I}$, is a diagonal matrix and so the core displacement field is simply $u_x = \alpha\, x$, $u_y = \alpha\, y$, $\alpha = p/(E + \nu)$, see Fig. 2.11. This is also the influence function for the average value of $\sigma_{xx} + \sigma_{yy}$ when the factor $E/|\Omega_e|(1+\nu)$ replaces p.

Core displacement field means that an arbitrary vector $\boldsymbol{u}_0$ can be added to the solution. This has no effect on the average value of the stresses because if the plate is freely floating, if it is untethered, then the applied load must satisfy the equilibrium condition

$$\int_\Omega p \cdot \boldsymbol{r}\, d\Omega + \int_\Gamma \boldsymbol{t} \cdot \boldsymbol{r}\, ds = 0 \tag{2.229}$$

where $\boldsymbol{r} = \boldsymbol{a} + \boldsymbol{x} \times \boldsymbol{\omega}$ is a rigid body motion of the plate and this property guarantees that the effect of $\boldsymbol{u}_0$ is nil when the influence function is evaluated.

Remark 2.7 The fluid pressure is a pure Neumann problem for the submerged plate but because the integral of the normal vector is zero along any closed contour

$$\int_\Gamma n_x\, ds = 0 \quad \int_\Gamma n_y\, ds = 0 \tag{2.230}$$

the problem is well posed, are the equilibrium conditions satisfied.

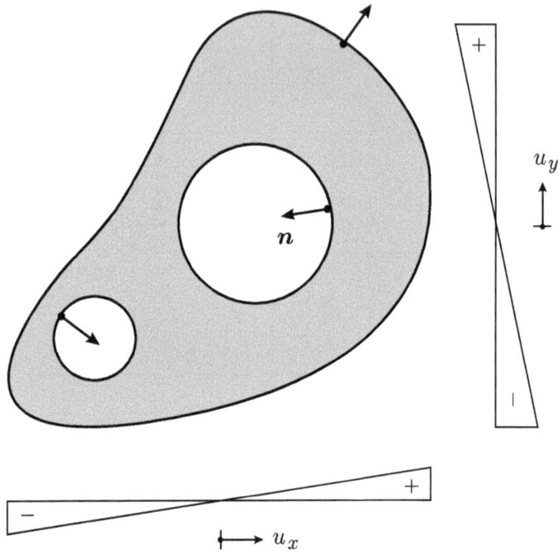

Fig. 2.11 Edge forces in the direction of the normal vector, as water pressure, generate a very simple displacement field in any plate; the field is determined up to arbitrary constants

Remark 2.8 The average values of the stresses (derivatives) must be zero if the edge of a plate is fixed all around because the forces $E/|\Omega_e|\, n_x$ and $E/|\Omega_e|\, n_y$ respectively applied to a fixed edge will effect nothing.

Remark 2.9 In beams the bending stresses σ_x typically exhibit an antisymmetric linear distribution in vertical direction (y) so that the average value of these stresses over the length of the beam (= plate)

$$\int_0^l \int_0^h \sigma_x(y, x)\, dy\, dx = 0 \tag{2.231}$$

is zero. This is easily understood by looking at the forces which generate the influence function for this integral, see Fig. 2.12. If $\nu = 0$ then only lateral loads E will act on the beam but because of $\nu = 0$ they will have no effects on the horizontal edges. If $\nu \neq 0$ then the stretching of the beam by the lateral forces $\pm E$ translates into a constriction of the beam but this is compensated by an opposite displacement produced by the vertical forces $\pm\nu\, E$ so that again the displacement of the horizontal edges is zero.

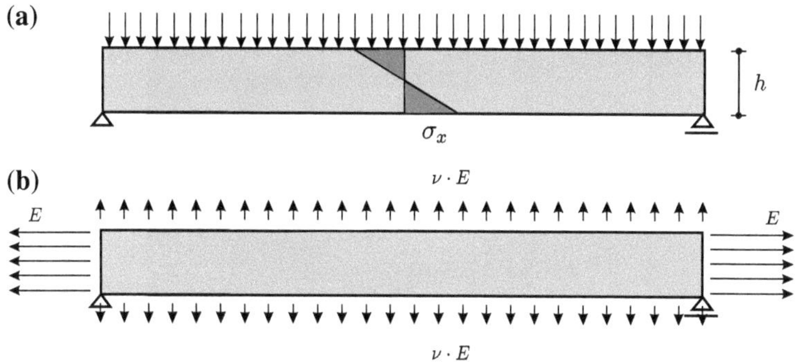

Fig. 2.12 Beam **a** bending stresses σ_x, **b** these forces generate the influence function for the average value of σ_x in the beam

2.5 Properties of Green's Functions

What is a point load mathematically, how do we define a Dirac delta? The engineer may help who defines point loads via the equilibrium conditions.

A point load $P = 1$ placed on a rope at a point x makes that the shear force V jumps by one unit

$$\lim_{\varepsilon \to 0} [V(x + 0.5\,\varepsilon) - V(x - 0.5\,\varepsilon)] = 1\,. \tag{2.232}$$

When the same force is placed on a membrane, see Fig. 2.13, then the *integral* of the slope $\partial u/\partial n$ over concentric circles Γ_{N_ε} must be 1 in the limit

$$\lim_{\varepsilon \to 0} \int_{\Gamma_{N_\varepsilon}} \frac{\partial u}{\partial n}(\boldsymbol{y})\, ds_y = \lim_{\varepsilon \to 0} \int_{\Gamma_{N_\varepsilon}} \frac{\partial u}{\partial n}(\boldsymbol{x} + \varepsilon\, \nabla r)\, ds_y = 1 \tag{2.233}$$

where

$$\nabla r = \nabla(\boldsymbol{y} - \boldsymbol{x}) = \begin{bmatrix} \cos\,\varphi \\ \sin\,\varphi \end{bmatrix} \tag{2.234}$$

is a "pointer" of length one which—like a compass needle—shows the direction to the points $\boldsymbol{y} = (\varepsilon, \varphi)$ (polar coordinates) on the edge Γ_ε.

Because the size of Γ_ε shrinks like $2\,\pi\,\varepsilon$ the slope of u must counterbalance this tendency by increasing as ε^{-1}. This is indeed the case: the displacement of a membrane near a unit point load has the form

$$u = -\frac{1}{2\,\pi}\,\ln\, r + \text{regular terms} \tag{2.235}$$

and the slope of the singular term on a concentric circle Γ_r with radius r is

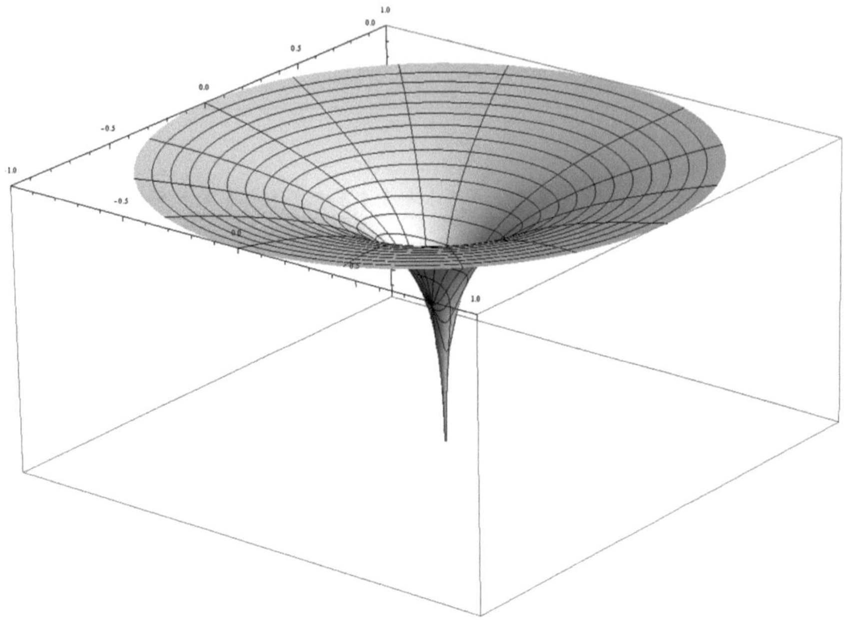

Fig. 2.13 A point load acting at the center of a circular membrane, Poisson equation

$$-\frac{1}{2\,\pi}\frac{\partial \ln r}{\partial n} = -\frac{1}{r}\nabla r \cdot \boldsymbol{n} = \frac{1}{r}\nabla r \cdot \nabla r = \frac{1}{2\,\pi\,r} \cdot 1 \qquad (2.236)$$

Note that the outward normal on the edge Γ_ε points towards the center $\boldsymbol{x}$, that is it has the opposite direction of ∇r, and this makes that $\nabla r \cdot \boldsymbol{n} = -\nabla r \cdot \nabla r$.

So by taking the limit of the singular term the point load at the center of the shrinking circle is recovered, see Fig. 2.14a,

$$\lim_{\varepsilon \to 0} \int_{\Gamma_{N_\varepsilon}} \frac{\partial u}{\partial n}(\mathbf{y})\, ds_{\mathbf{y}} = \lim_{\varepsilon \to 0} \int_0^{2\,\pi} \frac{1}{2\,\pi\,\varepsilon}\,\varepsilon\, d\varphi = 1\,. \qquad (2.237)$$

The contribution of the regular terms to this integral is zero because a bounded function integrated over a shrinking circle gives zero in the limit.

This is the engineering approach to Dirac deltas. If there is a "black hole" then the neighborhood of the black hole must show traces of its presence. When you circle the singularity once then you get a lift of one unit in the relevant quantity, the integral of the vertical forces or the integral of the deflection.

Note that the dimension n of the space, 2-D or 3-D, determines the strength $O(r^{-n})$ of the singularity. In 2-D the circle that surrounds the point load is of size $2\,\pi\,r$ and in 3-D it is a sphere of size $4\,\pi\,r^2$. This dimension argument is the reason

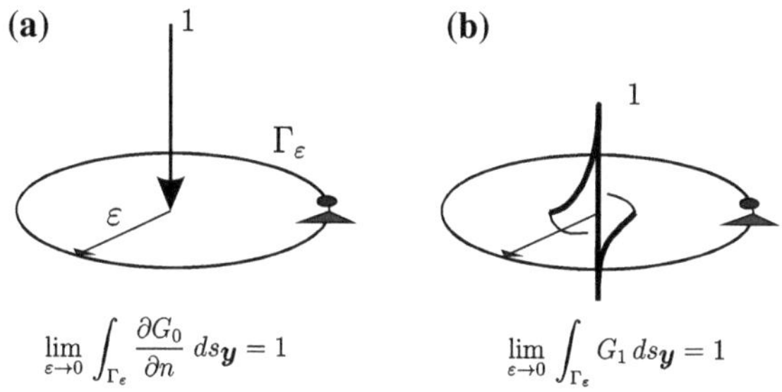

Fig. 2.14 A point load, solution G_0, and a dislocation, solution G_1

why the gravitational forces or electrostatic forces near a point mass or a point charge have the same order r^{-2}.

To summarize the main properties of Green's function let $\partial_i u,\ i = 0, 1, 2, 3$ be the boundary values of an abstract fourth-order equation $L\,u = p$ as for example of a Kirchhoff plate

$$\partial_0 u = u \quad \partial_1 u = \nabla u \cdot \boldsymbol{n} \quad \partial_2 u = m_n(u) \quad \partial_3 u = v_n(u)\,. \tag{2.238}$$

What is needed for an influence function of $\partial_1 u(\boldsymbol{x})$ (e.g.) is

- an integral identity such as $\mathscr{B}(u, v)$ which contains a boundary integral

$$\int_\Gamma \partial_1 u\, \partial_2 v\, ds \quad 1 + 2 = 2\,m - 1 = 3 \tag{2.239}$$

  with $\partial_1 u$ and the term conjugated to it, $\partial_2 v$,
- and a homogeneous solution $G(\boldsymbol{y}, \boldsymbol{x})$ of the governing equation with the property

$$\lim_{\varepsilon\to 0}\int_{\Gamma_{N_\varepsilon}} \partial_2 G(\boldsymbol{y}, \boldsymbol{x})\, ds_{\boldsymbol{y}} = 1 \tag{2.240}$$

  because this provides in the limit the Dirac delta effect

$$\partial_1 u(\boldsymbol{x}) = \lim_{\varepsilon\to 0}\int_{\Gamma_{N_\varepsilon}} \partial_2 G(\boldsymbol{y}, \boldsymbol{x})\, \partial_1 u(\boldsymbol{y})\, ds_{\boldsymbol{y}}\,. \tag{2.241}$$

Second order differential equations possess two Green's functions

$$G_0(\boldsymbol{y}, \boldsymbol{x}) = \text{single force} \quad G_1(\boldsymbol{y}, \boldsymbol{x}) = \text{dislocation} \tag{2.242}$$

and fourth order equations four such functions,

$$G_0(\boldsymbol{y}, \boldsymbol{x}) = \text{single force} \qquad G_1(\boldsymbol{y}, \boldsymbol{x}) = \text{moment} \tag{2.243}$$

$$G_2(\boldsymbol{y}, \boldsymbol{x}) = \text{twist} \qquad G_3(\boldsymbol{y}, \boldsymbol{x}) = \text{dislocation}\,. \tag{2.244}$$

2.5.1 Modern Approach

What we presented here was the classical theory. In FE-analysis things are much simpler, we must not be concerned with limits over shrinking circles and alike, we have more freedom and more room to invent a nearly infinite variety of Green's functions. It is only required that the functional $J(u)$ is linear. The nodal values g_i of the Green's function

$$G_h(\boldsymbol{y}, \boldsymbol{x}) = \sum_i g_i(\boldsymbol{x})\, \varphi_i(\boldsymbol{y}) \tag{2.245}$$

are the solution of the system $\boldsymbol{K}\,\boldsymbol{g} = \boldsymbol{j}$ where $j_i = J(\varphi_i)$ and once we have solved this system we can map p onto the value $J(u_h)$ of the FE-solution with ease

$$J(u_h) = \int_\Omega G_h(\boldsymbol{y}, \boldsymbol{x})\, p(\boldsymbol{y})\, d\Omega_{\boldsymbol{y}} = \boldsymbol{g}^T\, \boldsymbol{f}\,. \tag{2.246}$$

2.5.2 Maxwell

If a unit point load applied at one point $\boldsymbol{x}_1$ of an elastic structure results in a given deflection at another point $\boldsymbol{x}_2$, then the same load applied at $\boldsymbol{x}_2$ will result in the same deflection at $\boldsymbol{x}_1$.

This is Maxwell's theorem and the underlying principle[4]

$$\int_0^l G_0(y, x_2)\, \delta_0(y - x_1)\, dy = \int_0^l G_0(y, x_1)\, \delta_0(y - x_2)\, dy \tag{2.247}$$

can be extended to any pair $\{i, j\}$ of Green's functions $G_i(y, x_1)$ and $G_j(y, x_2)$

$$\int_0^l G_i(y, x_2)\, \delta_j(y - x_1)\, dy = \int_0^l G_j(y, x_1)\, \delta_i(y - x_2)\, dy\,. \tag{2.248}$$

[4] The indices on the Green's functions and the Dirac deltas are to distinguish the different functions, see Sect. 2.4.1.

Say $G_1(y, x_2)$ is the influence function for the shear force $V = H\,u'(x)$ in a rope at the point x_2

$$-H\frac{d^2}{dy^2}G_1(y, x_2) = \delta_1(y - x_2) \quad G_1(0, x_2) = G_1(l, x_2) = 0 \tag{2.249}$$

and $G_0(y, x_1)$ is the influence function for the deflection $u(x)$ of the rope at another point x_1

$$-H\frac{d^2}{dy^2}G_0(y, x_1) = \delta_0(y - x_1) \quad G_0(0, x_1) = G_0(l, x_1) = 0 \tag{2.250}$$

then

$$\int_0^l G_1(y, x_2)\,\delta_0(y - x_1)\,dy = \int_0^l G_0(y, x_1)\,\delta_1(y - x_2)\,dy \tag{2.251}$$

or

$$G_1(x_1, x_2) = V(G_0(x_2, x_1)) \tag{2.252}$$

that is the magnitude of G_1 at x_1 is the same as the magnitude of the shear force belonging to G_0 at the point x_2.

Formally this allows to derive any influence function $G_i(y, \xi)$ from $G_0(y, x)$ by the maneuver

$$\int_0^l G_0(y, x)\,\delta_i(y - \xi)\,dy = \int_0^l G_i(y, \xi)\,\delta_0(y - x)\,dy = G_i(x, \xi)\,. \tag{2.253}$$

But this is a rather cumbersome approach. To plot the influence function for the shear force $G_1(y, x)$ in a rope you would subdivide the rope into ten equally spaced points x_i and place first a unit force $P = 1$ at the point x_1, calculate the shear force V at x, then place the force at the point x_2, again calculate the shear force at x, etc. A linear plot of all these values would be a—for engineering purposes often good enough—first approximation of the influence function.

2.5.3 Modes of Decay

Generating an influence function can be compared to throwing a stone into a water basin and watching the waves slowly ebb away. If the basin is an intricate system of ponds and channels then the energy transport will be influenced by the width of the channels and the curvature of the channels.

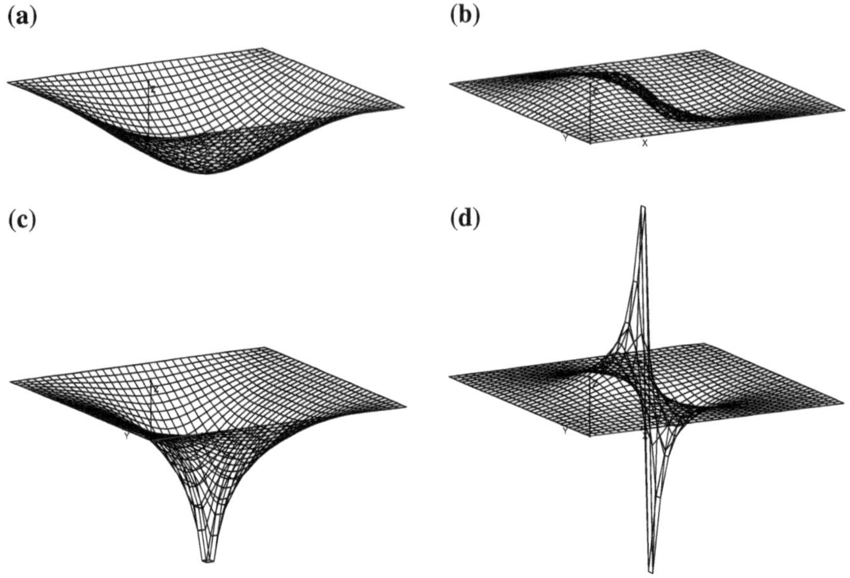

Fig. 2.15 Influence functions on a slab ($K\,\Delta\Delta w = p$), **a** deflection, **b** rotation, **c** moment m_{xx}, **d** shear force q_x

But the mode of decay, s. Fig. 2.15, primarily depends on the character of the influence function, and this in turn depends on the function value (here of a beam)

$$u(x) \quad u'(x) \quad u''(x) \quad u'''(x) \tag{2.254}$$

which the influence function extracts from the right-hand side, the fourth order derivative, $EIu^{IV} = p$,

$$\frac{d^i u(x)}{dx^i} = \int_0^l G_i(y, x)\, p(y)\, dy \quad i = 0, 1, 2, 3\,. \tag{2.255}$$

That is an influence function can be considered an *integral operator*. The more a kernel $G_i(y, x)$ integrates the more "volume" it has, the more it extends in all directions and the slower the rate of its decay, s. Fig. 2.15.

The kernel $G_0(y, x)$ maps p onto $u(x)$, it integrates four times! The opposite is the Dirac delta δ_0 which is a so-called *reproducing kernel*. It neither integrates nor differentiates; it has zero spread. It measures at one point only and the output is equal to the input

$$\int_0^l \delta_0(y, x)\, p(y)\, dy = p(x)\,. \tag{2.256}$$

In the list of the Green's functions of the beam, see (2.255), the Dirac delta would be the fifth kernel, $G_4 \equiv \delta_0$ which maps p onto itself—no gain.

2.5.4 Dipoles and Monopoles

Consider two opposite electric point charges a distance Δx apart and with a strength $\pm 1/\Delta x$ which is inversely proportional to their distance Δx. If the distance between the two charges shrinks to zero, $\Delta x \to 0$, they become a *dipole*.

In solid mechanics point charges are point loads $\pm 1/\Delta x$ and while an electric field needs no carrier, is "invisible" (to our limited senses), a dipole in a solid makes its presence felt by a gap, a unit dislocation.[5] Such a dislocation generates the influence function for a stress in a solid or—in the one-dimensional case—the normal force N in a bar.

Of a somewhat different type is the influence function for a displacement because it is generated by a point load $P = 1$, a *monopole*. Influence functions which are generated by monopoles sum, they add, while influence functions generated by dipoles measure differences, they are sensitive to imbalances.[6] Each of the influence functions in Fig. 2.15 belongs to either one of these two types:

- G.F. for displacements and bending moments *sum*.
- G.F. for rotations, stresses and shear forces *differentiate*

A monopole generates a dent, a pit. Anything that falls into the pit makes that the displacement increases. The pit sums, it integrates. A dipole instead represents a shear deformation and these two opposite movements differentiate.

The influence function for the slope u' in a beam comes from a dipole, a couple $M = 1$, which is the limit of two opposite point forces

$$M = \lim_{\Delta x \to 0} \frac{1}{\Delta x} \Delta x = 1\,. \tag{2.257}$$

If the resulting shear deformation is perfectly antisymmetric as it is in the middle of a hinged beam and if the load happens to be symmetric then the slope is zero; no need for the cross-section of the beam to rotate to counterbalance the effect of uneven forces. Also the influence function for a shear force V is of dipole-type, is a *"high-pass filter"* while the influence function for the bending moment M is of monopole-type is a *"low-pass filter"*. The latter influence function is generated by two moments $M = \pm 1/\Delta x$ which rotate inward—towards the bent—and so generate a symmetric deflection but with a sharp bent, a jump in the first derivative, at the source point.

[5] You see a gap only in 1-D. In 2-D you can only sense it if you circle the gap once. Then you will experience a displacement shift in the direction of the dislocation.

[6] We restrain at this point from introducing quadrupoles or even octupoles [5].

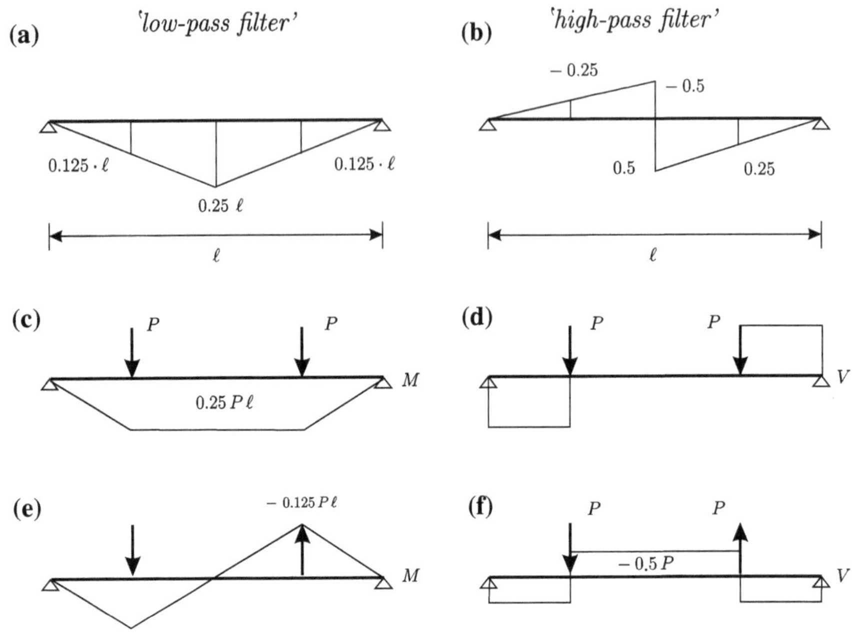

Fig. 2.16 Influence functions (uppermost row) for **a** the bending moment and **b** the shear force at the center of the beam, **c** and **d** bending moments and shear forces of a symmetric load and of an antisymmetric load, (**e**) and (**f**)

Now it is clear what happens: the maximum effect is observed if the loading and the influence are of the same type (symmetric—symmetric or antisymmetric—antisymmetric) and the minimum effect if the two have opposite "signs" are of opposite type, s. Fig. 2.16.

The difference between monopoles and dipoles also explains why displacements and bending moments are easier to approximate than stresses and shear forces. It is the difference between numerical integration and numerical differentiation, see Fig. 2.17 and also Figs. 2.18 and 2.19.

Remark 2.10 All influence functions for support reactions integrate though the support reactions are normal forces (stresses) or shear forces respectively and so we would expect that the influence functions differentiate. But at a fixed support one half of the shear deformation (= influence function) is hindered by the foundation. So with one half-wave being zero and the other half-wave making a full swing to produce the required dislocation $[[u]] = 1$ the influence function turns into an one-sided integration formula.

Remark 2.11 Not all influence functions decay! If parts of the released structure (released = after an N-, V- or M-hinge is built in) can perform rigid body movements then the opposite may be true, s. Fig. 2.20b.

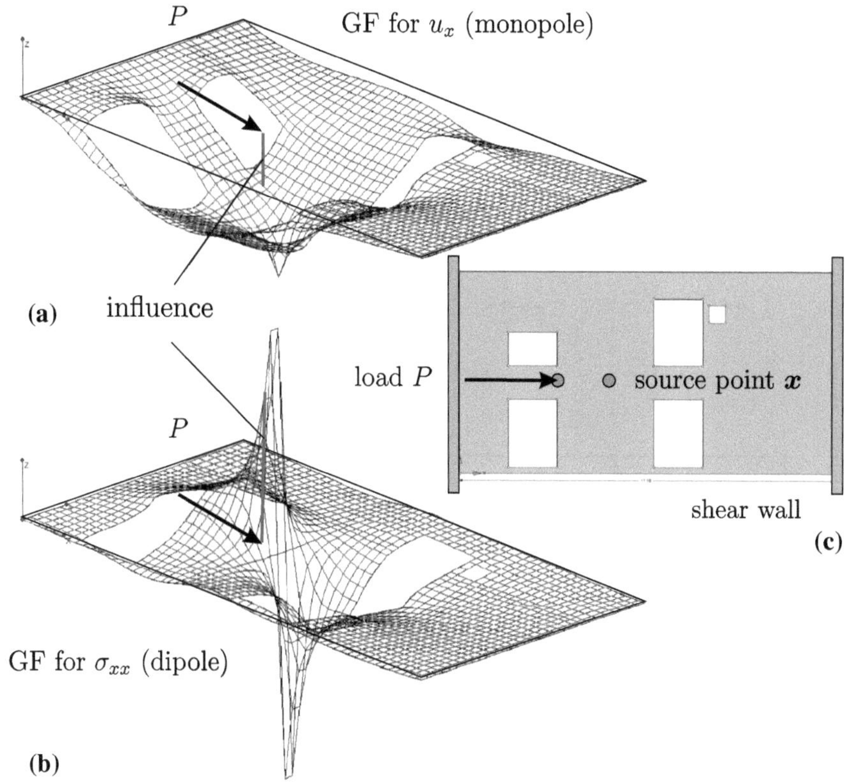

Fig. 2.17 Monopoles and dipoles, **a** influence function for the horizonal displacement and **b** the stress σ_{xx} in a shear wall **c** displayed are only the horizontal components of the two influence functions

2.5.5 Multipole Expansion

To be complete we mention the *multipole expansion* of influence functions whereby large complex problems in acoustics or in electromagnetic and gravitational field theory can be solved in reasonable time.

The gravitational potential generated by a body Ω with mass density $\rho(\boldsymbol{x})$ is the integral

$$u(\boldsymbol{x}) = -G \int_{\Omega} \frac{1}{r}\, \rho(\boldsymbol{y})\, d\Omega_{\boldsymbol{y}} \quad r = |\boldsymbol{y} - \boldsymbol{x}| \tag{2.258}$$

where the kernel $1/r$ is the free-space Green's function (fundamental solution) of the Laplacian

$$-\Delta \frac{1}{r} = 4\,\pi\,\delta(\boldsymbol{y} - \boldsymbol{x}) \tag{2.259}$$

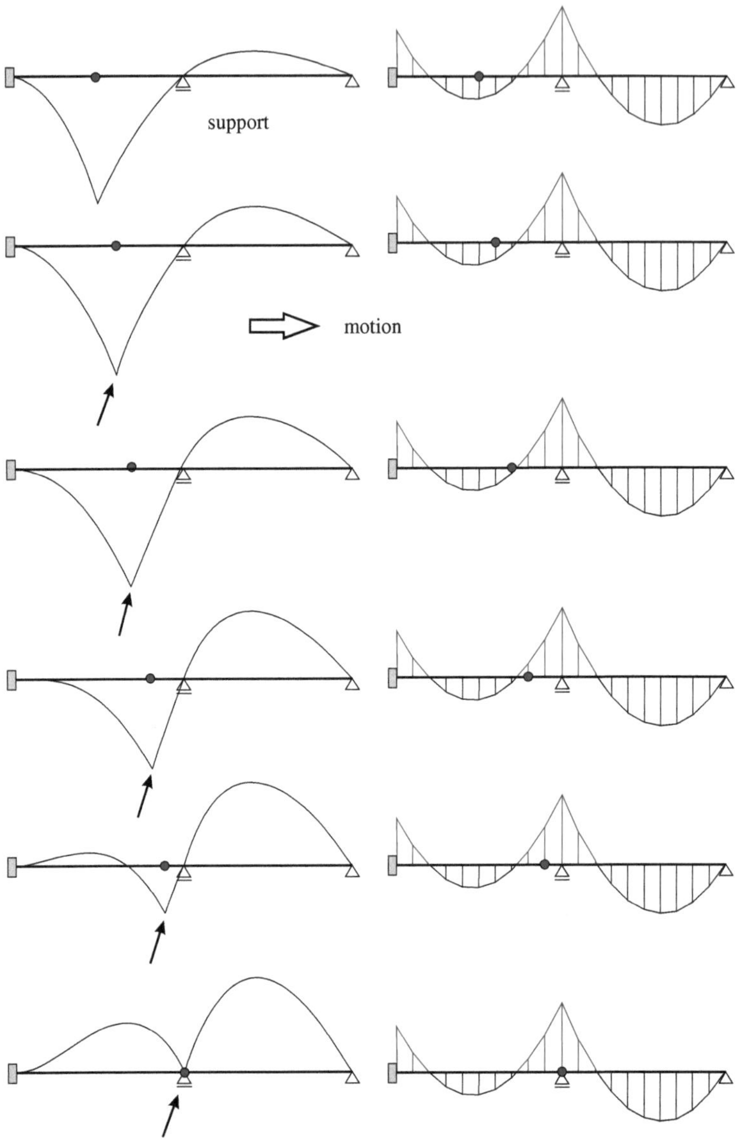

Fig. 2.18 How the influence functions for the bending moment $M = -EI\,u''$ wanders along the beam and keeps its shape

and G is the gravitational constant.[7]

By doing a Taylor expansion of the kernel

[7] Actually the complete Green's function is $1/(4\,\pi\,r)$. But in the literature the potential is always given in the form (2.258), that is the $1/4\,\pi$ must be contained in the constant G.

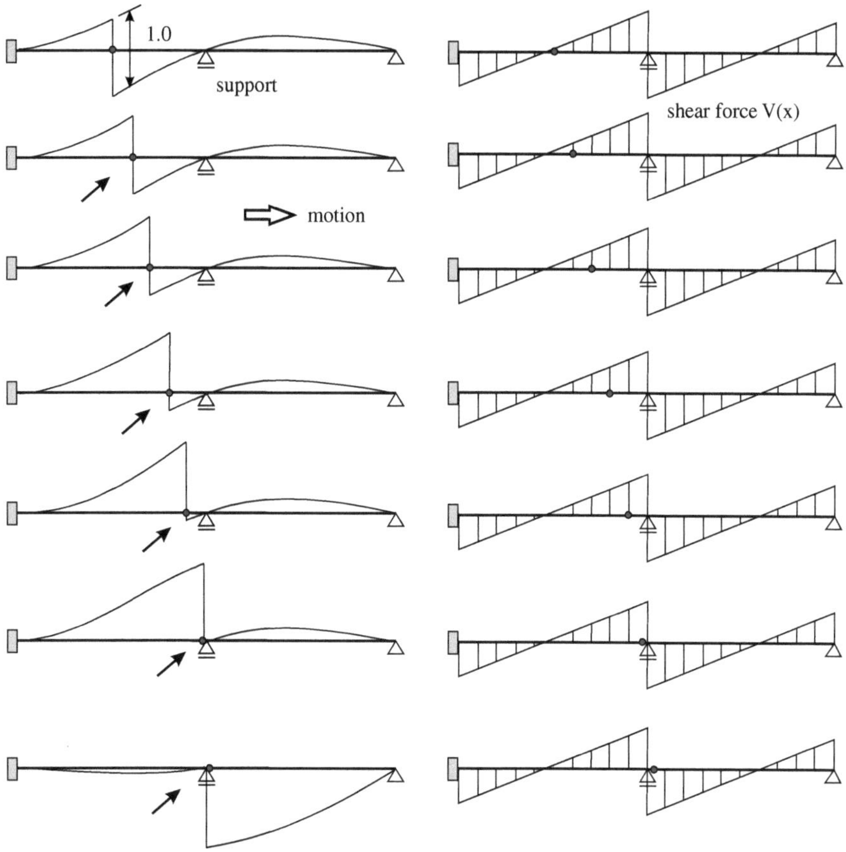

Fig. 2.19 How the influence function for the shear force $V = -EI\,u'''$ wanders along the beam but essentially stays the same. Note that near the fixed support the two-sided dislocation becomes a one-sided swing and so produces the maximal effect

$$\frac{1}{|\boldsymbol{y}-\boldsymbol{x}|} = \frac{1}{r}\left[1 + \frac{1}{r}\,\boldsymbol{e}_x \cdot \boldsymbol{y} + \frac{1}{2\,r^2}(3\,(\boldsymbol{e}_x \cdot \boldsymbol{y})^2 - |\boldsymbol{y}|^2) + O\left(\frac{|\boldsymbol{y}|}{|\boldsymbol{x}|}\right)^3\right], \quad (2.260)$$

where the unit vector $\boldsymbol{e}_x$ signals the direction to $\boldsymbol{x} = r\,\boldsymbol{e}_x$, the potential can be written as

$$u(\boldsymbol{x}) = u_{mon}(\boldsymbol{x}) + u_{dip}(\boldsymbol{x}) + u_{quad}(\boldsymbol{x}) + \ldots \quad (2.261)$$

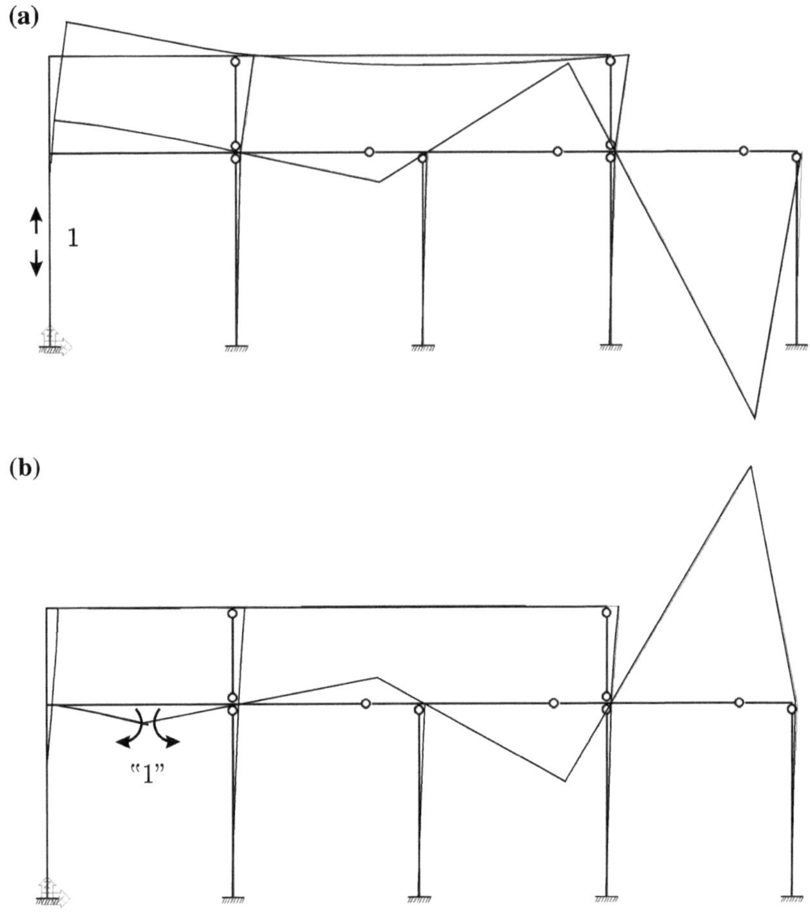

Fig. 2.20 **a** Influence function for the normal force N and **b** the bending moment M in a frame. Not all influence functions decay!

The potentials

$$u_{mon}(\boldsymbol{x}) = -\frac{G}{r} \int_{\Omega} \rho(\boldsymbol{y})\, d\Omega \tag{2.262}$$

$$u_{dip}(\boldsymbol{x}) = -\frac{G}{r^2} \int_{\Omega} \rho(\boldsymbol{y})\, \boldsymbol{e}_x \cdot \boldsymbol{y}\, d\Omega \tag{2.263}$$

$$u_{quad}(\boldsymbol{x}) = -\frac{G}{2\,r^3} \int_{\Omega} \rho(\boldsymbol{y})(3\,(\boldsymbol{e}_x \cdot \boldsymbol{y})^2 - |\boldsymbol{y}|^2)\, d\Omega \tag{2.264}$$

represent monopoles (+), dipoles (+−) and quadrupoles (+ − +−) respectively. If the Earth were a perfect sphere with a uniform density then its higher potentials would be zero, $u(\boldsymbol{x}) = u_{mon}(\boldsymbol{x})$. This expansion is often coupled with a multi-level clustering of the boundary elements, that is the cells or panels into which the surface of a vibrating machine is subdivided, and so the two techniques combined allow to solve exterior problems at "quasi-linear" costs.

2.5.6 *Infinite Energy*

The strain energy product $a(u, v)$ of two functions is an integral. On the diagonal, $v = u$, it is, up to the factor $1/2$, the internal energy

$$W_i = \frac{1}{2}\, a(u, u) \tag{2.265}$$

of the function u. If the integral does not exist, if the integral is infinite, it is said that the function has infinite energy. Most Green's functions have infinite energy.

Apply for example a point load $P = 1$ at the center, $\boldsymbol{x} = 0$, of a circular membrane with radius $R = 1$, see Fig. 2.13,

$$-\Delta G = P \cdot \delta(\boldsymbol{y} - \boldsymbol{0}) \quad G(\boldsymbol{y}, \boldsymbol{0}) = 0 \quad \boldsymbol{y} \in \Gamma\,. \tag{2.266}$$

The solution to this problem, the Green's function for $u(\boldsymbol{0})$

$$G(\boldsymbol{y}, \boldsymbol{0}) = -P\,\frac{1}{2\pi}\ln r \tag{2.267}$$

has a singularity at $\boldsymbol{y} = \boldsymbol{0}$. This means that the point force once placed on the center of the membrane will sink and sink and not stop before it has reached the point ∞, so that the exterior work done by the load is infinite

$$W_e = \int_\Omega -\Delta G \cdot G\, d\Omega_{\boldsymbol{y}} = P \int_\Omega \delta(\boldsymbol{y} - \boldsymbol{0})\, G(\boldsymbol{y}, \boldsymbol{0})\, d\Omega_{\boldsymbol{y}} = P \cdot \infty\,. \tag{2.268}$$

According to the principle of conservation of energy $W_e = W_i$ which is identical with Green's first identity

$$\mathscr{G}(u, u) = \underbrace{\int_\Omega -\Delta u\, u\, d\Omega + \int_\Gamma \frac{\partial u}{\partial n}\, u\, ds}_{2\,W_e} - \underbrace{a(u, u)}_{2\,W_i} = 0\,, \tag{2.269}$$

the internal energy ($d\Omega = r\, dr\, d\varphi$) must be infinite as well, which indeed it is

$$a(G,G) = \int_\Omega \nabla G \cdot \nabla G \, d\Omega = \frac{1}{2\pi} \int_\Omega \frac{1}{r^2}(r,_x^2 + r,_y^2)\, d\Omega$$
$$= \frac{1}{2\pi} \int_0^1 \frac{1}{r}\, dr \int_0^{2\pi} (\cos^2 \varphi + \sin^2 \varphi)\, d\varphi = \int_0^1 \frac{1}{r}\, dr = \infty\,, \tag{2.270}$$

and so we have in the end the—fitting—result

$$\mathscr{G}(G,G) = \infty - \infty = 0\,. \tag{2.271}$$

But when a guitar string is plugged with a finger

$$-\,H\,G'' = \delta(y-x) \quad G(0,x) = G(l,x) = 0 \tag{2.272}$$

which produces a triangular shape G, then the internal energy of the string is bounded

$$a(G,G) = \int_0^l H\,(G')^2\, dy < \infty \tag{2.273}$$

because G' is piecewise constant and finite.

The important point is that if the energy of the Green's functions G is bounded then the functional $J(u)$ which is associated with a Green's function

$$J(u) = \int_\Omega G(\boldsymbol{y}, \boldsymbol{x})\, p(\boldsymbol{y})\, d\Omega_{\boldsymbol{y}} \tag{2.274}$$

is also bounded and vice versa.

In the case of linear functionals bounded and continuous is the same. A functional on a space $\mathcal{V}$ is continuous if there exists a constant c, which is independent of the single u, such that for all u

$$|J(u)| < c\, \|u\|\,. \tag{2.275}$$

2.5.7 Genealogy of Influence Functions

Theoretically it suffices to know the Green's function of the functional

$$J(u) = u(x) = \int_0^l G(y,x)\, p(y)\, dy \tag{2.276}$$

because the Green's function $\tilde{G}(y, x)$ of any other linear functional $\tilde{J}(u)$ is simply

$$\tilde{G} = \tilde{J}(G) \tag{2.277}$$

as follows by direct substitution

$$\tilde{J}(u) = \int_0^l \tilde{J}(G)(y, x)\, p(y)\, dy = \int_0^l \tilde{G}(y, x)\, p(y)\, dy\,. \tag{2.278}$$

Though one must be careful because applying a functional $\tilde{J}$ under the integral sign is not a trivial task. There are some rules which apply to such maneuvers.

2.6 Sobolev's Embedding Theorem

The question whether a Green's functions has finite energy is answered by the following theorem.

Theorem 2.4 (Sobolev's Embedding Theorem) *Assume Ω is a bounded domain of $\mathbb{R}^n$ with a piecewise smooth edge (as in engineering applications). For $2(m-i) > n$ we have the inclusion*

$$C^i(\bar{\Omega}) \subset H^{i+m}(\Omega) \tag{2.279}$$

and there exist constants $c_i < \infty$ such that for all $u \in H^{i+m}(\Omega)$ the following estimate holds

$$\|u\|_{C^i(\bar{\Omega})} \leq c_i\, \|u\|_{H^{i+m}(\Omega)}\,. \tag{2.280}$$

The norm of a function u

$$\|u\|_{C^i(\bar{\Omega})} := \max_{0 \leq |j| \leq i} \left| \frac{\partial^{|j|} u(\boldsymbol{x})}{\partial x^j} \right| \tag{2.281}$$

is the maximum value of $|u|$ and its derivatives up to the order i on $\bar{\Omega}$.

This theorem implies that the strain energy induced by the point load is bounded and the conjugated quantity (the effect the point load produces) is finite and continuous if the three indices satisfy the inequality [6],

$$m - i > \frac{n}{2}\,. \tag{2.282}$$

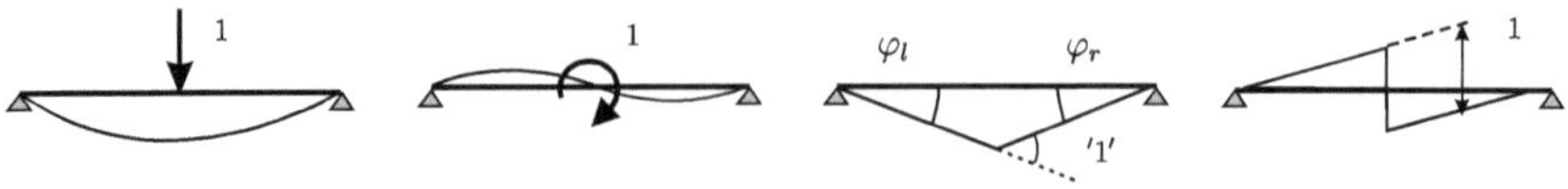

Fig. 2.21 The four singularities of a beam, '1' $\equiv \tan\varphi_l + \tan\varphi_r = 1$

Table 2.1 Energy is bounded (Yes) or (No)

singularity $m = 1$	$n = 1$ rope, bar, Timoshenko beam	$n = 2$ plate, Reissner–Mindlin	$n = 3$ 3-D
$i = 0$: 	Yes	No	No
$i = 1$: 	No	No	No
singularity $m = 2$	Euler–Bernoulli beam	Kirchhoff plate	
$i = 0$: 	Yes	Yes	
$i = 1$: 	Yes	No	
$i = 2$: 	No	No	
$i = 3$: 	No	No	

The order of the energy is

$$\begin{array}{ll} m = 1 & \text{Timoshenko beams, Reissner–Mindlin plates, elastic solids} \\ m = 2 & \text{Euler–Bernoulli beams, Kirchhoff plates} \end{array}$$

and the singularities in a second order equation ($2\,m = 2$) have the indices

$$i = 0 \quad \text{force} \quad i = 1 \quad \text{dislocation}$$

and in a fourth order equation ($2\,m = 4$), see Fig. 2.21,

$$\begin{array}{llll} i = 0 & \text{force} & i = 1 & \text{moment} \\ i = 2 & \text{bend} & i = 3 & \text{dislocation.} \end{array}$$

The index $n = 1, 2, 3$ corresponds to the space dimension. Table 2.1 summarizes the inequality (2.282).

What makes this theorem so important is the link between integrals and point values that is which index m must a Sobolev space $H^m(\Omega)$ have for it to contain $C(\Omega)$

$$C(\Omega) \subset H^m(\Omega) \tag{2.283}$$

or else when does $\|u\|_m < \infty$ imply that u is continuous (or to be precise: equivalent to a continuous function) on Ω. This is the case if $m > n/2$.

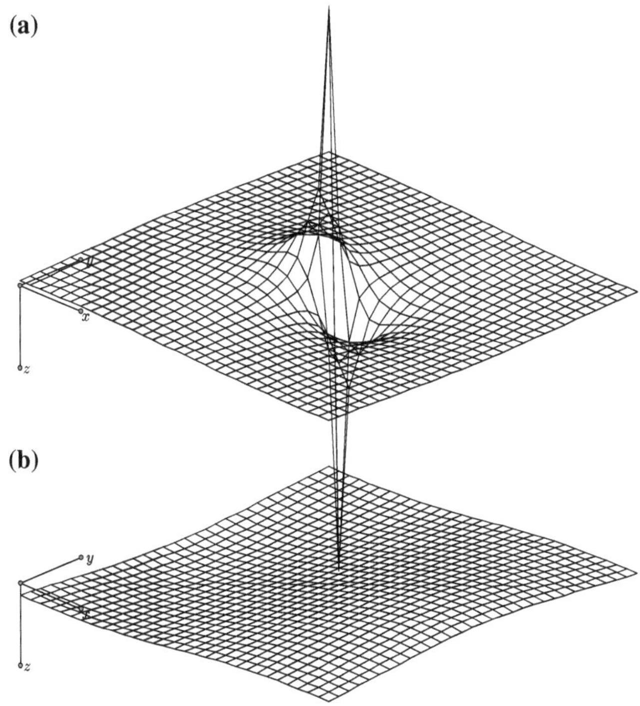

Fig. 2.22 Split of the Green's function for the first derivative $u_{,x}$ on a rectangular domain into **a** the fundamental solution $g_0(\boldsymbol{y}, \boldsymbol{x})$ and **b** the regular part $u_R(\boldsymbol{y}, \boldsymbol{x})$

If for example Ω is a (Kirchhoff) plate with energy space $H^2(\Omega)$, corresponding to $\Delta\Delta u = p$, and a function u lies in $H^2(\Omega)$—if all its derivatives up to the order 2 are square integrable on Ω—then u must be continuous, no sudden jumps are allowed, and it must also be bounded, $|u(\boldsymbol{x})| < \infty$, on Ω. There is guaranteed to be an upper limit to the deflection.

The space $H^1(\Omega)$ does have this property only in 1-D, $1 > 1/2$, but not in 2-D, $1 \not> 2/2$, and not in 3-D, $1 \not> 3/2$.

Remark 2.12 In 2-D the H^1-functions can have at most singularities at isolated points but they cannot be discontinuous along whole lines while in 3-D the H^1-functions can have singularities both at isolated points and along curves [7] A4.

2.7 Fundamental Solutions

Each Green's function

$$G(\boldsymbol{y}, \boldsymbol{x}) = g(\boldsymbol{y}, \boldsymbol{x}) + u_R(\boldsymbol{y}, \boldsymbol{x}) \tag{2.284}$$

can be split into a fundamental solution—in the case of the 2-D Poisson equation this would be the function

$$g(\boldsymbol{y}, \boldsymbol{x}) = -\frac{1}{2\pi} \ln r \tag{2.285}$$

and a regular solution $u_R(\boldsymbol{y}, \boldsymbol{x})$. The fundamental solution exhibits the particular feature which is characteristic for the Green's function, in this case

$$-\Delta g(\boldsymbol{y}, \boldsymbol{x}) = \delta(\boldsymbol{y} - \boldsymbol{x}) \tag{2.286}$$

while the regular part, which is a homogeneous solution of the governing equation,

$$-\Delta u_R(\boldsymbol{y}, \boldsymbol{x}) = 0\,, \tag{2.287}$$

makes that the sum $G = g + u_R$ satisfies the boundary conditions, see Fig. 2.22.

Because of their disregard for boundary conditions fundamental solutions are also called *free-space Green's functions*.

2.7.1 Influence Function

When Green's second identity is formulated with the fundamental solution (2.285) and a sufficiently smooth function u

$$\lim_{\varepsilon \to 0} B(g[\boldsymbol{x}], u) = 0 \tag{2.288}$$

and this expression is solved for $u(\boldsymbol{x})$

$$u(\boldsymbol{x}) = \int_\Gamma [g(\boldsymbol{y}, \boldsymbol{x}) \frac{\partial u}{\partial n}(\boldsymbol{y}) - \frac{\partial}{\partial n} g(\boldsymbol{y}, \boldsymbol{x})\, u(\boldsymbol{y})]\, ds_{\boldsymbol{y}} + \int_\Omega g(\boldsymbol{y}, \boldsymbol{x})\, p(\boldsymbol{y})\, d\Omega_{\boldsymbol{y}} \tag{2.289}$$

then this is an alternative influence function for $u(\boldsymbol{x})$. The boundary element method is based on this type of influence functions [8, 9]. It operates with fundamental solutions instead of Green's functions because fundamental solutions are problem independent.

But given a boundary value problem such as

$$-\Delta u = p \quad u = 0 \quad \text{on } \Gamma \tag{2.290}$$

the influence function (2.289) cannot be applied directly

$$u(\boldsymbol{x}) = \int_\Gamma g(\boldsymbol{y}, \boldsymbol{x}) \underbrace{\frac{\partial u}{\partial n}}_{?}(\boldsymbol{y})\, ds_{\boldsymbol{y}} + \int_\Omega g(\boldsymbol{y}, \boldsymbol{x})\, p(\boldsymbol{y})\, d\Omega_{\boldsymbol{y}} \tag{2.291}$$

because the slope is unknown on Γ and must first be determined by an integral equation (apply the previous equation to points on the boundary)

$$\int_{\Gamma} g(\boldsymbol{y}, \boldsymbol{x})\, s_h(\boldsymbol{y})\, ds_y = -\int_{\Omega} g(\boldsymbol{y}, \boldsymbol{x})\, p(\boldsymbol{y})\, d\Omega_y \quad \boldsymbol{x} \in \Gamma \tag{2.292}$$

where $s_h(\boldsymbol{y})$ is a boundary element approximation of the slope.

2.8 Ill-Posed Problems

The Green's function of the Laplace operator

$$-\Delta G(\boldsymbol{y}, \boldsymbol{x}) = \delta(\boldsymbol{y} - \boldsymbol{x}) \quad G(\boldsymbol{y}, \boldsymbol{x}) = 0 \quad \boldsymbol{y} \in \Gamma\,, \tag{2.293}$$

has—as most Green's functions do—infinite energy. This poses a problem because it is no longer possible to claim that the FE-solution minimizes the distance in the energy to the exact solution

$$\|G - G_h\|_E \le \|G - v_h\|_E \quad \forall\, v_h \in \mathcal{V}_h \tag{2.294}$$

because the distance to the point $\|G\| = \infty$ is always infinite—regardless of which point $G_h \in \mathcal{V}_h$ is picked as best approximation to G.

In the literature on goal-oriented adaptive refinement the Dirac deltas, the point loads $P = 1$, are therefore always replaced by equivalent surface loads $p = 1/|\Omega_\varepsilon|$ spread over a tiny disc Ω_ε centered at the source point so that the response of the medium, let us call this the function $\tilde{G}$, is almost the Green's function G. Numerically there is of course no difference between nodal forces f_i coming from P and p. The authors do this only to stay inside the bounds of the theory and to justify the estimate, see (4.20),

$$|u(\boldsymbol{x}) - u_h(\boldsymbol{x})| \le \|\tilde{G} - \tilde{G}_h\|_E \, \|u - u_h\|_E \tag{2.295}$$

which (theoretically) breaks down when instead of $\tilde{G}$ the exact Green's function is substituted because G has infinite energy.

To a large extent mathematical analysis is about inequalities about bounds. When you can provide bounds for an error term then you are in a much better position because you can then *control* the error and this is why the inequality above is invaluable. If you can control something then this means that you understand on what this something depends.

From the perspective of the theory of weak boundary value problems the search for an approximation G_h is an ill-posed problem because (1) the exact solution G does not lie in the energy space $H^1(\Omega)$ of the Laplace operator and (2) the functional $J(v)$ is not bounded on $H^1(\Omega)$.

But the finite element method simply disregards such warnings and operates with Green's functions just as if they happened to have finite energy and the success proves the FE-method right. Which certainly is a curious situation: *In an FE-program the input is mostly processed by kernel functions which are solutions of ill-posed problems.*

Remark 2.13 The functional $J(u) = u(\boldsymbol{x})$ would be a bounded functional on $H^1(\Omega)$ if there would exist a constant $c < \infty$, independent of the single u, such that

$$|J(u)| \leq c\,\|u\|_1 \quad \forall u \in H^1(\Omega)\,. \tag{2.296}$$

This would be true if $C(\Omega)$ can be "continuously embedded" into $H^1(\Omega)$

$$C(\Omega) \subset H^1(\Omega) \tag{2.297}$$

which is another way of saying that there exists a constant $c < \infty$ such that

$$\max_{\boldsymbol{x}\in\Omega} |u(\boldsymbol{x})| \leq c\,\|u\|_1 \tag{2.298}$$

for all functions u in $H^1(\Omega)$. Think of it this way: if the integral of u squared and its first-order derivatives squared, ($\sim$ *energy*), tends to zero then u uniformly tends to zero—pointwise (!). When the stored heat (= energy) in a small box Ω is nearly zero then the temperature $u(\boldsymbol{x})$ at each point of Ω must be near absolute zero. Or what is the same: if two functions u and v are close in the H^1-norm then they are also close in the C-norm which means

$$\max_{\boldsymbol{x}\in\Omega} |u(\boldsymbol{x}) - v(\boldsymbol{x})| \leq c\,\|u - v\|_1\,. \tag{2.299}$$

But the function $u = -\ln(-\ln^{-1} r)$ has a finite H^1-norm on a disc Ω_ρ with radius $\rho = 0.5$ and centered at $\boldsymbol{x}$

$$\|u\|_1^2 = \int_\Omega (u^2 + u_{,x}^2 + u_{,y}^2)\,d\Omega = \int_0^{2\pi}\int_0^{\rho} \left(u^2 + \frac{4}{r^2\,\ln^2 r^2}\right) r\,dr\,d\varphi < \infty \tag{2.300}$$

but becomes infinite at the center, $r = 0$ and so the functional $J(u) = u(0)$ cannot be a bounded functional on $H^1(\Omega_\rho)$.

2.9 Nonlinear Problems

Green's functions are only applicable to linear problems because scalar products are linear and so the superposition principle is limited to linear problems

$$u_1(\boldsymbol{x}) + u_2(\boldsymbol{x}) = \int_\Omega G(\boldsymbol{y}, \boldsymbol{x})\,(p_1(\boldsymbol{y}) + p_2(\boldsymbol{y}))\,d\Omega_{\boldsymbol{y}}\,. \tag{2.301}$$

But two observations suggest to extend at least the concept of a Green's function to nonlinear problems:

- Functionals $J(u)$ are not limited to linear problems.
- In linear problems Green's functions are identical with the Lagrange multiplier λ

$$\mathcal{L}(u, \lambda) := J(u) - (a(u, \lambda) - (p, \lambda)) \tag{2.302}$$

because

$$\mathcal{L}_{,u} = J(v) - a(v, \lambda) = 0 \qquad \forall\, v \in \mathcal{V} \quad \Rightarrow \lambda = G \tag{2.303}$$

$$\mathcal{L}_{,\lambda} = a(u, v) - (p, v) = 0 \qquad \forall\, v \in \mathcal{V} \tag{2.304}$$

and in the nonlinear case λ corresponds to the Green's function at the linearization point. So Lagrange multipliers inherently carry over the concept of a Green's function also to nonlinear problems although one must be fully aware that a nonlinear solution does *not* allow such a representation as in (2.301). But the very same idea which has been applied so successfully to linear problems, *goal-oriented adaptive refinement*, see Sect. 4.7, can also be applied to nonlinear problems and in this regard the extension to nonlinear problems is important.

2.9.1 Lagrange Multiplier

Assume a function $f(x, y)$ is to be minimized under the side condition that the point (x, y) satisfies an equation $g(x, y) = 0$. With the two functions f and g and the Lagrange multiplier λ we form the expression

$$\mathcal{L}(x, y, \lambda) = f(x, y) + \lambda\, g(x, y)\,, \tag{2.305}$$

and the point $\{x, y, \lambda\}$ at which L becomes stationary

$$d\mathcal{L} = \mathcal{L}_{,x}\; dx + \mathcal{L}_{,y}\; dy + \mathcal{L}_{,\lambda}\; d\lambda = 0 \tag{2.306}$$

is also the point x, y at which $f(x, y)$ attains its minimum value.

Lagrange multipliers are not restricted to optimization problems. Imagine a function $f(x)$ is to be evaluated at a point x and x is subject to the constraint $g(x) = 0$ as in

$$f(x) = \sin^2(x) \quad g(x) = x - 1 = 0\,. \tag{2.307}$$

Then the expression

$$\mathcal{L} = f(x) + \lambda\, g(x) \tag{2.308}$$

is stationary for a certain value of λ at the point $\{x, \lambda\}$, that is

$$d\mathcal{L} = \mathcal{L}_{,x}\, dx + \mathcal{L}_{,\lambda}\, d\lambda = 0\,. \tag{2.309}$$

The point $\{x, \lambda\}$ is found by solving the two equations

$$\mathcal{L}_{,x} = f'(x) + \lambda\, g'(x) = 0 \qquad \text{(this determines } \lambda) \tag{2.310}$$
$$\mathcal{L}_{,\lambda} = g(x) = 0 \qquad \text{(this determines } x)\,. \tag{2.311}$$

In the following the function f will be a functional $J(\boldsymbol{u})$ and the vector $\boldsymbol{u}$ is subject to the constraint $\boldsymbol{K}\,\boldsymbol{u} = \boldsymbol{f}$.

2.9.2 Lagrange Multiplier and Linear Algebra

Lagrange multipliers are denoted by λ or $\boldsymbol{\lambda}$ and we follow this convention here but in essence they are identical with the nodal vector $\boldsymbol{g}$ of the Green's function, so $\boldsymbol{\lambda} \equiv \boldsymbol{g}$.

Let

$$J(\boldsymbol{u}) = \boldsymbol{j}^T\,\boldsymbol{u} \tag{2.312}$$

a linear functional on $\mathbb{R}^n$ which is to be evaluated under the side condition that the vector $\boldsymbol{u}$ satisfies the symmetric $(n \times n)$ system

$$\boldsymbol{K}\,\boldsymbol{u} = \boldsymbol{f}\,. \tag{2.313}$$

With a third vector $\boldsymbol{\lambda} \in \mathbb{R}^n$ we form the Lagrange functional

$$\mathcal{L}(\boldsymbol{u}, \boldsymbol{\lambda}) = J(\boldsymbol{u}) - \boldsymbol{\lambda}^T\,(\boldsymbol{K}\,\boldsymbol{u} - \boldsymbol{f}) \tag{2.314}$$

and the point $\{\boldsymbol{u}, \boldsymbol{\lambda}\}$ at which the functional $\mathcal{L}$ is stationary

$$d\mathcal{L} = \mathcal{L}_{,u_i}\; du_i - \mathcal{L}_{,\lambda_i}\; d\lambda_i = 0 \tag{2.315}$$

is determined by the two equations

$$\boldsymbol{K}\,\boldsymbol{u} = \boldsymbol{f} \quad \boldsymbol{K}\,\boldsymbol{\lambda} = \boldsymbol{j} \tag{2.316}$$

and so it follows that

$$J(\boldsymbol{u}) = \boldsymbol{\lambda}^T\,\boldsymbol{f} \tag{2.317}$$

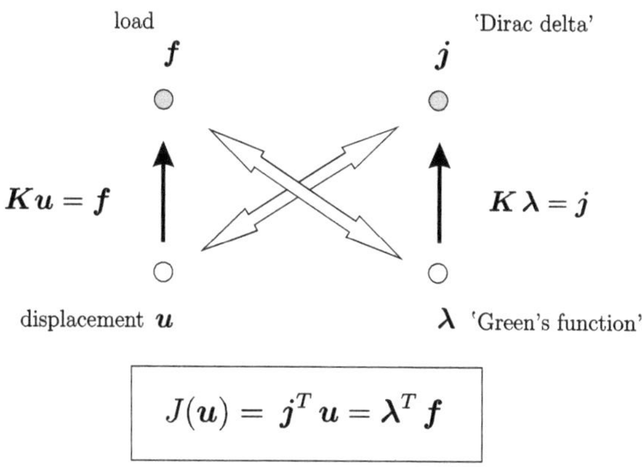

Fig. 2.23 The Lagrange multiplier method is an application of Betti's theorem that is of the duality inherent in the scalar product, $\pi^T = \pi$

which means that the Lagrange multiplier plays the same role als the nodal vector $\boldsymbol{g}$ of the Green's function in FE-analysis, see Fig. 2.23.

The Lagrange multiplier method as presented here is identical with Betti's theorem that is it is a simple application of the identity ($\pi^T = \pi$)

$$\mathscr{B}(\boldsymbol{u}, \boldsymbol{\lambda}) = \boldsymbol{\lambda}^T \boldsymbol{K}\, \boldsymbol{u} - \boldsymbol{u}^T \boldsymbol{K}\, \boldsymbol{\lambda} = 0 \tag{2.318}$$

which holds true for any two vectors $\boldsymbol{u}$, $\boldsymbol{\lambda}$ and a symmetric matrix $\boldsymbol{K}$. This immediately implies that if the two vectors are solutions of (2.316) then

$$\mathscr{B}(\boldsymbol{u}, \boldsymbol{\lambda}) = \boldsymbol{\lambda}^T \boldsymbol{f} - \boldsymbol{u}^T \boldsymbol{j} = 0 \tag{2.319}$$

or $J(\boldsymbol{u}) = \boldsymbol{\lambda}^T \boldsymbol{f}$.

The stationarity condition $d\mathcal{L} = 0$ only serves to derive the two equations (2.316).

2.9.3 Nonlinear Functionals

Next let $J(u)$ a nonlinear functional. We only presuppose that $J(0) = 0$. The expression

$$J'(u; v) := \frac{d}{d\varepsilon} J(u + \varepsilon\, v)|_{\varepsilon=0} \tag{2.320}$$

is the Gateaux derivative of the functional $J(u)$. Because of the chain rule of the Calculus this expression is linear in the second argument, in v. Let $f(s)$ be the function

$$f(s) := J(u + s\, e) \quad e = u - u_h\,, \quad 0 \leq s \leq 1\,. \tag{2.321}$$

According to the Fundamental Theorem of the Calculus we have

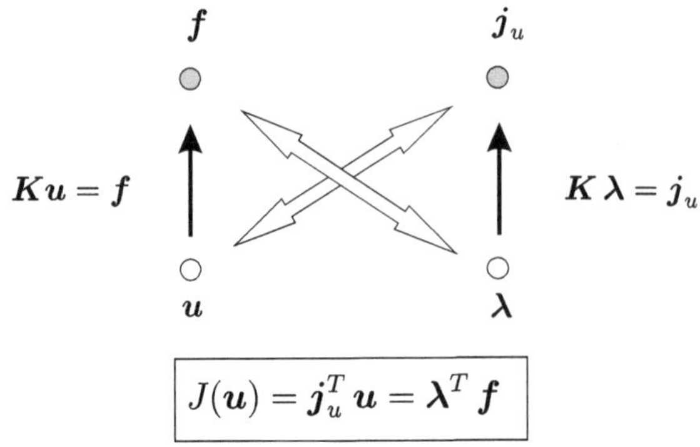

Fig. 2.24 The Lagrange multiplier method applied to a non-linear functional

$$f(1) - f(0) = J(u) - J(u_h) = \int_0^1 f'(s)\,ds = \int_0^1 J'(u_h + s\,e; e)\,ds \qquad (2.322)$$

and in particular, if $u = 0$,

$$-J(u_h) = \int_0^1 J'(u_h - s\,u_h; -u_h)\,ds = -\int_0^1 J'(u_h - s\,u_h; u_h)\,ds\,. \qquad (2.323)$$

In FE-analysis the function u_h is a weighted sum of the shape functions

$$u_h(\boldsymbol{x}) = \sum_{i=1}^{n} u_i\,\varphi_i(\boldsymbol{x}) \qquad (2.324)$$

and so, see Fig. 2.24,

$$J(u_h) = J(\boldsymbol{u}) := \boldsymbol{j}_u^T\,\boldsymbol{u} = \sum_{i=1}^{n} j_{u\,i}\,u_i \quad j_{ui} := \int_0^1 J'(\boldsymbol{u}_h - s\,\boldsymbol{u}_h; \varphi_i)\,ds\,. \qquad (2.325)$$

The functional $J(\boldsymbol{u})$ ($\boldsymbol{u}$ is a vector) has the same value as $J(u_h)$ (u_h is a function) if $\boldsymbol{u}$ is the nodal vector of u_h as in (2.324). The point $\{\boldsymbol{u}, \boldsymbol{\lambda}\}$ at which the Lagrange functional

$$\mathcal{L}(\boldsymbol{u}, \boldsymbol{\lambda}) = J(\boldsymbol{u}) - \boldsymbol{\lambda}^T\,(\boldsymbol{K}\,\boldsymbol{u} - \boldsymbol{f}) \qquad (2.326)$$

becomes stationary is determined by the two equations

$$\boldsymbol{K}\,\boldsymbol{u} = \boldsymbol{f} \quad \boldsymbol{K}\,\boldsymbol{\lambda} = \boldsymbol{j}_u\,, \qquad (2.327)$$

and we have

$$J(\boldsymbol{u}) = \boldsymbol{j}_u^T \boldsymbol{u} = \boldsymbol{\lambda}^T \boldsymbol{K}\,\boldsymbol{u} = \boldsymbol{\lambda}^T \boldsymbol{f}\,. \tag{2.328}$$

This looks like magic: seemingly a nonlinear functional can be given the form of a scalar product. But the vector $\boldsymbol{\lambda}$ which multiplies $\boldsymbol{f}$ depends on the argument u_h. Only in linear problems is $\boldsymbol{\lambda}$ truly a constant vector. The (pseudo) linearity in (2.328) is due to the chain rule of the Calculus.

Example 2.7 Let

$$J(u) = \int_0^l u^2\,dx \tag{2.329}$$

then

$$J'(u;\,v) = \int_0^l 2\,u\,v\,dx \quad j_{ui} = \int_0^l u_h\,\varphi_i\,dx \tag{2.330}$$

and

$$J(u_h) = \boldsymbol{\lambda}^T \boldsymbol{f} = \boldsymbol{j}_u^T \boldsymbol{K}^{-1} \boldsymbol{f} = \boldsymbol{j}_u^T \boldsymbol{u}\,. \tag{2.331}$$

A direct evaluation of $J(u_h)$ gives

$$J(u_h) = J(u_1\,\varphi_1 + \ldots + u_n\,\varphi_n) = \boldsymbol{u}^T \boldsymbol{M}\,\boldsymbol{u} \quad m_{ij} = \int_0^l \varphi_i\,\varphi_j\,dx \tag{2.332}$$

where $\boldsymbol{M} = [m_{ij}]$ is the "mass matrix" and we see that $\boldsymbol{j}_u = \boldsymbol{M}\,\boldsymbol{u}$.

Because $\boldsymbol{\lambda} = \boldsymbol{K}^{-1}\,\boldsymbol{j}_u$ depends on $\boldsymbol{u}$ the direct evaluation (2.332) is probably always faster than (2.331).

The real value of the vector $\boldsymbol{\lambda} = \boldsymbol{K}^{-1}\boldsymbol{j}_u$ is that it can be plotted. It represents the sensitivity of the functional $J(\boldsymbol{u})$ with respect to the vector $\boldsymbol{f}$. Imagine a planar mesh with two degrees of freedom at each node. At each node $\boldsymbol{x}_i$ the two corresponding values of $\boldsymbol{\lambda}$ form a small vector $\boldsymbol{\lambda}_i$ as for example in Figs. 3.20, 3.21 and 3.22 in Sect. 3.10.2. Nodes where $\boldsymbol{\lambda}_i \equiv \boldsymbol{g}_i$ is orthogonal to the vector $\boldsymbol{f}_i$ (which is the local representative of $\boldsymbol{f}$ at node $\boldsymbol{x}_i$) do not contribute to $J(\boldsymbol{u})$.

2.9.4 Nonlinear Problems

Next let us assume that the boundary value problem itself is nonlinear

$$a(u_h;\varphi_i) = (p,\varphi_i) \quad i = 1, 2, \ldots n\,, \tag{2.333}$$

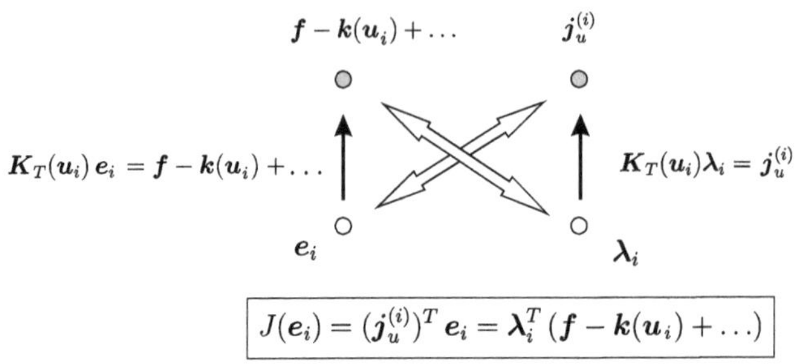

Fig. 2.25 The Lagrange multiplier method applied to a nonlinear problem

and so the vector $\boldsymbol{u} = \{u_i\}$ of nodal values of the FE-solution

$$u_h(\boldsymbol{x}) = \sum_{i=1}^{n} u_i\, \varphi_i(\boldsymbol{x}) \tag{2.334}$$

satisfies the nonlinear vector-valued equation

$$\boldsymbol{k}(\boldsymbol{u}) = \boldsymbol{f} \tag{2.335}$$

where

$$k_i(\boldsymbol{u}) := a(u_h; \varphi_i) \quad f_i = (p, \varphi_i) \quad i = 1, 2, \ldots n\,. \tag{2.336}$$

We allow, as before, that $J(u)$ is a nonlinear functional

$$J(\boldsymbol{u}) = \boldsymbol{j}_u^T\, \boldsymbol{u}\,. \tag{2.337}$$

Hence the Lagrange functional is

$$\mathcal{L}(\boldsymbol{u}, \boldsymbol{\lambda}) = \boldsymbol{u}^T\, \boldsymbol{j}_u - \boldsymbol{\lambda}^T (\boldsymbol{k}(\boldsymbol{u}) - \boldsymbol{f}) \tag{2.338}$$

and we conclude that the stationary point $\{\boldsymbol{u}, \boldsymbol{\lambda}\}$ is determined by the two equations

$$\boldsymbol{k}(\boldsymbol{u}) = \boldsymbol{f} \quad \boldsymbol{K}_T(\boldsymbol{u})\, \boldsymbol{\lambda} = \boldsymbol{j}_u \tag{2.339}$$

where $\boldsymbol{K}_T$ is the tangent stiffness matrix at the point $\boldsymbol{u}$.

This result is no longer symmetric and not what we had hoped for. If the first equation were of the same form as the second

$$\boldsymbol{K}_T(\boldsymbol{u})\, \boldsymbol{u} = \boldsymbol{f} \tag{2.340}$$

then we could apply immediately the previous logic and write $J(\boldsymbol{u}) = \boldsymbol{\lambda}^T\, \boldsymbol{f}$.

So let us assume for a moment it were symmetric. With an arbitrary vector $\boldsymbol{u}_i$ we can construct the matrix $\boldsymbol{K}_T(\boldsymbol{u}_i)$ and therewith the identity ($\pi^T = \pi$)

$$\mathscr{B}(\boldsymbol{e}, \boldsymbol{\lambda}) = \boldsymbol{\lambda}^T \boldsymbol{K}_T(\boldsymbol{u}_i)\, \boldsymbol{e} - \boldsymbol{e}^T \boldsymbol{K}_T(\boldsymbol{u}_i)\, \boldsymbol{\lambda} = 0 \tag{2.341}$$

which holds true for all vectors $\boldsymbol{e}$ and $\boldsymbol{\lambda}$.

The solution of $\boldsymbol{k}(\boldsymbol{u}) = \boldsymbol{f}$ is a zero of the vector-valued function

$$\boldsymbol{s}(\boldsymbol{u}) := \boldsymbol{k}(\boldsymbol{u}) - \boldsymbol{f} \tag{2.342}$$

and so if we follow the logic of Newton's algorithm

$$\boldsymbol{s}(\boldsymbol{u}_{i+1}) = \boldsymbol{0} = \boldsymbol{s}(\boldsymbol{u}_i) + \boldsymbol{s}'(\boldsymbol{u}_i)\,(\boldsymbol{u}_{i+1} - \boldsymbol{u}_i) + \cdots \tag{2.343}$$

and if we let

$$\boldsymbol{e}_{i+1} = \boldsymbol{u}_{i+1} - \boldsymbol{u}_i \tag{2.344}$$

then we are led to

$$\boldsymbol{K}_T(\boldsymbol{u}_i)\, \boldsymbol{e}_{i+1} \simeq \boldsymbol{f} - \boldsymbol{k}(\boldsymbol{u}_i)\,. \tag{2.345}$$

Denoting

$$\boldsymbol{K}_T(\boldsymbol{u}_i)\, \boldsymbol{\lambda}_i = \boldsymbol{j}_u^{(i)} \tag{2.346}$$

the identity (4.47) implies

$$\mathscr{B}(\boldsymbol{e}_{i+1}, \boldsymbol{\lambda}_i) = \underbrace{\boldsymbol{\lambda}_i^T \boldsymbol{K}_T(\boldsymbol{u}_i)}_{(\boldsymbol{j}_u^{(i)})^T}\, \boldsymbol{e}_{i+1} - \boldsymbol{e}_{i+1}^T \boldsymbol{K}_T(\boldsymbol{u}_i)\, \boldsymbol{\lambda}_i = 0 \tag{2.347}$$

or

$$J(\boldsymbol{e}_{i+1}) = (\boldsymbol{j}_u^{(i)})^T \boldsymbol{e}_{i+1} = \boldsymbol{e}_{i+1}^T \boldsymbol{K}_T(\boldsymbol{u}_i)\, \boldsymbol{\lambda}_i \simeq (\boldsymbol{f} - \boldsymbol{k}(\boldsymbol{u}_i))^T \boldsymbol{\lambda}_i \tag{2.348}$$

which means that the vector $\boldsymbol{\lambda}_i$ and the residual $\boldsymbol{k}(\boldsymbol{u}_i) - \boldsymbol{f}$ at the linearization point allow to approximate the error $J(\boldsymbol{e}_{i+1})$, see Fig. 2.25.

This basically is the algebra by which goal-oriented adaptive refinement can be extended to nonlinear problems. For more details see Sect. 4.7.

Remark 2.14 We assumed the tangent stiffness matrix $\boldsymbol{K}_T$ to be symmetric but the extension to non-symmetric matrices is obvious: replace the second matrix $\boldsymbol{K}_T$ in (4.47) by $\boldsymbol{K}_T^T$.

2.10 Mixed Problems

The boundary value problem

$$-\,\Delta u = p \quad \text{on } \Omega \quad u = 0 \quad \text{on } \Gamma \tag{2.349}$$

can be split into a coupled system

$$\nabla u - \boldsymbol{\sigma} = \boldsymbol{\sigma}_0 \tag{2.350}$$

$$-\text{div}\, \boldsymbol{\sigma} = p \tag{2.351}$$

for the two functions u and $\boldsymbol{\sigma}$ or $\boldsymbol{v} = \{u, \boldsymbol{\sigma}\}^T$. We have added an additional term $\boldsymbol{\sigma}_0$ on the right-hand side to account for possible initial stresses.

To this system belongs the identity

$$\begin{aligned}\mathscr{G}(\boldsymbol{v}, \hat{\boldsymbol{v}}) = &\int_\Omega \underbrace{\left[(\nabla u - \boldsymbol{\sigma}) \cdot \hat{\boldsymbol{\sigma}} - \text{div}\, \boldsymbol{\sigma}\, \hat{u}\right]}_{\boldsymbol{L}\,\boldsymbol{v}\cdot\hat{\boldsymbol{v}}} d\Omega + \int_\Gamma \boldsymbol{\sigma} \cdot \boldsymbol{n}\, \hat{u}\, ds \\ &- \int_\Omega (\nabla u \cdot \hat{\boldsymbol{\sigma}} + \nabla \hat{u} \cdot \boldsymbol{\sigma} - \boldsymbol{\sigma} \cdot \hat{\boldsymbol{\sigma}})\, d\Omega = 0\end{aligned} \tag{2.352}$$

and Green's second identity is the expression

$$\mathscr{B}(\boldsymbol{v}, \hat{\boldsymbol{v}}) = \int_\Omega \boldsymbol{L}\,\boldsymbol{v}\cdot\hat{\boldsymbol{v}}\, d\Omega + \int_\Gamma \boldsymbol{\sigma}\cdot\boldsymbol{n}\, \hat{u}\, ds - \int_\Gamma \hat{\boldsymbol{\sigma}}\cdot\boldsymbol{n}\, u\, ds - \int_\Omega \boldsymbol{v}\cdot\boldsymbol{L}(\hat{\boldsymbol{v}})\, d\Omega = 0\,. \tag{2.353}$$

Let G be the Green's function for the value of $u(\boldsymbol{x})$ at some interior point of Ω

$$-\,\Delta G(\boldsymbol{y}, \boldsymbol{x}) = \delta(\boldsymbol{y} - \boldsymbol{x}) \quad G(\boldsymbol{y}, \boldsymbol{x}) = 0 \quad \boldsymbol{y} \in \Gamma \tag{2.354}$$

and let the "mixed representation" of this function be

$$\boldsymbol{g} = \{G, \boldsymbol{\sigma}_G\}^T\,, \quad \boldsymbol{\sigma}_G = \nabla G \tag{2.355}$$

which is a homogeneous solution of $\boldsymbol{L}\,\boldsymbol{v} = \boldsymbol{0}$ at all points except at $\boldsymbol{x}$, then in the limit

$$\lim_{\varepsilon \to 0} \mathscr{B}(\boldsymbol{g}, \boldsymbol{v})_{\Omega_\varepsilon} = 0 \tag{2.356}$$

out of

$$\lim_{\varepsilon \to 0} \int_{\Gamma_{N_\varepsilon(\boldsymbol{x})}} \boldsymbol{\sigma}_G(\boldsymbol{y}, \boldsymbol{x}) \cdot \boldsymbol{n}\, u\, ds = u(\boldsymbol{x}) \tag{2.357}$$

would pop $u(\boldsymbol{x})$ and so we have found an influence function

$$u(\boldsymbol{x}) = \int_{\Omega} (G(\boldsymbol{y}, \boldsymbol{x})\, p(\boldsymbol{y}) + \boldsymbol{\sigma}_G(\boldsymbol{y}, \boldsymbol{x}) \cdot \boldsymbol{\sigma}_0(\boldsymbol{y}))\, d\Omega_{\boldsymbol{y}}\,. \tag{2.358}$$

The kernels in the influence function for σ_1 and σ_2

$$G^{(1)}(\boldsymbol{y}, \boldsymbol{x}) = \frac{\partial G}{\partial x_1}(\boldsymbol{y}, \boldsymbol{x}) \quad G^{(2)} = \frac{\partial G}{\partial x_2}(\boldsymbol{y}, \boldsymbol{x}) \tag{2.359}$$

represent unit dislocations in x_1 and x_2 direction respectively and the σ_i are limit values of the integrals

$$\lim_{\varepsilon \to 0} \int_{\Gamma_{N_\varepsilon(\boldsymbol{x})}} \boldsymbol{\sigma} \cdot \boldsymbol{n}\, G^{(i)}\, ds = \sigma_i(\boldsymbol{x}) \tag{2.360}$$

and so

$$\sigma_i(\boldsymbol{x}) = \int_{\Omega} (G^{(i)}(\boldsymbol{y}, \boldsymbol{x})\, p(\boldsymbol{y}) + \boldsymbol{\sigma}_{G^{(i)}}(\boldsymbol{y}, \boldsymbol{x}) \cdot \boldsymbol{\sigma}_0(\boldsymbol{y}))\, d\Omega_{\boldsymbol{y}} \tag{2.361}$$

where

$$\boldsymbol{\sigma}_{G^{(i)}} = \nabla G^{(i)}\,. \tag{2.362}$$

References

1. Hartmann F, Katz C (2007) Structural analysis with finite elements, 2nd edn. Springer, Berlin
2. Gockenbach M (2006) Understanding and implementing the finite element method. SIAM, Philadelphia
3. Strang G (2007) Computational science and engineering. Wellesley-Cambridge Press, Wellesley
4. Barton G (1989) Elements of Green's Functions and Propagation, Potentials, Diffusion and Waves. Oxford Scientific Publications, Oxford
5. Snieder R (2004) A guided tour of mathematical methods for the physical sciencсs. Cambridge University Press, Cambridge
6. Hartmann F (1985) The mathematical foundation of structural mechanics. Springer, Berlin
7. Šolin P (2006) Partial differential equations and the finite element method. Wiley, Hoboken
8. Hisao GC, Wendland WL (2008) Boundary integral equations. Springer, Berlin
9. Hartmann F (1989) Introduction to boundary elements. Springer, Berlin

Chapter 3
Finite Elements and Green's Functions

In this chapter we discuss the various features of FE-solutions in so far these features can be traced back to the nature of FE-influence functions, that is to the ways an FE-program approximates Green's functions. Evidence of this close connection between Green's functions and the FE-method is the fact that the columns of the inverse of a stiffness matrix are the discrete Green's functions of the nodal values u_i.

As in classical analysis where the solution of an equation can be written as the integral of the Green's function and the right-hand side, the program input is processed by these discrete Green's functions. They provide the output the engineer sees on the screen. In linear analysis (and to some extent also in nonlinear analysis) an FE-program can be seen as an exercise in approximating the Green's functions of an engineering problem with piecewise polynomial shape functions.

The FE-solution is the exact solution of an equivalent load case p_h, whereby equivalent is to be interpreted in the sense of the principle of virtual displacements: $(p, \varphi_i) = (p_h, \varphi_i)$. This property allows to extend Betti's theorem $(p_1, u_2) = (p_2, u_1)$ to FE-solutions in a particular way: keeping the loads in the equation but replacing the exact solutions by their FE-approximations, and this extension makes Tottenham's equation possible which establishes that FE-results are the scalar product between the approximate Green's functions and the original right-hand side of the differential equation.

While the concept of a Green's function is a classical and beautiful idea but with limited applicability because it puts as much strain on the analyst to calculate a Green's functions as to solve the original problem, it is in the realm of weak formulations that Green's functions really shine because in a weak formulation the Green's function is the Riesz element of the functional and approximating a Green's functions becomes as easy as solving an engineering problem with finite elements. This makes it possible to extend the concept of Green's function to arbitrary linear and bounded functionals.

F. Hartmann, *Green's Functions and Finite Elements*,
DOI: 10.1007/978-3-642-29523-2_3, © Springer-Verlag Berlin Heidelberg 2013

3.1 Poisson Equation

Let Ω be a regular planar domain with boundary Γ. On Ω we consider the boundary value problem

$$-\Delta u := -(u_{,xx} + u_{,yy}) = p \quad \text{on } \Omega \quad u = 0 \quad \text{on } \Gamma \tag{3.1}$$

where p is assumed to be a sufficiently regular function, so that

$$\mathcal{V} := \{u \in H^1(\Omega)\,|u = 0\} \tag{3.2}$$

can be considered the appropriate solution space. Green's first identity (2.86) formulated with the solution u and a test function $v \in \mathcal{V}$

$$\mathcal{G}(u, v) = \int_\Omega p\, v\, d\Omega - \int_\Omega \nabla u \cdot \nabla v\, d\Omega = 0 \tag{3.3}$$

shows that u has the property

$$\int_\Omega \nabla u \cdot \nabla v\, d\Omega = \int_\Omega p\, v\, d\Omega \qquad \forall\, v \in \mathcal{V} \tag{3.4}$$

or

$$a(u, v) = (p, v) \qquad \forall\, v \in \mathcal{V} \tag{3.5}$$

for short.

Let the edge of Ω be polygonal in shape so that Ω can be partitioned into a mesh of triangular or rectangular elements, see Fig. 3.1. To the n nodes $\boldsymbol{x}_j$, $j = 1, 2, \ldots n$ of the mesh belongs a set of n (piecewise linear or quadratic) shape functions $\varphi_i(\boldsymbol{x})$

$$\varphi_i(\boldsymbol{x}_j) = \begin{cases} 1 & i = j \\ 0 & i \neq j \end{cases} \qquad \boldsymbol{x}_j = \text{node} \tag{3.6}$$

which form a partition of unity and which constitute a "nodal basis" for the trial space $\mathcal{V}_h \subset \mathcal{V}$.

Given a function v in $\mathcal{V}$ its projection v_h onto the trial space $\mathcal{V}_h$ is the function $v_h \in \mathcal{V}_h$ which satisfies the n equations

$$a(v - v_h, \varphi_i) = 0 \qquad i = 1, 2, \ldots n \tag{3.7}$$

that is the error $e = v - v_h$ is orthogonal—in the sense of the strain energy product—to all $\varphi_i \in \mathcal{V}_h$. The FE-solution

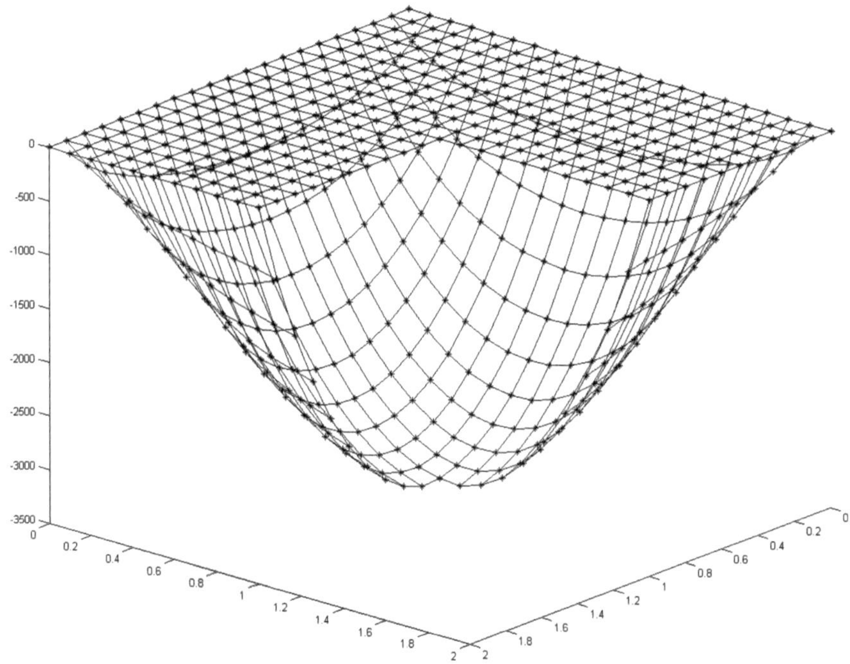

Fig. 3.1 Membrane

$$u_h(\boldsymbol{x}) = \sum_i u_i\, \varphi_i(\boldsymbol{x}) \tag{3.8}$$

is the projection of u onto $\mathcal{V}_h$ in this metric and in this particular case the orthogonality condition

$$a(u - u_h, \varphi_i) = \int_\Omega \nabla(u - u_h) \cdot \nabla\varphi_i \, d\Omega = 0 \qquad \forall\, \varphi_i \in \mathcal{V}_h \tag{3.9}$$

is known as the *Galerkin orthogonality*.

Green's first identity

$$\mathcal{G}(u, \varphi_i) = \int_\Omega p\, \varphi_i \, d\Omega + \underbrace{\int_\Gamma \frac{\partial u}{\partial n}\, \varphi_i \, ds}_{=0} - a(u, \varphi_i) = 0 \tag{3.10}$$

allows to substitute for the virtual internal energy the virtual exterior work of the right-hand side and so (3.9) is equivalent to the n equations

$$a(u_h, \varphi_i) = (p, \varphi_i) \qquad i = 1, 2, \ldots, n \tag{3.11}$$

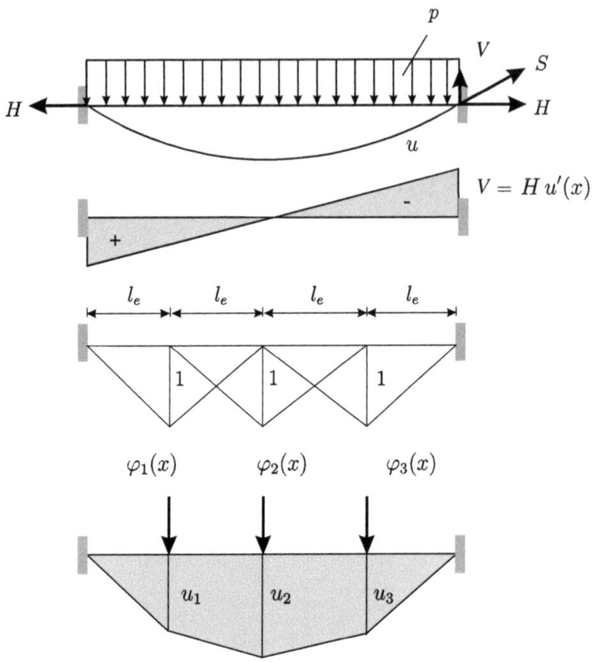

Fig. 3.2 FE-analysis of a taut rope

or the system

$$\boldsymbol{K}\,\boldsymbol{u} = \boldsymbol{f} \tag{3.12}$$

with entries

$$k_{ij} = a(\varphi_i, \varphi_j) \qquad f_i = (p, \varphi_i). \tag{3.13}$$

3.2 The FE-Load Case p_h

Given an equation such as

$$3\,x = 7.7 \tag{3.14}$$

and an approximate solution, $x_h = 2.5$, by doing a *backward error analysis*—simply substitute x_h into the equation—it is found that x_h is the exact solution to the slightly perturbed equation

$$3\,x_h = 7.5. \tag{3.15}$$

The same can be done with the FE-solution u_h: substitute the approximate solution u_h into the original differential equation

$$- \Delta u_h = p_h \tag{3.16}$$

and out pops the right-hand side p_h, the "FE-load case" p_h, which is solved by the FE-solution. If for example u_h is the function $u_h(\boldsymbol{x}) = x^3\, y^2$ then p_h is the function

$$- \Delta\, u_h = -\frac{\partial^2 u_h}{\partial x^2} - \frac{\partial^2 u_h}{\partial y^2} = -(6\, x\, y^2 + 2\, x^3) =: p_h(\boldsymbol{x}). \tag{3.17}$$

Complications arise because the typical FE-solutions are only piecewise smooth and so the differential operator is only applicable inside the elements. To quantify the jumps in the derivatives across the element boundaries Green's first identity is formulated with the FE-solution u_h and the function $v = 1$ on each element

$$\mathcal{G}(u_h, 1)_{\Omega_e} = \int_{\Omega_e} -\Delta\, u_h \cdot 1\, d\Omega + \int_{\Gamma_e} \frac{\partial u_h}{\partial n} \cdot 1\, ds - \underbrace{a(u_h, 1)_{\Omega_e}}_{=0} = 0 \tag{3.18}$$

and this string of zeros is added

$$\mathcal{G}(u_h, 1) := \sum_e \mathcal{G}(u_h, 1)_{\Omega_e} = 0 + 0 + \ldots + 0 = 0 \tag{3.19}$$

which makes

$$\mathcal{G}(u_h, 1) := \sum_e \int_{\Omega_e} -\Delta\, u_h \cdot 1\, d\Omega + \sum_k \int_{\Gamma_k} l_k \cdot 1\, ds = 0 \tag{3.20}$$

that automatically jump terms along the mesh lines Γ_k emerge

$$l_k = \frac{\partial u_h}{\partial n}_+ - \frac{\partial u_h}{\partial n}_- \tag{3.21}$$

which can be interpreted as vertical line forces l_k which produce kinks in the smooth fabric of the membrane. If an element borders on the outer edge Γ then l_k is simply identical with the slope $\partial u_h / \partial n$ of the element at the edge Γ.

Together with the smooth element forces inside each element, $p_h^{(e)} := -\Delta u_h$ (restricted to element Ω_e), these line forces represent the FE-load case p_h. For a 1-D example see Fig. 3.2.

In the case of the plate in Fig. 3.3 the element force $\boldsymbol{p}_h$ is a vector field with two components p_x and p_y.

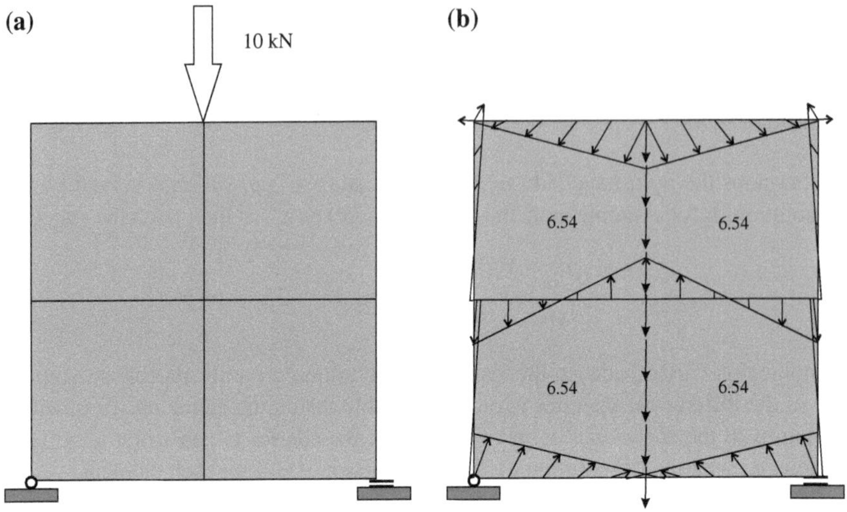

Fig. 3.3 How an FE-program approximates a point force **a** original load case, **b** FE-load case p_h, the numbers are the Euclidean sum $p_h = \sqrt{(p_x, p_x) + (p_y, p_y)}$ of the integrals of the distributed forces within the elements (bilinear elements)

3.2.1 Distributions

In most cases p_h is a rather odd looking function if at all and so we take the liberty to call the FE-load case a *distribution*. A distribution $d(\boldsymbol{x})$ is something which can be sensed by a test function $\varphi \in C_0^\infty(\Omega)$ that is when we shake this "something" with φ then we register virtual work

$$\delta W = \int_0^l d(\boldsymbol{x})\, \varphi(\boldsymbol{x})\, d\Omega \tag{3.22}$$

which allows even very wild distributions $d(\boldsymbol{x})$, see Fig. 3.4, to be measured. This virtual work approach is in agreement with the FE-method which is not interested in the shape of things, how something looks on the screen, but in the virtual work this something contributes on acting through virtual displacements.

3.2.2 Notation

Because eventually a long list of element loads p_e and line loads l_k is necessary to detail a distribution p_h the virtual work statement can become very lengthy

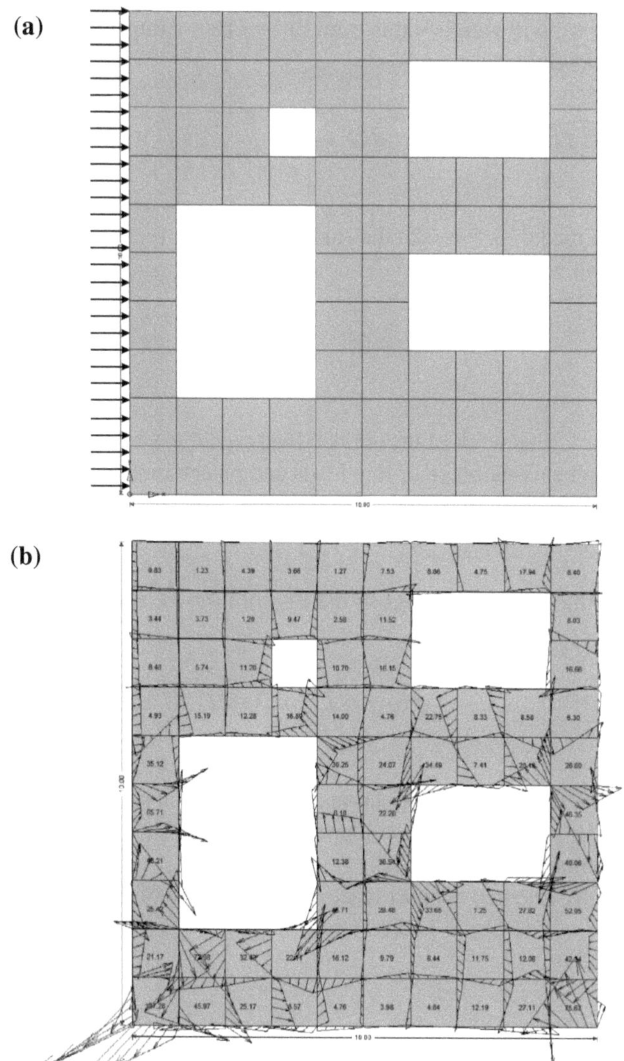

Fig. 3.4 The work done by **a** the original load p and **b** the FE-loads p_h are the same with respect to the shape functions φ_i, $\delta W_e(p_h, \varphi_i) = \delta W_e(p, \varphi_i)$

$$\int_\Omega p(u_h)\, v\, d\Omega := \sum_e \int_{\Omega_e} -\Delta\, u_h \cdot v\, d\Omega + \sum_k \int_{\Gamma_k} l_k \cdot v\, ds\,, \tag{3.23}$$

so that the integral on the left serves as a short hand. We understand the operator $p(.)$ applied to a function u_h to be the list of all the force terms of u_h when Green's first identity $\mathscr{G}(u_h, 1)$ is formulated as in (3.19) and $p(u_h)\, v$ is then the appropriate pairing. Often we will simplify this even further by writing p_h instead of $p(u_h)$ so that all the following expressions

$$\int_\Omega p_h\, v\, d\Omega := \int_\Omega p(u_h)\, v\, d\Omega = \sum_e \int_{\Omega_e} -\Delta u_h \cdot v\, d\Omega + \dots \tag{3.24}$$

have the same meaning. We take the same liberty with regard to the original right-hand side p and write

$$\int_\Omega p(u)\, v\, \Omega \qquad \text{or} \qquad \int_\Omega p\, v\, d\Omega \tag{3.25}$$

for the exterior virtual work of the applied load regardless of whether p is a uniform load or a complex assemblage of line loads and piecewise smooth distributed loads (checker-board loads) or else.

We hope that this sacrifice in exactness will be compensated by a gain in readability.

The distribution which belongs to the FE-solution u_h constitutes a virtual work functional

$$J(v) = \int_\Omega p(u_h)\, v\, d\Omega \tag{3.26}$$

and in particular is (3.20) the application of this functional to the displacement $v = 1$. The definition of $p(u_h)$ includes the support reactions and this is why the forces $p(u_h)$ are self-equilibrated, $J(1) = 0$.

The vertically oriented support reactions on the edge of the membrane are (as in a rope, $V = H\, u'$, see Fig. 1.1) the H-fold slope

$$V = H\, \frac{\partial u}{\partial n} \tag{3.27}$$

where H is the uniform prestress which we set $H = 1$ for simplicity. Green's first identity, $\mathcal{G}(u_h, 1) = 0$, guarantees that the edge forces balance the applied load (Γ_k are now only internal mesh lines)

$$-\int_\Gamma \frac{\partial u_h}{\partial n} \cdot 1\, ds = \sum_e \int_{\Omega_e} -\Delta u_h \cdot 1\, d\Omega + \sum_k \int_{\Gamma_k} l_k \cdot 1\, ds. \tag{3.28}$$

The slope on the edge of a membrane is usually negative because the deflection decreases towards the edge and so the left-hand side comes out positive.

The following theorem clarifies the role of the FE-load case p_h and why it is central for the finite element method.

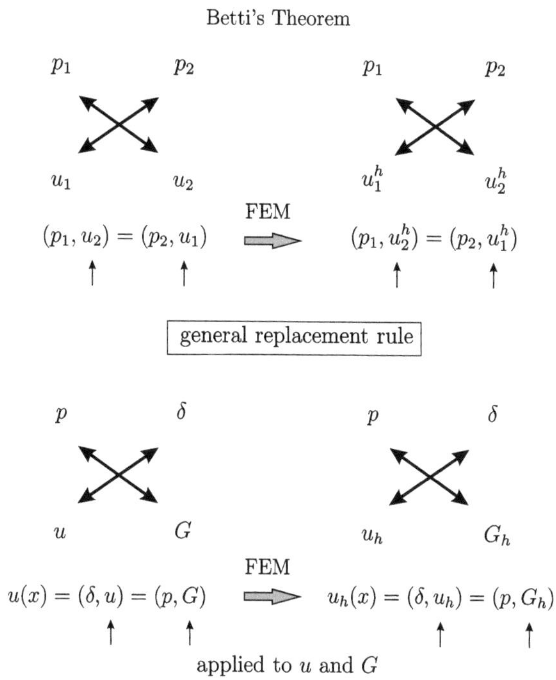

Fig. 3.5 Betti's theorem and its extension to FE-solutions

Theorem 3.1 (Equivalence) *The exact load case* p *and the FE-load case* p_h *are equivalent with respect to all* $\varphi_i \in \mathcal{V}_h$

$$\int_\Omega p(u)\,\varphi_i\,d\Omega = \int_\Omega p(u_h)\,\varphi_i\,d\Omega \qquad \forall\,\varphi_i \in \mathcal{V}_h \tag{3.29}$$

and therefore with respect to all functions $v_h \in \mathcal{V}_h$ *because the* φ_i *form a basis of* $\mathcal{V}_h$.

One cannot tell apart the original load case p from the substitute load case p_h by shaking the membrane with one of the shape functions φ_i. The response in terms of virtual exterior work is in both cases the same.

This theorem is simply a reformulation of the Galerkin orthogonality

$$0 = a(u - u_h, \varphi_i) = a(u, \varphi_i) - a(u_h, \varphi_i) = \int_\Omega p(u)\,\varphi_i\,d\Omega - \int_\Omega p(u_h)\,\varphi_i\,d\Omega \tag{3.30}$$

in terms of exterior—instead of interior—virtual work.

3.3 Extensions

Betti's theorem, which states that the reciprocal exterior work of two solutions u_1 and u_2 is the same, $(p_1, u_2) = (p_2, u_1)$, can be extended to FE-solutions in the following sense $(p_1, u_2^h) = (p_2, u_1^h)$.

3.3.1 Betti's Theorem: Extended

Betti's theorem, discovered by Enrico Betti in 1872, states that for a linear elastic structure subjected to two sets of forces $\{P_i\}, i = 1, ..., m$ and $\{Q_j\}, j = 1, 2, ..., n$, the work done by the set $\{P\}$ through the displacements produced by the set $\{Q\}$ is equal to the work done by the set $\{Q\}$ through the displacements produced by the set $\{P\}$.

This result, also known as the Maxwell–Betti reciprocal work (or reciprocity) theorem, applies to any linear self-adjoint operator and its solutions. It is a verbal assessment of Green's second identity.

If the membrane is subjected to two different load distributions, p_1 and p_2,

$$-\Delta u_1 = p_1 \quad \text{on } \Omega \quad u_1 = 0 \quad \text{on } \Gamma \tag{3.31}$$

$$-\Delta u_2 = p_2 \quad \text{on } \Omega \quad u_2 = 0 \quad \text{on } \Gamma \tag{3.32}$$

then Green's identity

$$\mathscr{B}(u_1, u_2) = \int_\Omega -\Delta u_1\, u_2\, d\Omega + \int_\Gamma \frac{\partial u_1}{\partial n}\, u_2\, ds - \int_\Gamma \frac{\partial u_2}{\partial n}\, u_1\, ds - \int_\Omega -\Delta u_1\, u_2\, d\Omega = 0 \tag{3.33}$$

implies that the reciprocal exterior work is the same

$$\mathscr{B}(u_1, u_2) = \int_\Omega p_1\, u_2\, d\Omega - \int_\Omega u_1\, p_2\, d\Omega = 0 \tag{3.34}$$

which is Betti's theorem. This theorem can be extended, see Fig. 3.5, to FE-solutions in the following sense:

Theorem 3.2 (Betti's theorem—extended) *If u_1^h is the FE-approximation of (3.31) and if u_2^h is the FE-approximation of (3.32) then Betti's theorem (3.34) remains valid if u_1 and u_2 are replaced by u_1^h and u_2^h*

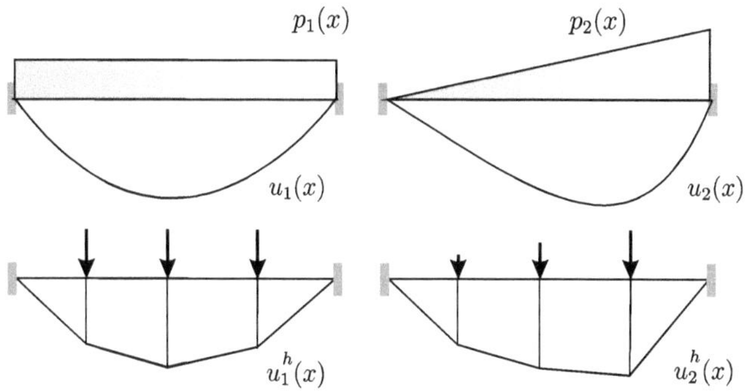

$$\int_0^l p_1\, u_2\, dx = \int_0^l p_2\, u_1\, dx \qquad \text{Betti}$$
$$\int_0^l p_1\, u_2^h\, dx = \int_0^l p_2\, u_1^h\, dx \qquad \text{Betti-extended}$$

Fig. 3.6 Betti and Betti—extended

$$\int_\Omega p_1\, u_2^h\, d\Omega = \int_\Omega p_2\, u_1^h\, d\Omega. \tag{3.35}$$

The proof is based on the *Equivalence Theorem*

$$\int_\Omega p_1^h\, u_2^h\, d\Omega = \int_\Omega p_1\, u_2^h\, d\Omega \int_\Omega p_2^h\, u_1^h\, d\Omega = \int_\Omega p_2\, u_1^h\, d\Omega \tag{3.36}$$

and Betti's theorem

$$\mathscr{B}(u_1^h, u_2^h) = \int_\Omega p_1^h\, u_2^h\, d\Omega - \int_\Omega p_2^h\, u_1^h\, d\Omega = 0 \tag{3.37}$$

so that, see Fig. 3.6,

$$\int_\Omega p_1^h\, u_2^h\, d\Omega = \int_\Omega p_1\, u_2^h\, d\Omega = \int_\Omega p_2^h\, u_1 \varphi\, d\Omega = \int_\Omega p_2\, u_1^h\, d\Omega. \tag{3.38}$$

3.3.2 Tottenham's Equation

The Green's function of the boundary value problem (3.1) is the solution to the boundary value problem

$$-\,\Delta G(\boldsymbol{y},\boldsymbol{x}) = \delta(\boldsymbol{y}-\boldsymbol{x}) \quad \text{on } \Omega \qquad G(\boldsymbol{y},\boldsymbol{x}) = 0 \quad \boldsymbol{y} \in \Gamma \tag{3.39}$$

where the Dirac delta has the properties

$$\delta(\boldsymbol{y}-\boldsymbol{x}) = 0 \qquad \boldsymbol{y} \neq \boldsymbol{x} \tag{3.40}$$

$$\int_\Omega \delta(\boldsymbol{y}-\boldsymbol{x})\, v(\boldsymbol{y})\, d\Omega_{\boldsymbol{y}} = v(\boldsymbol{x}) \qquad \boldsymbol{x} \in \Omega\,,\, v \in C^0(\Omega). \tag{3.41}$$

With the help of the Green's function the exact solution can be written as

$$u(\boldsymbol{x}) = \int_\Omega G(\boldsymbol{y},\boldsymbol{x})\, p(\boldsymbol{y})\, d\Omega_{\boldsymbol{y}}. \tag{3.42}$$

The approximate Green's function $G_h(\boldsymbol{y},\boldsymbol{x})$, the projection of $G(\boldsymbol{y},\boldsymbol{x})$ onto the subset $\mathcal{V}_h$, has the form

$$G_h(\boldsymbol{y},\boldsymbol{x}) = \sum_i g_i(\boldsymbol{x})\, \varphi_i(\boldsymbol{y}) \tag{3.43}$$

where the notation $g_i(\boldsymbol{x})$ is to indicate that the nodal values g_i depend on the location of the source point $\boldsymbol{x}$. This "separation of variables" pattern is typical for FE-Green's functions. The vector $\boldsymbol{g}(\boldsymbol{x})$ is the solution of the system

$$\boldsymbol{K}\, \boldsymbol{g}(\boldsymbol{x}) = \boldsymbol{j}(\boldsymbol{x}) \qquad j_i(\boldsymbol{x}) = \int_\Omega \delta(\boldsymbol{y}-\boldsymbol{x})\, \varphi_i(\boldsymbol{y})\; d\Omega_{\boldsymbol{y}} = \varphi_i(\boldsymbol{x}). \tag{3.44}$$

The function $G_h(\boldsymbol{y},\boldsymbol{x})$ is the Green's function of the FE-solution $u_h(\boldsymbol{x})$ as the following theorem explains [1].

Theorem 3.3 (Tottenham's equation) *If u is the solution to the boundary value problem (3.1), u_h its FE-approximation and $G_h(\boldsymbol{y},\boldsymbol{x})$ the FE-approximation of the Green's function for the point value $u(\boldsymbol{x})$ then holds*

$$u_h(\boldsymbol{x}) = \int_\Omega G_h(\boldsymbol{y},\boldsymbol{x})\, p(\boldsymbol{y})\, d\Omega_{\boldsymbol{y}}. \tag{3.45}$$

According to Betti's theorem the reciprocal exterior work of the two solutions of the two load cases

$$-\Delta u = p \qquad u = 0 \qquad \text{on } \Gamma \tag{3.46}$$

$$-\Delta G = \delta \qquad G = 0 \qquad \text{on } \Gamma \tag{3.47}$$

is the same and so

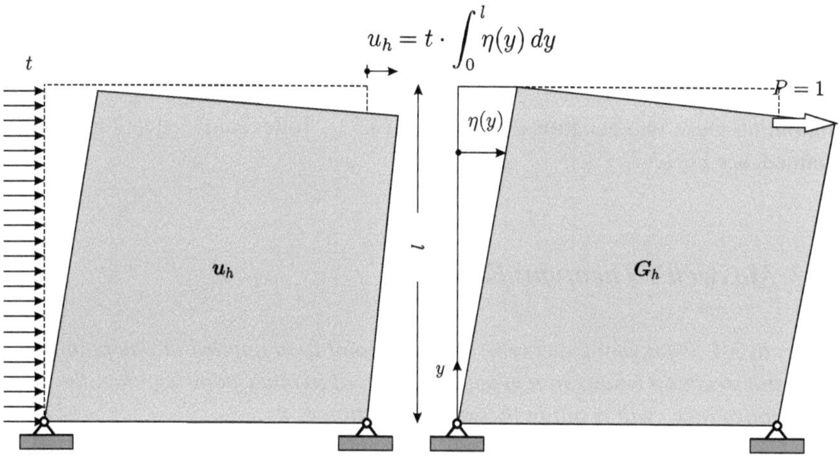

Fig. 3.7 Tottenham's equation: the FE-corner displacement u_h is equal to the work done by the load t on acting through the approximate influence function

$$u(\boldsymbol{x}) = \int\limits_{\Omega} \delta(\boldsymbol{y} - \boldsymbol{x})\, u(\boldsymbol{y})\, d\Omega_{\boldsymbol{y}} = \int\limits_{\Omega} G(\boldsymbol{y}, \boldsymbol{x})\, p(\boldsymbol{y})\, d\Omega_{\boldsymbol{y}}. \tag{3.48}$$

According to *Betti's Theorem-Extended* we may, (set $p_1 = p$ and $p_2 = \delta$), replace in (3.45) the exact solutions u and G by their FE-approximations and so

$$u_h(\boldsymbol{x}) = \int\limits_{\Omega} \delta(\boldsymbol{y} - \boldsymbol{x})\, u_h(\boldsymbol{y})\, d\Omega_{\boldsymbol{y}} = \int\limits_{\Omega} G_h(\boldsymbol{y}, \boldsymbol{x})\, p(\boldsymbol{y})\, d\Omega_{\boldsymbol{y}} \tag{3.49}$$

which is (3.45). □

The following arguments lead to the same result: Green's first identity implies

$$\mathscr{G}(u_h, G_h) = \int\limits_{\Omega} G_h\, p_h\, d\Omega - a(u_h, G_h) = 0 \tag{3.50}$$

and because of $(p_h, G_h) = (p, G_h)$ this is identical with

$$a(u_h, G_h) = \int\limits_{\Omega} G_h\, p\, d\Omega \tag{3.51}$$

where G_h plays the role of a test function, a virtual displacement. Then the roles are interchanged: G_h becomes the FE-approximation to $-\Delta\, G = \delta$ and u_h is interpreted as a virtual displacement and so it follows

$$u_h(\boldsymbol{x}) = \int_\Omega \delta(\boldsymbol{y} - \boldsymbol{x})\, u_h(\boldsymbol{y})\, d\Omega_y = a(G_h, u_h). \tag{3.52}$$

Combining these two equations, (3.51) and (3.52), Tottenham's equation (3.45) is obtained, see Fig. 3.7.

3.3.3 Maxwell's Theorem: Extended

Theorem 3.4 (Maxwell's theorem) *If a unit point load applied at one point $\boldsymbol{x}_1$ of an elastic structure results in a given deflection at another point $\boldsymbol{x}_2$, then the same load applied at $\boldsymbol{x}_2$ will result in the same deflection at $\boldsymbol{x}_1$.*

Maxwell's theorem is a particular instance of Betti's theorem

$$\int_\Omega p_1\, u_2\, d\Omega = \int_\Omega p_2\, u_1\, d\Omega. \tag{3.53}$$

Namely if $p_1 = \delta(\boldsymbol{y} - \boldsymbol{x}_1)$ and $p_2 = \delta(\boldsymbol{y} - \boldsymbol{x}_2)$ are two point loads then this implies

$$u_2(\boldsymbol{x}_1) = \int_\Omega \delta(\boldsymbol{y} - \boldsymbol{x}_1)\, u_2(\boldsymbol{y})\, d\Omega = \int_\Omega \delta(\boldsymbol{y} - \boldsymbol{x}_2)\, u_1(\boldsymbol{y})\, d\Omega = u_1(\boldsymbol{x}_2) \tag{3.54}$$

which is

$$u_2(\boldsymbol{x}_1) = u_1(\boldsymbol{x}_2) \qquad \mathit{Maxwell's\ Theorem}. \tag{3.55}$$

Does Maxwell's Theorem also apply to FE-solutions? The symmetry of the stiffness matrix implies that Maxwell's Theorem holds true with regard to the nodal displacements: a unit load at a node $\boldsymbol{x}_i$ effects at a node $\boldsymbol{x}_j$ the same displacement as a unit load at node $\boldsymbol{x}_j$ will at node $\boldsymbol{x}_i$

$$\boldsymbol{e}_j^T\, \boldsymbol{K}\, \boldsymbol{e}_i = k_{ij} = k_{ji} = \boldsymbol{e}_i^T\, \boldsymbol{K}\, \boldsymbol{e}_j. \tag{3.56}$$

But does this also hold true for arbitrary other points of the mesh? Yes, as will be shown in the following.

Imagine that a membrane Ω is subjected to two different unit point loads $\delta(\boldsymbol{y} - \boldsymbol{x}_1)$ and $\delta(\boldsymbol{y} - \boldsymbol{x}_2)$ respectively. Let $u_1(\boldsymbol{x})$ and $u_2(\boldsymbol{x})$ be the deflections of the membrane in these two load cases.

Formulating Green's second identity with these two solutions on Ω minus two small circular neighborhoods $N_\varepsilon(\boldsymbol{x}_1)$ and $N_\eta(\boldsymbol{x}_2)$ of the two source points $\boldsymbol{x}_1$ and $\boldsymbol{x}_2$ and taking the limit gives

$$\lim_{\varepsilon\to 0,\eta\to 0} \mathscr{B}(u_1, u_2)_{\Omega-N_\varepsilon-N_\eta} = u_2(\boldsymbol{x}_1) - u_1(\boldsymbol{x}_2) = 0 \tag{3.57}$$

which is Maxwell's theorem: the echoes are the same. A unit heat source (e.g.) at $\boldsymbol{x}_1$ produces at the point $\boldsymbol{x}_2$ the same temperature as a unit heat source placed at the point $\boldsymbol{x}_2$ will at the point $\boldsymbol{x}_1$.

So in the terminology of Betti's theorem Maxwell's theorem is the expression

$$\int_\Omega \delta(\boldsymbol{y} - \boldsymbol{x}_1)\, u_2(\boldsymbol{y})\, d\Omega_{\boldsymbol{y}} = \int_\Omega \delta(\boldsymbol{y} - \boldsymbol{x}_2)\, u_1(\boldsymbol{y})\, d\Omega_{\boldsymbol{y}} \tag{3.58}$$

and according to *Betti's theorem—extended* the exact solutions may be replaced by their FE-approximations that is

$$\int_\Omega \delta(\boldsymbol{y} - \boldsymbol{x}_1)\, u_2^h(\boldsymbol{y})\, d\Omega_{\boldsymbol{y}} = \int_\Omega \delta(\boldsymbol{y} - \boldsymbol{x}_2)\, u_1^h(\boldsymbol{y})\, d\Omega_{\boldsymbol{y}} \tag{3.59}$$

or

$$u_2^h(\boldsymbol{x}_1) = u_1^h(\boldsymbol{x}_2) \qquad \textit{Maxwell's Theorem—extended}. \tag{3.60}$$

And this holds true for arbitrary pairs of points. So also in FE-analysis the symmetry of the echoes is preserved. And this symmetry can be extended to arbitrary linear functionals.

Theorem 3.5 (Maxwell for functionals) *If $J_1(u)$ and $J_2(u)$ are linear functionals on $\mathcal{V}$ and if G_1 and G_2 are the corresponding Riesz elements respectively then*

$$J_1(G_2) = J_2(G_1) \tag{3.61}$$

and the FE-approximations obey the same law

$$J_1(G_2^h) = J_2(G_1^h). \tag{3.62}$$

The displacement field generated by a horizontal dislocation of the point $\boldsymbol{x}$ in Fig. 3.8b is the Green's function $\boldsymbol{G}_1$ for the stress σ_{xx} at this point. To this field belong certain shear stresses σ_{xy} in the cross section $(0, l)$. And the integral of theses stresses has the same value as the stress σ_{xx} at the point $\boldsymbol{x}$ if the two halves of the cross section $(0, l)$ slide in opposite direction by one unit length (= influence function $\boldsymbol{G}_2$ for the resultant shear force) so that

$$J_1(\boldsymbol{G}_2) = \sigma_{xx}(\boldsymbol{G}_2)(\boldsymbol{x}) = \int_0^l \sigma_{xy}(\boldsymbol{G}_1)(s)\, ds = J_2(\boldsymbol{G}_1). \tag{3.63}$$

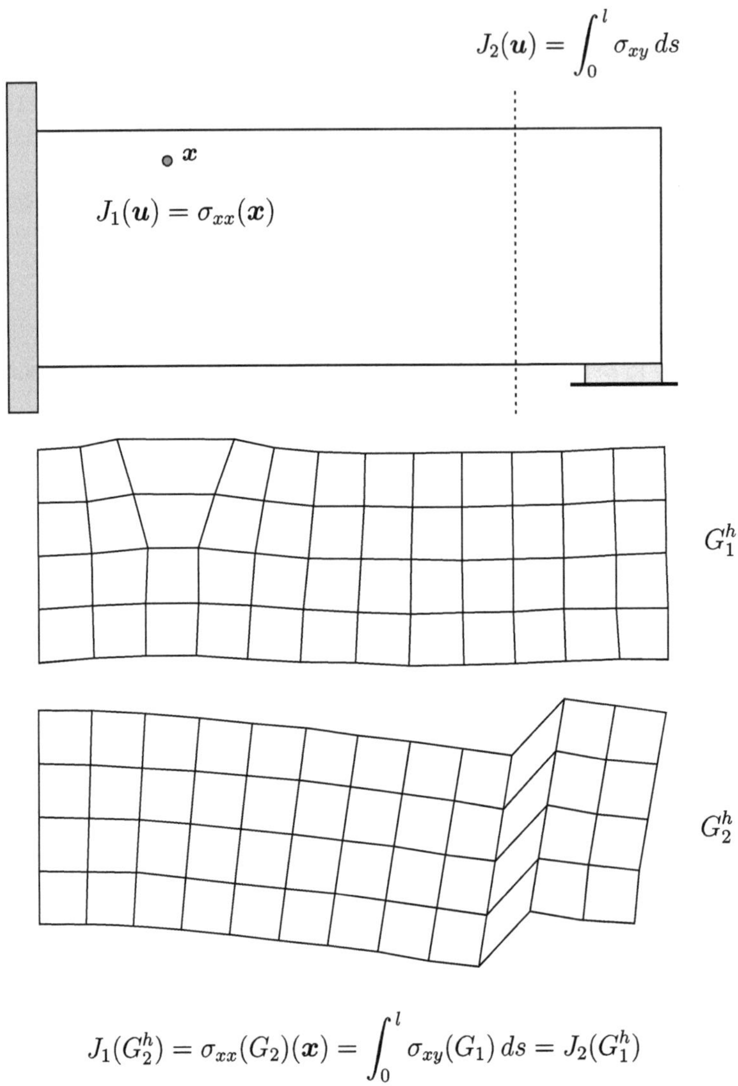

Fig. 3.8 The Green's functions (Riesz elements) of the two functionals satisfy $J_1(G_2) = J_2(G_1)$ and this also holds true for the FE-solutions, $J_1(G_2^h) = J_2(G_1^h)$

3.4 Proxies

The FE-Green's function $G_h(\boldsymbol{y}, \boldsymbol{x})$ tries to mimic the actions of the original Green's function

$$u(\boldsymbol{x}) = \int_{\Omega} G(\boldsymbol{y}, \boldsymbol{x})\, p(\boldsymbol{y})\, d\Omega_{\boldsymbol{y}} \qquad \text{orig.} \tag{3.64}$$

$$u_h(\boldsymbol{x}) = \int_{\Omega} G_h(\boldsymbol{y}, \boldsymbol{x})\, p(\boldsymbol{y})\, d\Omega_{\boldsymbol{y}} \qquad \text{subst.} \tag{3.65}$$

and the FE-Dirac delta $\delta_h(\boldsymbol{y}, \boldsymbol{x})$ tries the same with the original Dirac delta $\delta(\boldsymbol{y} - \boldsymbol{x})$

$$u(\boldsymbol{x}) = \int_{\Omega} \delta(\boldsymbol{y} - \boldsymbol{x})\, u(\boldsymbol{y})\, d\Omega_{\boldsymbol{y}} \qquad \text{orig.} \tag{3.66}$$

$$u_h(\boldsymbol{x}) = \int_{\Omega} \delta_h(\boldsymbol{y}, \boldsymbol{x})\, u(\boldsymbol{y})\, d\Omega_{\boldsymbol{y}} \qquad \text{subst.} \tag{3.67}$$

but they both fail. What they produce is only an approximation $u_h(\boldsymbol{x})$ to the exact value $u(\boldsymbol{x})$.

But on the space $\mathcal{V}_h$ they perform flawlessly. On the space $\mathcal{V}_h$ they are as good as the originals as will be explained in the following.

Remark 3.1 We write $\delta_h(\boldsymbol{y}, \boldsymbol{x})$ and not $\delta_h(\boldsymbol{y} - \boldsymbol{x})$ because δ_h represents a certain distribution of exterior forces or loads which is specific for each point $\boldsymbol{x}$ and which is not invariant to shifts; to different pairs of points $\boldsymbol{y}, \boldsymbol{x}$ and $\hat{\boldsymbol{y}}, \hat{\boldsymbol{x}}$ having the same distance $r = |\hat{\boldsymbol{y}} - \hat{\boldsymbol{x}}| = |\boldsymbol{y} - \boldsymbol{x}|$ belong different sets of forces δ_h.

3.4.1 $G_h = G$ on $\mathcal{V}_h^*$

Let us assume that the solution to the boundary value problem

$$-\Delta u = p \qquad \text{on } \Omega \qquad u = 0 \qquad \text{on } \Gamma \tag{3.68}$$

lies in $\mathcal{V}_h$. Green's first identity

$$\mathcal{G}(u, \varphi_i) = \int_{\Omega} p\, \varphi_i\, d\Omega - a(u, \varphi_i) = 0 \qquad \forall \varphi_i \in \mathcal{V}_h \tag{3.69}$$

then implies that the FE-solution u_h

$$a(u_h, \varphi_i) = (p, \varphi_i) \qquad \forall \varphi_i \in \mathcal{V}_h \tag{3.70}$$

is identical with u. But if u lies in $\mathcal{V}_h$ then the approximate kernel G_h—though it is not identical with the exact kernel G—must map p onto the exact solution $u = u_h$

$$u(\boldsymbol{x}) - u_h(\boldsymbol{x}) = \int_\Omega [G(\boldsymbol{y}, \boldsymbol{x}) - G_h(\boldsymbol{y}, \boldsymbol{x})]\, p(\boldsymbol{y})\, d\Omega_{\boldsymbol{y}} = 0 \tag{3.71}$$

and so the error $G - G_h$—which is not zero (!)—must be orthogonal to the right-hand side p if the solution lies in $\mathcal{V}_h$.

Proof To see this it suffices to show for one φ_i that

$$\varphi_i(\boldsymbol{x}) - \varphi_i(\boldsymbol{x}) = \int_\Omega [G(\boldsymbol{y}, \boldsymbol{x}) - G_h(\boldsymbol{y}, \boldsymbol{x})]\, p(\varphi_i)(\boldsymbol{y})\, d\Omega_{\boldsymbol{y}} = 0 \tag{3.72}$$

where $p(\varphi_i)$ is the distribution that belongs to the shape function φ_i. The first part

$$\varphi_i(\boldsymbol{x}) = \int_\Omega G(\boldsymbol{y}, \boldsymbol{x})\, p(\varphi_i)(\boldsymbol{y})\, d\Omega_{\boldsymbol{y}} \tag{3.73}$$

is evident. To prove the second part

$$\varphi_i(\boldsymbol{x}) = \int_\Omega G_h(\boldsymbol{y}, \boldsymbol{x})\, p(\varphi_i)(\boldsymbol{y})\, d\Omega_{\boldsymbol{y}} \tag{3.74}$$

which in short-hand notation is

$$(\delta[\boldsymbol{x}], \varphi_i) = (G_h, p(\varphi_i)) \tag{3.75}$$

we recall *Betti's Theorem-Extended* which asserts that in the equation

$$W_{12} = (\delta[\boldsymbol{x}], \varphi_i) = (G, p(\varphi_i)) = W_{21} \tag{3.76}$$

the functions φ_i and G may be replaced by their projections onto $\mathcal{V}_h$. Because φ_i lies in $\mathcal{V}_h$ its projection is identical with φ_i and so (3.75) is established. □

This result applies of course to *any* linear functional.

3.4.2 $\delta_h = \delta$ on $\mathcal{V}_h$

On $\mathcal{V}_h$ the blurred Dirac delta δ_h finds its target as well as the true Dirac delta.

The Dirac delta $\delta(\boldsymbol{y} - \boldsymbol{x})$ is not a proper function. But the approximate Dirac delta $\delta_h(\boldsymbol{y}, \boldsymbol{x})$ is real, it can be displayed on the screen and the work done by δ_h on acting through a virtual displacement, see Fig. 3.9, can be calculated.

The approximate Dirac delta $\delta_h(\boldsymbol{y}, \boldsymbol{x})$ is the right-hand side, the distribution, that belongs to the approximate Green's function $G_h(\boldsymbol{y}, \boldsymbol{x})$ in the sense of (3.19)

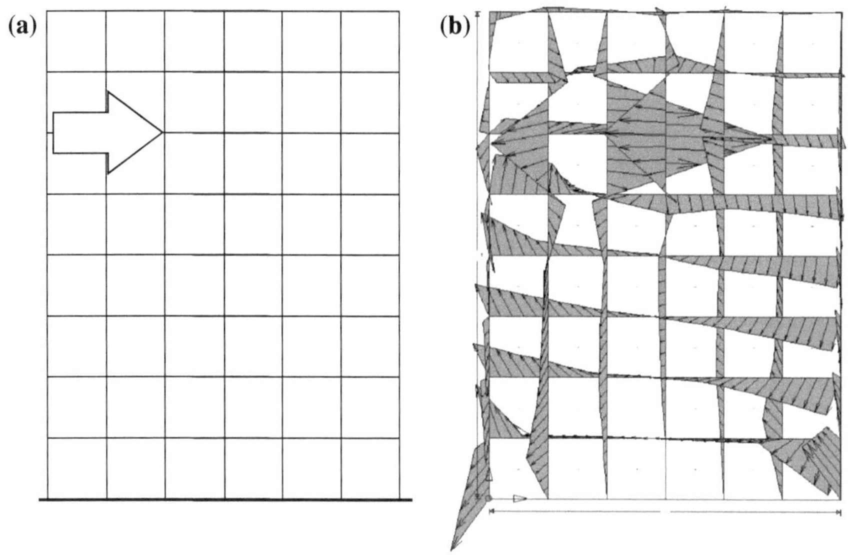

Fig. 3.9 **a** Dirac Delta δ and **b** Dirac Delta δ_h

$$\mathscr{G}(G_h, 1) := \sum_e \int_{\Omega_e} -\Delta\, G_h \cdot 1\, d\Omega + \sum_k \int_{\Gamma_k} l_k \cdot 1\, ds = 0. \tag{3.77}$$

That is $\delta_h(\boldsymbol{y}, \boldsymbol{x})$ is the assemblage of all the element loads $p_h^e := -\Delta G_h(\boldsymbol{y}, \boldsymbol{x})$ (on an element Ω_e) and the collection of all the jumps l_k of the normal derivative in between the elements and on the exterior edge (lines Γ_k). On the exterior edge the jump is identical with the slope of the FE-solution.

Because the two load cases $\delta(\boldsymbol{y} - \boldsymbol{x})$ and $\delta_h(\boldsymbol{y}, \boldsymbol{x})$ are equivalent with respect to all $\varphi_i \in \mathcal{V}_h$ it follows

$$\varphi_i(\boldsymbol{x}) = \int_{\Omega} \delta(\boldsymbol{y} - \boldsymbol{x})\, \varphi_i(\boldsymbol{y})\, d\Omega_y = \int_{\Omega} \delta_h(\boldsymbol{y}, \boldsymbol{x})\, \varphi_i(\boldsymbol{y})\, d\Omega_y \qquad \forall\, \varphi_i \in \mathcal{V}_h \tag{3.78}$$

and so for any function $u_h \in \mathcal{V}_h$ as well

$$u_h(\boldsymbol{x}) = \int_{\Omega} \delta_h(\boldsymbol{y}, \boldsymbol{x})\, u_h(\boldsymbol{y})\, d\Omega_y. \tag{3.79}$$

These substitute Dirac deltas definitely are no point sources. But on $\mathcal{V}_h$ their actions produce the same effect as the true Dirac deltas. This is the meaning of (3.78).

In Fig. 3.9 are displayed the true Dirac delta and the substitute delta. Both are work-equivalent on $\mathcal{V}_h$. The work done by the true point load $P = 1$ on acting through

$v_h(\boldsymbol{x})$ and the work done by the element loads and line loads of the substitute δ_h on acting through v_h

$$v_h(\boldsymbol{x}) = \begin{cases} \int\limits_\Omega \delta(\boldsymbol{y} - \boldsymbol{x})\, v_h(\boldsymbol{y})\, d\Omega_{\boldsymbol{y}} \\ \int\limits_\Omega \delta_h(\boldsymbol{y}, \boldsymbol{x})\, v_h(\boldsymbol{y})\, d\Omega_{\boldsymbol{y}} \end{cases} \tag{3.80}$$

is the same.

The integral (3.79) basically is the interpolation operator which reconstructs $u_h(\boldsymbol{x})$ by weighting its nodal values with the nodal forces j_i of the approximate Green's function

$$u_h(\boldsymbol{x}) = \int\limits_\Omega \delta_h(\boldsymbol{y}, \boldsymbol{x})\, u_h(\boldsymbol{y})\, d\Omega_{\boldsymbol{y}} = \boldsymbol{j}(\boldsymbol{x})^T\, \boldsymbol{u} \tag{3.81}$$

and these nodal forces in turn are simply the values of the different shape functions at the point $\boldsymbol{x}$

$$u_h(\boldsymbol{x}) = \boldsymbol{j}(\boldsymbol{x})^T\, \boldsymbol{u} = \sum_i j_i\, u_i = \sum_i \varphi_i(\boldsymbol{x})\, u_i. \tag{3.82}$$

Remark 3.2 The side by side comparison of δ with δ_h in Fig. 3.9 may look impressive—the approximate Dirac delta seems to shoulder a heavy load in contrast to the slender point-like original Dirac delta—but one may not forget that because of the Galerkin orthogonality—we let the source point $\boldsymbol{x}$ a node $\boldsymbol{x}_j$ and δ_{ij} is the *Kronecker delta*

$$\begin{aligned} a(G[\boldsymbol{x}_j] - G_h[\boldsymbol{x}_j], \varphi_i) &= (\delta[\boldsymbol{x}_j] - \delta_h[\boldsymbol{x}_j], \varphi_i) \\ &= \varphi_i(\boldsymbol{x}_j) - \int\limits_\Omega \delta_h(\boldsymbol{y}, \boldsymbol{x}_j)\, \varphi_i(\boldsymbol{y})\, d\Omega_{\boldsymbol{y}} \\ &= \delta_{ij} - \int\limits_\Omega \delta_h(\boldsymbol{y}, \boldsymbol{x}_j)\, \varphi_i(\boldsymbol{y})\, d\Omega_{\boldsymbol{y}} = 0 \end{aligned} \tag{3.83}$$

almost all shape functions $\varphi_i(\boldsymbol{y})$ are orthogonal to δ_h, so most of δ_h is "non-existent" in the weak sense, is a "weak" zero.

But the solution u senses the presence of δ_h and of each part of it! The work done by δ_h on acting through u is the value of the FE-solution at the source point $\boldsymbol{x}_j$

$$u_h(\boldsymbol{x}_j) = \int\limits_\Omega \delta_h(\boldsymbol{y}, \boldsymbol{x}_j)\, u(\boldsymbol{y})\, d\Omega_{\boldsymbol{y}} \tag{3.84}$$

and in evaluating this integral each element load p_e^h and each line load l_k that contributes to δ_h counts.

3.4.3 $J_h(u) = J(u)$ on $\mathcal{V}_h$

Each linear and bounded functional $J(u)$ can be identified with the action of a certain Dirac delta $\delta(\boldsymbol{y}-\boldsymbol{x})$ or a Green's function $G(\boldsymbol{y},\boldsymbol{x})$ respectively

$$J(u) = \int_\Omega \delta(\boldsymbol{y}-\boldsymbol{x})\, u(\boldsymbol{y})\, d\Omega_{\boldsymbol{y}} = \int_\Omega G(\boldsymbol{y},\boldsymbol{x})\, p(\boldsymbol{y})\, d\Omega_{\boldsymbol{y}} \tag{3.85}$$

where the Green's function is the Riesz element of the functional $J(u)$.

When G is projected onto $\mathcal{V}_h$, $G \to G_h$, then the new kernel $G_h(\boldsymbol{y},\boldsymbol{x})$ constitutes a new functional

$$J_h(u) = \int_\Omega \delta_h(\boldsymbol{y},\boldsymbol{x})\, u(\boldsymbol{y})\, d\Omega_{\boldsymbol{y}} = \int_\Omega G_h(\boldsymbol{y},\boldsymbol{x})\, p(\boldsymbol{y})\, d\Omega_{\boldsymbol{y}} \tag{3.86}$$

which in general is different from $J(u)$

$$J_h(u) \neq J(u) \tag{3.87}$$

that is the two functionals map the same function u in general onto different points on the real axis. But when we restrict the functionals to $\mathcal{V}_h$ then the results agree.

Theorem 3.6 (Equivalence) *On $\mathcal{V}_h$ the two functionals agree*

$$J(\varphi_i) = J_h(\varphi_i) \qquad \forall\, \varphi_i \in \mathcal{V}_h. \tag{3.88}$$

An alternative version of this theorem is the following.

Theorem 3.7 (Dual Galerkin orthogonality) *If G_h is the projection of the Green's function G onto the subspace $\mathcal{V}_h$ and $p(v_h)$ the right-hand side (= distribution) of a trial function $v_h \in \mathcal{V}_h$ then holds*

$$\int_\Omega [G(\boldsymbol{y},\boldsymbol{x}) - G_h(\boldsymbol{y},\boldsymbol{x})]\, p(v_h)(\boldsymbol{y})\, d\Omega_{\boldsymbol{y}} = 0 \qquad \forall\, v_h \in \mathcal{V}_h. \tag{3.89}$$

We call it *dual Galerkin orthogonality* because it is conjugate to the classical Galerkin orthogonality

$$a(u-u_h, v_h) = \int_\Omega (p - p_h)\, v_h\, d\Omega = 0 \qquad \forall\, v_h \in \mathcal{V}_h \tag{3.90}$$

which states that the error in the right-hand side is orthogonal to $\mathcal{V}_h$. Here the statement is that the error in the Green's function is orthogonal to the functions in $\mathcal{V}_h^*$, which are the right-hand sides, the distributions or load cases, which belong to the trial functions $v_h \in \mathcal{V}_h$. Galerkin tests with the trial functions v_h while here the test is performed with the loads or distributions $p_h(v_h)$ which belong to the v_h.

This result implies that if the exact solution lies in $\mathcal{V}_h$ then

$$J(u) - J_h(u) = \int_\Omega [G(\boldsymbol{y}, \boldsymbol{x}) - G_h(\boldsymbol{y}, \boldsymbol{x})]\, p(\boldsymbol{y})\, d\Omega_{\boldsymbol{y}} = 0. \tag{3.91}$$

On the larger space $\mathcal{V}$ the two functionals differ in general

$$J(v) \neq J_h(v) \qquad v \notin \mathcal{V}_h \tag{3.92}$$

because the Galerkin orthogonality only holds true on the subset $\mathcal{V}_h \subset \mathcal{V}$. But the following theorem applies:

Theorem 3.8 (Dual Reciprocity) *If u_h is the projection of u onto the subspace $\mathcal{V}_h$ and G_h the projection of G then*

$$J(u_h) = J_h(u). \tag{3.93}$$

Proof The proof is easy

$$J(u_h) = (\delta, u_h) = a(G, u_h) = a(G_h, u_h) = (p, G_h) = J_h(u) \tag{3.94}$$

and basically is an application of *Betti's theorem—extended*. Or it can be read as Tottenham's equation

$$J(u_h) = (\delta, u_h) = u_h(\boldsymbol{x}) = \int_0^l G_h\, p\, dy = J_h(u). \quad \square \tag{3.95}$$

Imagine there is a traffic jam on a two-span bridge. The pier in the middle of the bridge has to withstand a certain portion of the total weight of the stalled cars and trucks. When this problem is worked out with an FE-program then (1) the original load p gets replaced by a work-equivalent load p_h

$$\text{(load case } p\text{)}\ u \ \rightarrow\ u_h \ \text{(load case } p_h\text{)} \tag{3.96}$$

and (2) the pier reaction $J(u)$ is calculated with the substitute FE-influence function

$$J_h(u_h) = \text{FE pier reaction}. \tag{3.97}$$

So it seems that two errors are committed: wrong load and wrong "yardstick". But in truth it is only one because

$$J_h(u_h) = J(u_h). \tag{3.98}$$

What the FE-influence function and the exact influence function measure is the same because u_h lies in $\mathcal{V}_h$. Error #2 never materializes in an FE-code.

We can now also establish when the error

$$J(u_h) - J(u) = 0 \tag{3.99}$$

of an FE-solution is zero.

Theorem 3.9 (Exact values)
Sufficient conditions

1. *If the Riesz element G of a functional $J()$ lies in $\mathcal{V}_h$ then it is identical with its projection, $G_h = G$, that is then holds*

$$J_h(u) = J(u) \qquad \forall\, u \in \mathcal{V} \tag{3.100}$$

and so also

$$J(u_h) = J_h(u) = J(u). \tag{3.101}$$

2. *If the exact solution lies in $\mathcal{V}_h$, $u = u_h$ (its projection) then the error in any Green's functions is orthogonal to the right-hand side p*

$$J(u) - J(u_h) = \int_\Omega [G(\boldsymbol{y}, \boldsymbol{x}) - G_h(\boldsymbol{y}, \boldsymbol{x})]\, p(\boldsymbol{y})\, d\Omega_y = 0. \tag{3.102}$$

Necessary condition

1. *If a value is exact, $J(u_h) = J(u)$ then the error in the Green's function must be orthogonal to the right-hand side p*

$$J(u) - J(u_h) = \int_\Omega [G(\boldsymbol{y}, \boldsymbol{x}) - G_h(\boldsymbol{y}, \boldsymbol{x})]\, p(\boldsymbol{y})\, d\Omega_y = 0. \tag{3.103}$$

Consider for example the functional

$$J(u) := \int_\Omega -\Delta\, u\, d\Omega + \int_\Gamma \frac{\partial u}{\partial n}\, ds \tag{3.104}$$

which is zero for any function $u \in C^2(\Omega)$ because it is simply the integration by parts formula

$$J(u) = \mathscr{G}(u, 1) = \int_\Omega -\Delta\, u \cdot 1\, d\Omega + \int_\Gamma \frac{\partial u}{\partial n} \cdot 1\, ds - \underbrace{a(u, 1)}_{=0} = 0. \tag{3.105}$$

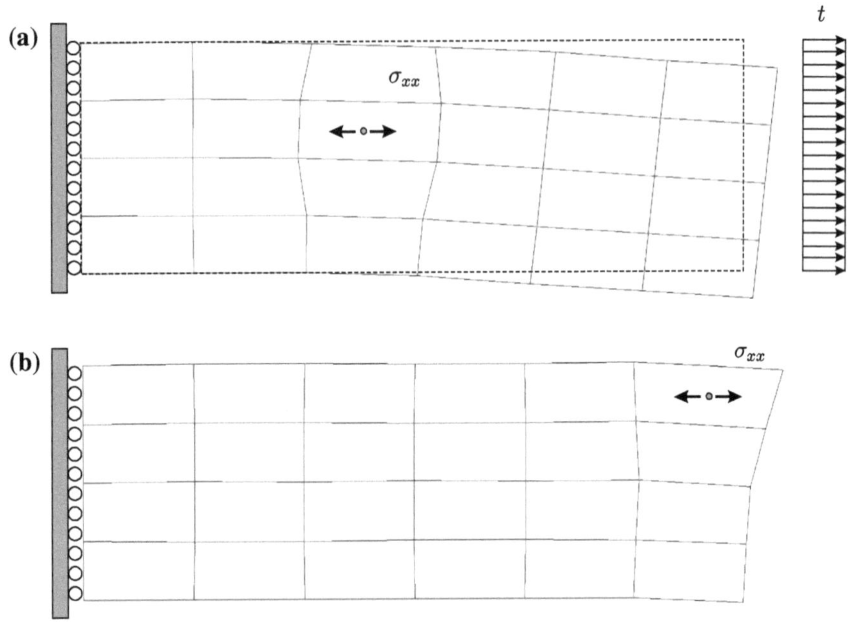

Fig. 3.10 FE-influence functions for σ_{xx}. In the vicinity of the source point the functions certainly are not correct but they can predict the correct value σ_{xx} when constant forces pull at the edge. All FE-influence function for $\sigma_{xx}(\boldsymbol{x})$, where $\boldsymbol{x}$ can be any point, have this property!

In mechanical terms it expresses the fact that the tractions $\partial u/\partial n$ on the boundary balance the pressure $p = -\Delta u$ applied to a membrane. Any smooth function satisfies this (natural) equilibrium condition.

The Riesz element of the functional (3.104) is the function $G(\boldsymbol{y}, \boldsymbol{x}) = 1$ and because the shape functions φ_i form a partition of unity the unit-function $G(\boldsymbol{y}, \boldsymbol{x}) = 1$ lies in $\mathcal{V}_h^+$ (the full or "unrestrained" space $\mathcal{V}_h$ which knows nothing of boundary conditions) and therefore it is guaranteed that also the FE-solution u_h satisfies the equilibrium condition,

$$J(u_h) = \sum_e \int_{\Omega_e} -\Delta\, G_h\, d\Omega + \sum_k \int_{\Gamma_k} j_k\, ds = 0 \tag{3.106}$$

where the first sum extends over the interior of all elements Ω_e and the second sum measures the jumps j_k in the slope along all interelement boundaries Γ_k which here also include the edge of the domain.

In Fig. 3.10 are plotted different influence functions for the stress σ_{xx} in a plate. These influence functions certainly are not exact in the neighborhood of the source point. But their shape at the right end of the plate must be correct in an *integral sense* because we know from the *patch test* that such a simple problem is exactly solvable with bilinear elements and this means that when a uniform load t pulls on the edge then the influence functions

$$\sigma_{xx}(\boldsymbol{x}) = t \cdot \int_0^l G_h(\boldsymbol{x}, \boldsymbol{y})\, ds_y = t \cdot l \cdot G_{avg}(\boldsymbol{x}) \tag{3.107}$$

must evaluate to the exact value.

This means that the horizontal component $G_h(\boldsymbol{x}, \boldsymbol{y})$ of the FE-influence function obviously has the correct average value $G_{avg}(\boldsymbol{x})$ at the right end of the plate and this must be true, as we know from the patch test, for all points of the plate. The piecewise polynomial Green's functions, the influence functions for the stresses $\sigma_{xx}(\boldsymbol{x})$ at all the points $\boldsymbol{x}$ of the plate, are "crooked" and "flawed" and far from being perfect but they have one property in common: their average value $G_{avg}(\boldsymbol{x})$ over the right edge is *exact*. This is indeed a remarkable property.

Just imagine a point $\boldsymbol{x}$ very, very close to the edge. No hope for the shape functions to come close—even only in rudimentary form—to the exact Green's function but the average value is correct—definitely.

Remark 3.3 The patch test can be seen as a test of an FE-model in how far it is able to produce influence functions which are exact in an integral sense because most often the loads applied in a patch test are uniform. The outcome of a patch test in 2-D and 3-D with a point load applied is guaranteed to be negative.

3.4.4 Summary

So an FE-solution can be written in six different ways

$$\begin{aligned} u_h(\boldsymbol{x}) &= \int_\Omega G(\boldsymbol{y}, \boldsymbol{x}) p_h(\boldsymbol{y})\, d\Omega_y = \int_\Omega G_h(\boldsymbol{y}, \boldsymbol{x}) p_h(\boldsymbol{y})\, d\Omega_y \\ &= \int_\Omega G_h(\boldsymbol{y}, \boldsymbol{x}) p(\boldsymbol{y})\, d\Omega_y \\ &= \int_\Omega \delta(\boldsymbol{y} - \boldsymbol{x}) u_h(\boldsymbol{y})\, d\Omega_y = \int_\Omega \delta_h(\boldsymbol{y}, \boldsymbol{x}) u_h(\boldsymbol{y})\, d\Omega_y \\ &= \int_\Omega \delta_h(\boldsymbol{y}, \boldsymbol{x}) u(\boldsymbol{y})\, d\Omega_y\,, \end{aligned} \tag{3.108}$$

not counting the weak forms

$$u_h(x) = a(G, u_h) = a(G_h, u_h) = a(G_h, u). \tag{3.109}$$

Splines or Bezier curves are controlled by a (finite) set of control points or knots and in a similar way the deflection $u(x)$ of a rope is controlled by the Green's functions $G(y, x)$ of the rope

$$u(x) = \int_0^l G(y, x)\, p(y)\, dy. \tag{3.110}$$

They represent an infinite number of invisible weights on $u(x)$ which control the "fit" between u and the applied load p. The solution to the boundary value problem (1.30) is the only function which passes all tests

$$u(x) = \frac{p}{2}\, x\, (l - x) = \int_0^l G(y, x)\, p(y)\, dy \tag{3.111}$$

$$\textit{for all Green's functions } G(y, x) \textit{ of the rope} \tag{3.112}$$

while the FE-solution $u_h(x)$ fails in most of these tests

$$u_h(x) \neq \int_0^l G(y, x)\, p(y)\, dy. \tag{3.113}$$

The only exception being the Green's functions of the nodes because the piecewise linear FE-solution of the rope interpolates the exact curve at the nodes, $u_h(x_i) = u(x_i)$.

The amazing fact is that when we decide to forego the exact Green's functions and instead repeat the tests with the approximate Green's functions then the FE-solution passes *all* tests

$$u_h(x) = \int_0^l G_h(y, x)\, p(y)\, dy \tag{3.114}$$

$$\textit{for all FE-Green's functions } G_h(y, x) \textit{ of the rope.}$$

That is if we make our own rules, if the shape $u_h(x)$ is subjected to a different set of control points—figuratively speaking—it fits perfectly. In this sense the FE-method is consistent.

So according to (3.108) the FE-method can be interpreted in two ways:

- either as a modification of the right-hand side, $p \to p_h$
- or as a modification of the kernels, $G(y, x) \to G_h(y, x)$.

Remarkably both strategies produce the same error.

3.5 Dirac Energy

Influence functions express a balance, an energy balance. The work done by a unit point load $P = 1$ on acting through the deflection $u(x)$ of the rope

$$1 \cdot u(x) = \int_{\Omega} G(y, x)\, p(y)\, dy \tag{3.115}$$

is the same as the work done by the distributed load p on acting through $G(y, x)$, the reaction of the rope to the point load.

The factor 1 is essential because otherwise the units would not match

$$\begin{aligned} force \cdot length = 1 \cdot u(x_c) &= \int_0^l G(y, x)\, p(y)\, dy \\ &= length \cdot force/length \cdot length \end{aligned} \tag{3.116}$$

So the output of an influence function is an *energy*. We call this energy quantum the "Dirac energy".

It is the work done by the applied load on acting through the influence function.

This principle finds its simplest expression in a see-saw, see Fig. 3.11b: the work done by the two weights is zero for each turn of the see-saw

$$\begin{aligned} P_l\, u_l - P_r\, u_r &= P_l \tan\varphi\, h_l - P_r \tan\varphi\, h_r \\ &= (P_l\, h_l - P_r\, h_r) \tan\varphi = 0 \qquad \forall \varphi \end{aligned} \tag{3.117}$$

because the two loads balance, $P_l\, h_l = P_r\, h_r$.

In this sense each influence function is a see-saw. To calculate the shear force $V(x)$ of the frame at a point x, see Fig. 3.11a, we install a shear hinge at x and apply a unit dislocation to the hinge so that the shear force $V(x)$ does the work $-V(x) \cdot 1$,

$$-V(x)\, u(x_-) - V(x)\, u(x_+) = -V(x)\,(u(x_-) + u(x_+)) = -V(x) \cdot 1. \tag{3.118}$$

The work done by the load, the point force P, on acting through the displacement u, initiated by the dislocation, must be the opposite of this value because

$$\underbrace{-V(x) \cdot 1 + P\, u}_{W_{1,2}} = 0. \tag{3.119}$$

This is Betti's theorem, $W_{1,2} = W_{2,1}$. Betti's theorem is the logic of the see-saw and pulleys, see Fig. 3.12. (The work $W_{2,1}$ is zero, see the following remark).

So to each internal action, $V(x)$, $N(x)$, $M(x)$ etc., belongs a certain mechanism, a certain see-saw and when we activate this mechanism—unlock the hinge—and

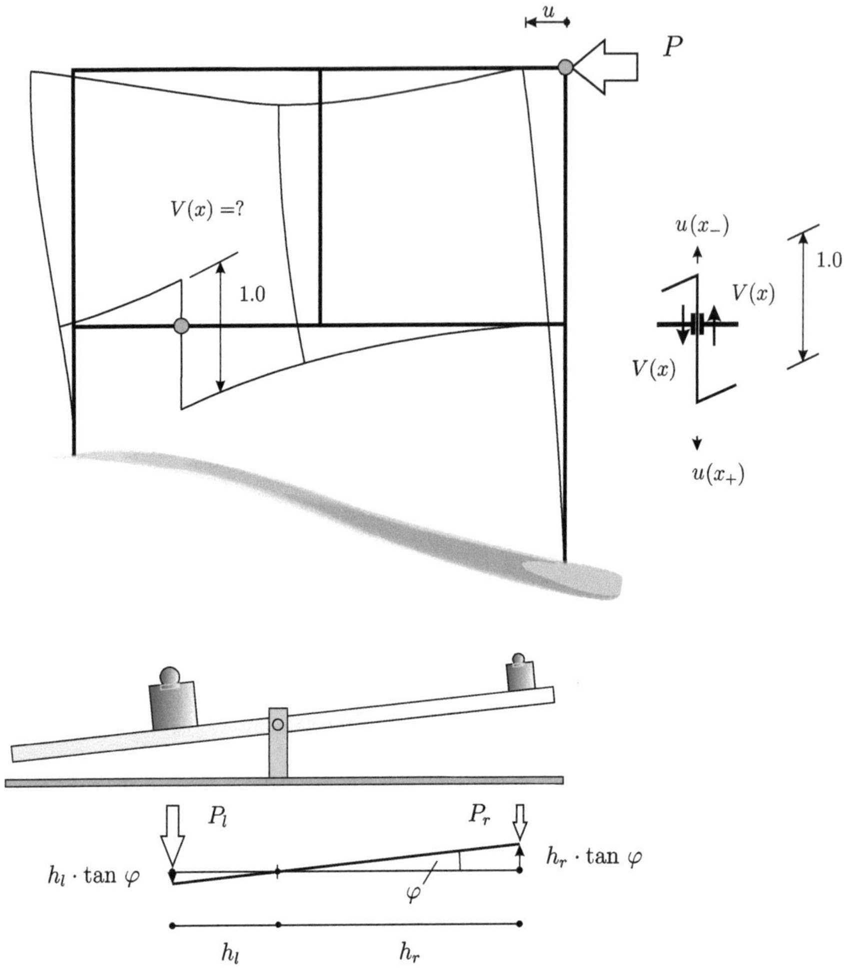

Fig. 3.11 An influence function resembles a see-saw

count the work done by the load on acting through the displacements initiated by spreading the hinge then we learn how large the internal action must be to balance the work of the load. In FE-analysis we hinder the free motions of the structure and so the mechanism gets a wrong signal in return, the effect of the unit spread which reaches P is u_h

$$-V_h(x) \cdot 1 + P\, u_h = 0 \tag{3.120}$$

and not the exact effect u

$$-V(x) \cdot 1 + P\, u = 0 \tag{3.121}$$

and so $V_h(x) \neq V(x)$. *An FE-program gets the Dirac energies wrong.*

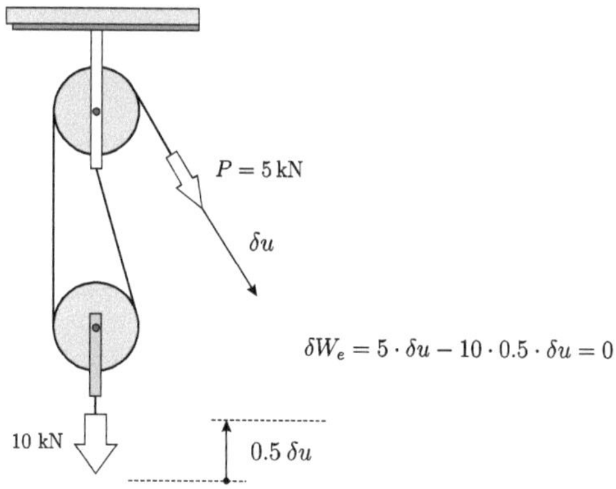

Fig. 3.12 A pulley demonstrates a basic principle of linear mechanics: equilibrium means equal work, means "symmetry" with respect to translations

We conclude that the kinematics of a mesh, the richness in details, determines the accuracy of an FE-solution, see Figs. 3.13 and 3.14.

- mesh = kinematics = accuracy of influence functions = quality of results.

In the Sleipner-platform accident the FE-bending moment was about half the exact value because of the poor kinematics of the mesh in the region of the tri-cells [2].

It can now also be stated what a good design is. The energy balance

$$V(x) = \frac{P\,u}{1} = P \cdot \frac{u}{1} \qquad \leftarrow \qquad \text{make } u \text{ small!} \tag{3.122}$$

signals that the aim must be a design where the displacement u which reaches the point load P due to the unit spread (the denominator) of the hinge is as small as possible because then $V(x)$ will only be a small fraction of the applied load P.

Throw a stone into the water and watch the ripples! The smaller the ripples that reach the load, the better. Archimedes' lever (willingly) is the opposite of a good design: a unit displacement of the tip of the lever on the left leads to a very large displacement u at the other end of the lever. (*Vice versa:* Archimedes has miles to go for to lift the Earth just one iota!).

Remark 3.4 Equation (3.122) shows that influence is a ratio, $u/1$, of two displacements and therefore it is of no concern whether the unit spread is 1 mm, 1 cm, 1 m or even 1 mile. It could be any value. A unit spread only makes calculations easier.

Remark 3.5 Influence functions for *displacements* are based on Betti's theorem

$$W_{1,2} = W_{2,1}. \tag{3.123}$$

The work done by a unit force $P = 1$ on acting through $u(x)$ is the same as the work done by the applied load on acting through the displacement initiated by the point load.

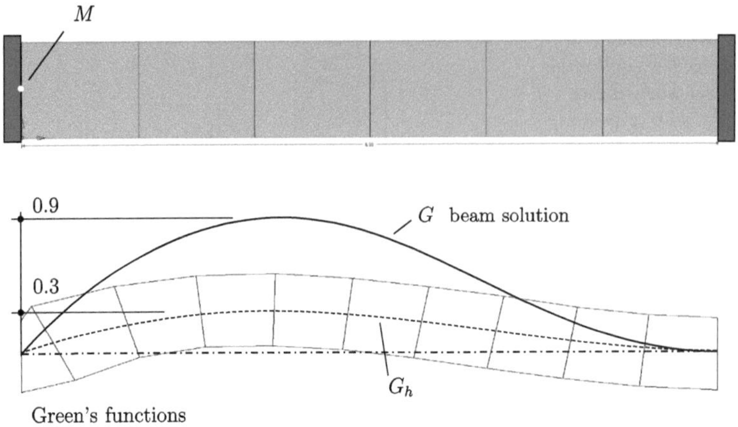

Fig. 3.13 Influence function for the bending moment M in a beam like structure; exact solution and FE-solution; the poor kinematics of the bilinear elements leads to a large error

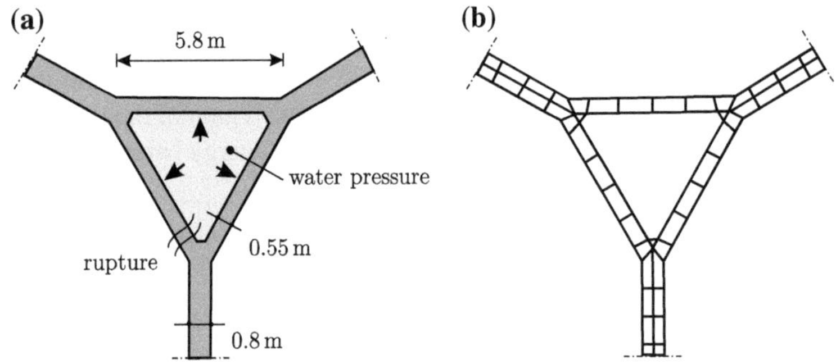

Fig. 3.14 The tri-cell of the Sleipner-platform and the FE-mesh which was too coarse to produce reliable results for the bending moments

Influence functions for *force terms* are based on the same equation

$$W_{1,2} = 0 \tag{3.124}$$

but $W_{2,1}$ is always zero. It is the work done by the two forces which effect the unit spread of the hinge. These forces $\pm F$, one on each side of the hinge, have opposite signs but are otherwise equal so that their total work is zero

$$W_{2,1} = F\, u(x_-) - F\, u(x_+) = F\,(u(x_-) - u(x_+)) = 0 \tag{3.125}$$

because the displacement u of the primary problem is continuous at x.

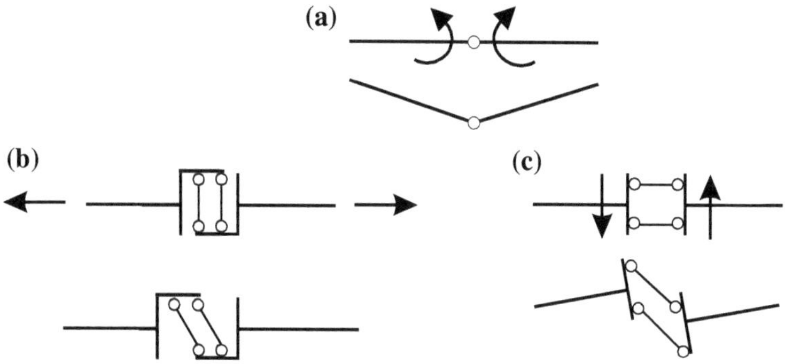

Fig. 3.15 Hinges allow to generate Green's function in frame analysis, **a** M-hinge, **b** N-hinge, **c** V-hinge

Remark 3.6 The technique to construct influence functions by unit spreads of $N-$, $M-$ or $V-$hinges, see Fig. 3.15, is called in structural mechanics an application of the *Müller–Breslau principle*.

3.6 Generalized Green's Functions

Virtual work, $(p, \delta u)$, is a functional and this is why functionals play such a central role in the theory of weak boundary value problems. When we make the transition from the formulation of the boundary value problem

$$- \Delta u = p \qquad \text{on } \Omega \qquad u = 0 \qquad \text{on } \Gamma \tag{3.126}$$

in a pointwise sense to a formulation in a variational sense

$$a(u, v) = \int_\Omega p\, v\, d\Omega \qquad \forall\, v \in \mathcal{V} \tag{3.127}$$

the right-hand side p becomes a linear functional

$$J(v) := \int_\Omega p\, v\, d\Omega\,, \tag{3.128}$$

that is (3.127) is equivalent to

$$a(u, v) = J(v) \qquad \forall\, v \in \mathcal{V} \tag{3.129}$$

and so—by definition—the variational solution u of (3.126) is also the Riesz element[1] of the functional $J(v)$. This can be verified by formulating Green's first identity with the solution u and a test function v

$$\mathscr{G}(v, u) = \int_{\Omega} -\Delta v\, u\, d\Omega - a(v, u) = \int_{\Omega} -\Delta v\, u\, d\Omega - J(v) = 0 \tag{3.130}$$

or

$$\int_{\Omega} -\Delta v\, u\, d\Omega = J(v) \tag{3.131}$$

which confirms that the work done by $-\Delta v = p$ on acting through the Green's function u is the value $J(v)$.

So each FE-solution is automatically also the Riesz element of the functional $J(v)$ on the right-hand side and—vice versa—given any linear functional $J(v)$ the solution to the variational problem (3.129) is the Riesz element of the functional.

We call the Riesz element of the functional $J(v)$ the *generalized Green's function* of $J(v)$. So both terms are synonyms.

Historically Green's functions came first and the concept of a Riesz element came later. The term generalized is to indicate that the existence of a generalized Green's function and of its properties is based on the Hilbert space theory of weak boundary value problems. The benefit of this modern approach is that to any (bounded and linear) functional $J(u)$ can be assigned a Green's function in the sense of (3.129).

And in FE-methods these ideas are extended to unbounded functionals like the derivatives

$$J(u) = u_{,x_1} \tag{3.132}$$

disregarding all warnings that this is not allowed because it transgresses the bounds of the Hilbert space theory but it is a huge gain in applicability and the success proves the FE-method right.

3.6.1 Arbitrary Deltas

Given linear functionals such as

$$J_a(u) = u(x) \qquad J_b(u) = \sigma(u)(x) \qquad J_c(u) = \int_0^l u(x)\, dx \tag{3.133}$$

each of these functionals can be identified with the action of a certain Dirac delta

[1] The *Lax–Milgram Theorem* is essentially a statement about the existence of the Riesz element.

$$J_a(u) = \int_0^l \delta_a(y - x)\, u(y)\, dy \tag{3.134}$$

$$J_b(u) = \int_0^l \delta_b(y - x)\, u(y)\, dy \tag{3.135}$$

$$J_c(u) = \int_0^l \delta_c(x)\, u(x)\, dx. \tag{3.136}$$

The first delta, δ_a, represents a point load and the second, δ_b, a dislocation while the third is the unit function, $\delta_c(x) = 1$, a simple translation. But such a physical interpretation of the Dirac deltas is not a necessary prerequisite and in the case of more complicated functionals it may even not be possible to provide such a physical interpretation. But this is if no concern because the Dirac delta plays only a symbolic role.

Consider the Poisson equation on a planar domain Ω with an edge Γ

$$-\,\Delta\, u(\boldsymbol{x}) = p(\boldsymbol{x}) \qquad u = 0 \qquad \text{on } \Gamma \tag{3.137}$$

and let the functional $J(u)$ be the average value of the gradient of u over a small circular region $\Omega_\rho(\boldsymbol{x})$ with radius ρ centered at a point $\boldsymbol{x} \in \Omega$

$$J(u) = \frac{1}{\pi\, c^2} \int_{\Omega_\rho(\boldsymbol{x})} (u,_{y_1}(\boldsymbol{y}) + u,_{y_2}(\boldsymbol{y}))\, d\Omega. \tag{3.138}$$

We postulate that this functional can be expressed as

$$J(u) = \int_\Omega \delta_X(\boldsymbol{y} - \boldsymbol{x})\, u(\boldsymbol{y})\, d\Omega_{\boldsymbol{y}} \tag{3.139}$$

where no word is said about what the Dirac delta $\delta_X(\boldsymbol{y} - \boldsymbol{x})$ looks like, which actions on the part of δ_X produce $J(u)$. We only claim that the scalar product (= integral) of δ_X with the function u provides the value $J(u)$.

Next it is only necessary to find the function G which has the delta function as its right-hand side, that is which solves the equation

$$-\,\Delta\, G(\boldsymbol{y}, \boldsymbol{x}) = \delta_X(\boldsymbol{y} - \boldsymbol{x}) \qquad G(\boldsymbol{y}, \boldsymbol{x}) = 0 \qquad \boldsymbol{y} \in \Gamma \tag{3.140}$$

because Green's second identity, see (2.93), then implies

$$\mathscr{B}(G, u) = \underbrace{\int_\Omega \delta_X(\mathbf{y} - \mathbf{x})\, u(\mathbf{y})\, d\Omega_y}_{J(u)} - \int_\Omega G(\mathbf{x}, \mathbf{y})\, p(\mathbf{y})\, d\Omega_y = 0 \tag{3.141}$$

or

$$J(u) = \int_\Omega G(\mathbf{x}, \mathbf{y})\, p(\mathbf{y})\, d\Omega_y. \tag{3.142}$$

This is the idea.

It remains to approximate G on $\mathcal{V}_h$ with the n shape functions

$$G_h(\mathbf{y}, \mathbf{x}) = \sum_{i=1}^{n} g_i(\mathbf{x})\, \varphi_i(\mathbf{y}) \tag{3.143}$$

and to determine the nodal values $g_i(\mathbf{x})$ by solving the n equations

$$a(G_h, \varphi_i) = (\delta_X(\mathbf{y} - \mathbf{x}), \varphi_i) = J(\varphi_i) \qquad i = 1, 2, \ldots n \tag{3.144}$$

or in matrix notation

$$\boldsymbol{K}\, \boldsymbol{g}(\mathbf{x}) = \boldsymbol{j}(\mathbf{x}) \qquad k_{ij} = a(\varphi_i, \varphi_j) \qquad j_i = J(\varphi_i). \tag{3.145}$$

So the extension of the FE-method to Green's functions is easy and straightforward.[2] The equivalent nodal forces are simply the values of the functional at each φ_i, that is $f_i = J(\varphi_i)$, or in our notation $f_i = j_i = J(\varphi_i)$.

Let us summarize this as follows:

Theorem 3.10 (Equivalent nodal forces for Green's functions) *The Riesz element or generalized Green's function G of a linear and bounded functional is the solution of the variational problem*

$$a(G, v) = J(v) \qquad \forall\, v \in \mathcal{V} \tag{3.146}$$

and its projection

$$G_h(\mathbf{y}, \mathbf{x}) = \sum_{i} g_i(\mathbf{x})\, \varphi_i(\mathbf{y}) \tag{3.147}$$

onto the space $\mathcal{V}_h$ is the solution of the variational problem

$$a(G_h, \varphi_i) = J(\varphi_i) \qquad \forall\, \varphi_i \in \mathcal{V}_h \tag{3.148}$$

[2] Strictly speaking it is not because most Green's function have no finite energy, $a(G, G) = \infty$, but at present we simply assume they do.

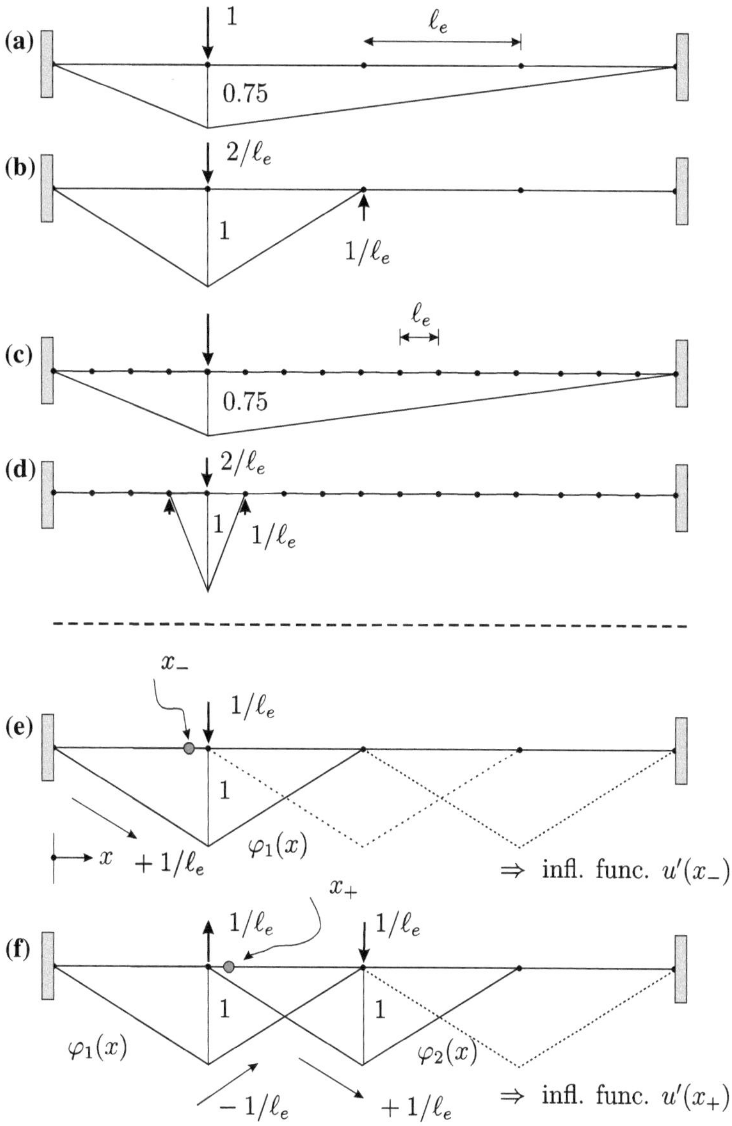

Fig. 3.16 Rope $-u'' = p$ **a** influence function for the nodal deflection **b** for the discontinuity in the slope $u'(x_-) - u'(x_+)$ at the same node **c** on the finer mesh the influence function for the nodal deflection stays the same while **d** the influence function for the discontinuity shrinks—the mesh improves; the nodal forces of **e** minus **f** are the nodal forces in **b**. Note that the peak 1 in **b** and **d** remains the same. It is not a discretization error

that is the equivalent nodal forces for a generalized Green's functions are

$$f_i = j_i = J(\varphi_i). \tag{3.149}$$

The notation j_i is the notation we prefer for these nodal forces.

Because the FE-shape functions are only piecewise smooth, functionals which differentiate must be handled with care. If for example $J(u) = u'(x)$ is the derivative at a node and the φ_i are hat functions then to each side of the node belongs a different influence function and different equivalent nodal forces j_i which generate these influence functions.

And if $J(u) = u''(x)$ is the second derivative at the center of a linear element then all the $j_i = J(\varphi_i) = \varphi_i'' = 0$ are zero—a piecewise linear FE-solution has no curvature. So that the Green's function $G_h = 0$ is zero and consequently $J_h(u) = 0$ for all $u \in \mathcal{V}$. Note that this result is in agreement with the *dual reciprocity* for functionals formulated earlier

$$J(u_h) = u_h''(x) = 0 = J_h(u). \tag{3.150}$$

The fact that the derivative of the FE-solution jumps at a node but the exact solution does not qualifies the functional

$$J_\Delta(u_h) := u_h'(x_-) - u_h'(x_+) \tag{3.151}$$

as *error functional* because it only reacts to the FE-solution and the functions $\varphi_i \in \mathcal{V}_h$. The aim in refining the mesh, see Fig. 3.16, must be that

$$\lim_{h \to 0} J_\Delta(u_h) = 0. \tag{3.152}$$

Remark 3.7 The influence function for the jump in the derivative is of even type (monopole) and so if the load is the same on both sides of the node then the discontinuity is pronounced while opposite loads (checkerboard loads) will produce no kink in $u_h(x)$.

3.7 Influence Functions for Integral Values

If $u(x)$ is the deflection of the rope

$$-u''(x) = p \qquad 0 < x < l \qquad u(0) = u(l) = 0 \tag{3.153}$$

then

$$u(x) = \int_0^l G(y,x)\, p(y)\, dy \tag{3.154}$$

and so the integral of the deflection is

$$J(u) = \int_0^l u(x)\,dx = \int_0^l \int_0^l G(y,x)\,p(y)\,dy\,dx$$
$$= \int_0^l \underbrace{[\int_0^l G(y,x)\cdot 1\,dx]}_{G_\Sigma(y)}\,p(y)\,dy = \int_0^l G_\Sigma(y)\,p(y)\,dy. \tag{3.155}$$

Obviously is the Green's function $G_\Sigma(x)$ of this functional the solution of the boundary value problem

$$-G''_\Sigma(x) = 1 \qquad 0 < x < l \qquad G_\Sigma(0) = G_\Sigma(l) = 0 \tag{3.156}$$

where the right-hand side, $p = 1$, resembles a series of tightly packed Dirac deltas—one Dirac delta at each point x of the rope.

If the integral extends only over a part of the rope

$$J(u) = \int_{x_a}^{x_b} u(x)\,dx \tag{3.157}$$

then the right-hand side vanishes outside the interval $[x_a, x_b]$

$$-G''_\Sigma(x) = p \qquad 0 < x < l \qquad p(x) = \begin{cases} 1 & x \in [x_a, x_b] \\ 0 & x \notin [x_a, x_b] \end{cases} \tag{3.158}$$

and in the limit, if $[x_a, x_b]$ shrinks to a mere point, p transforms again into a Dirac delta.

3.7.1 Nodal Forces

In agreement with the rule that $j_i = J(\varphi_i)$ the equivalent nodal forces (we write j_i instead of f_i) which generate the FE-approximation $G^h_\Sigma(x)$ of G_Σ are the integrals of the shape functions

$$j_i = J(\varphi_i) = \int_0^l \varphi_i(x)\,dx. \tag{3.159}$$

This rule applies to any integral functional as for example the following functionals

$$J_a(u) = \int_0^l u'(x)\,dx \qquad J_b(u) = \int_\Omega \sigma_{xx}(\boldsymbol{x})\,d\Omega \qquad J_c(u) = \int_\Omega m_{yy}\,d\Omega \tag{3.160}$$

to which belong the equivalent nodal forces

$$j_i = J_a(\varphi_i) = \int_0^l \varphi_i'(x)\,dx \qquad j_i = J_b(\varphi_i) = \int_\Omega \sigma_{xx}(\varphi_i)(\boldsymbol{x})\,d\Omega$$
$$j_i = J_c(\varphi_i) = \int_\Omega m_{yy}(\varphi_i)\,d\Omega. \tag{3.161}$$

Example 3.1 The influence function for the bending moment

$$J(\boldsymbol{u})(x) = \int_{-1}^{+1} \sigma_{xx}(y)\cdot y\,dy \qquad y = \text{lever arm} \tag{3.162}$$

in section $A - A$ (with horizontal coordinate x) of the cantilever plate in Fig. 3.17 is generated by the equivalent nodal forces

$$j_i(x) = \int_{-1}^{+1} \sigma_{xx}(\varphi_i)(x, y)\cdot y\,dy \tag{3.163}$$

which are the moments generated by the vector-valued shape functions $\varphi_i(x, y)$ (displacement fields) associated with the nodal degrees of freedom u_i.

Within a bilinear element of size $a \times b$ the stress distribution σ_{xx} is a function of the nodal values

$$\sigma_{xx}(x, y) = \frac{E}{a\,b\,(-1+\nu^2)}\Big[b\,(u_1 - u_3) + a\,\nu\,(u_2 - u_8)$$
$$+ x\,\nu\,(-u_2 + u_4 - u_6 + u_8) + y\,(-u_1 + u_3 - u_5 + u_7)\Big]. \tag{3.164}$$

Letting $u_1 = 1$ and all other $u_i = 0$ gives $\sigma_{xx}(\varphi_1)(x, y)$, etc..

Substituting this formula into (3.163) we obtain ($a = b = 1.0, \nu = 0.1$), the equivalent nodal forces j_i displayed in Fig. 3.17b. The shape these forces produce, see Fig. 3.17a, is not the perfect 45^0 rotation but at least the value at the upper corner point comes relatively close to the true value, 2.47 $\sim$ 2.5, though at other nodes the deviations are markedly larger.

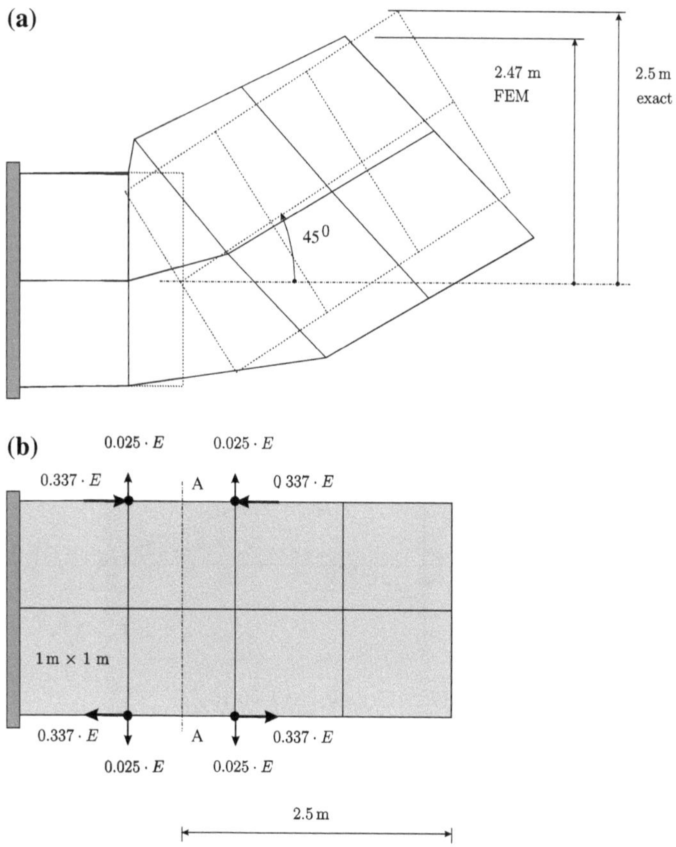

Fig. 3.17 Cantilever plate, influence function for the bending moment M in section $A - A$ **a** exact solution (*dotted lines*) and FE-solution **b** equivalent nodal forces which generate the FE-Green's function for M

3.8 Weak Influence Functions

There is a need to distinguish between—as we call them—*weak* and *strong* influence functions. Strong influence functions are based on Betti's theorem (classical duality) while weak influence functions are based on an evaluation of the strain energy product $a(G, u)$ between the Green's function and u. An engineer would say that weak influence functions are based on the *principle of virtual forces*; the Dirac delta being the virtual force.

A continuous beam, see Fig. 3.18, may exemplify the difference between weak and strong influence functions. The beam is subjected to a single moment at the intermediate support so that the exact solution is a piecewise cubic function and two elements suffice to produce the exact solution $u = u_h$ (the fact that the FE-solution is exact is not essentially in the following).

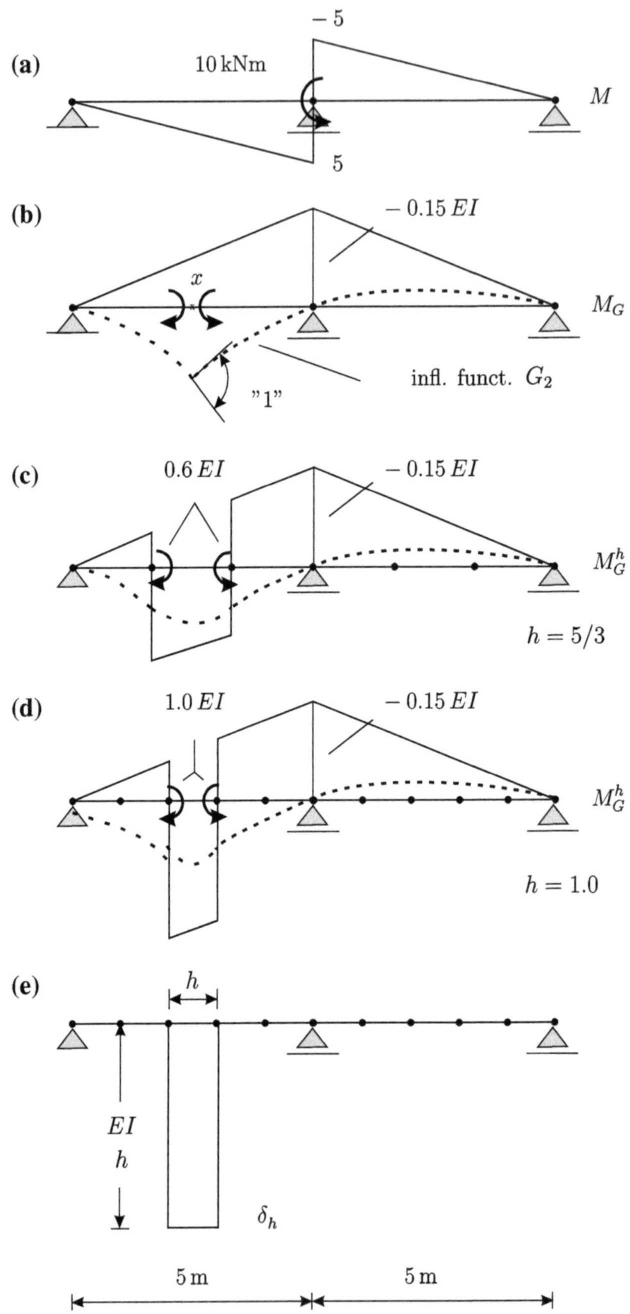

Fig. 3.18 Continuous beam, **a** bending moment, **b** bending moment of the exact influence function, **c** and **d** bending moment of the approximate influence function; Fig. **d** is the sum of Fig. **b** plus the approximate Dirac Delta in Fig. **e**

The couple produces at the center of the first span, $x = 2.5$, a moment of $M(x) = 2.5$ kNm as can be verified with the influence function for $M(x)$

$$M(x) = \int_0^l G_2(y, x)\, p(y)\, dy \equiv G_2'(y_2, x) \cdot 10\,\text{kNm} = 2.5\,\text{kNm}. \tag{3.165}$$

According to Betti the value $M(x)$ is equal to the work done by the load p on acting through the influence function $G_2(y, x)$. Because the load is a couple $M = 10$ kNm and a couple performs work on acting through rotations M must be multiplied with the slope $G_2'(y_2, x)$ of the Green's function at the point y_2, the location of M. This gives a value of 2.5 kNm.

Querying the weak influence function for the same result returns zero because the strain energy product between the Green's function $G_2(y, x)$ and u is zero

$$a(G_2, u) = \int_0^l EI\, G_2''\, u''\, dy = \int_0^l \frac{M\, M_G}{EI} = 0. \tag{3.166}$$

In this case the result is evident because M is antisymmetric and M_G is symmetric but this result is no coincidence, it is a rule. Any other result would mean that the derivative $u'(x)$ is discontinuous at x.

But when the same equation (3.166) is formulated with the FE-approximation G_2^h—instead of the exact function G_2—then the result is the bending moment at x

$$\begin{aligned} M_h(x) = a(G_2^h, u) &= \int_0^l \frac{M_G^h\, M}{EI} dy \\ &= \frac{1}{EI}\left[\frac{1}{6} \cdot 5 \cdot (0.2\,\text{EI} + 2 \cdot 0.05\,\text{EI}) \cdot 5 \right. \\ &\quad \left. + \frac{1}{3} \cdot (-0.15)\,\text{EI} \cdot (-5) \cdot 5\right] = 2.5. \end{aligned} \tag{3.167}$$

This seems strange. But technically it is easy to explain: the bending moment M_G^h is the sum of the exact moment M_G in Fig. 3.18b plus a small box of width h (= length of the element) and height EI/h, see Fig. 3.18e, and so

$$a(G_2^h, u) = \int_0^l \frac{M_G\, M}{EI} dy + \int_{x-h/2}^{x+h/2} \frac{EI}{h} \frac{M}{EI}\, dy = 0 + M(x). \tag{3.168}$$

The second integral is just the average value of the bending moment on the element that contains the source point x. But because $M(x) = M_h(x)$ is a linear function this value is just the value of $M(x)$ at the center of the element.

This result (3.168) means that

$$M_G^h = M_G + \delta_h \tag{3.169}$$

can be split into the exact kernel M_G and an error term which in its actions corresponds to an approximate Dirac delta δ_h, the box. On $\mathcal{V}_h$ the box is as good as δ because all shape functions (Hermite polynomials) have linear second order derivatives and the average value of a linear function is equal to the value at the center of the interval.

In Sect. 2.4.4 we constructed a sequence of functions $G_1^{\Delta x}$ which, as Δx tends to zero, converges towards G_1, the influence function for the first derivative at the center of the rope. It holds:

1. The strain energy product between one such function $G_1^{\Delta x}$ and the slope deflection $u_h(x)$ is the difference quotient at x

$$\mathcal{G}(G_1^{\Delta x}, u) = \frac{1}{\Delta x}\, u(x + 0.5\,\Delta x) - \frac{1}{\Delta x}\, u(x - 0.5\,\Delta x) - a(G_1^{\Delta x}, u) = 0.$$

2. And because $u_h(x) \in \mathcal{V}_h$ is piecewise linear and therefore linear at x (not a node) the difference quotient is identical with the derivative

$$\left[\frac{1}{\Delta x}\, u_h(x + 0.5\,\Delta x) - \frac{1}{\Delta x}\, u_h(x - 0.5\,\Delta x)\right] = u_h'(x). \tag{3.170}$$

This is the logic.

When force terms—essentially derivatives—are calculated in FE-analysis with the strain energy product then the result is a finite difference expression which on $\mathcal{V}_h$ is identical with the derivative(s). Stated otherwise: the exact Dirac delta δ is replaced by an approximate delta δ_h but on $\mathcal{V}_h$ the two are identical in their actions.

Remark 3.8 Given the shape functions of a beam of length h, see (1.91), the bending moments, $M_i = -EI\,\varphi_i''(x)$, are

$$\begin{aligned} M_1(x) &= -EI\left(-\frac{6}{h^2} + \frac{12\,x}{h^3}\right) & M_3(x) &= -EI\left(\frac{6}{h^2} - \frac{12\,x}{h^3}\right) \\ M_2(x) &= -EI\left(\frac{4}{h} - \frac{6x}{h^2}\right) & M_4(x) &= -EI\left(\frac{2}{h} - \frac{6\,x}{h^2}\right) \end{aligned} \tag{3.171}$$

so that the nodal forces $j_i = M(\varphi_i)(h/2)$ which generate the FE-influence function G_2^h are

$$\downarrow\ j_1 = j_3 = 0\ \downarrow \qquad \circlearrowleft\ j_2 = -\frac{EI}{h} \qquad j_4 = \frac{EI}{h}\ \circlearrowleft\,. \tag{3.172}$$

3.9 Weak Influence Functions Have More Choices

When we apply Betti's theorem then we usually assume that the primal solution, $-\Delta u = p$, and the dual solution, $-\Delta G = \delta(\boldsymbol{x}, \boldsymbol{y})$, come from the same system. But Betti's theorem remains valid if the two solutions, let us call these u_I and G_{II}, satisfy different sets of boundary conditions. But under these circumstances additional adjoint terms on the edge of the domain contribute to the statement $W_{1,2} = W_{2,1}$ and often these additional boundary terms are unknown a priori and must be determined by integral equations as in the *boundary element method*.

We have the same freedom with regard to weak influence functions. But because weak influence functions evaluate the strain energy product via the nodal displacements

$$J(u)(\boldsymbol{x}) = a(G_{II}, u_I) = \boldsymbol{g}_{II}^T \boldsymbol{K}_{II}\, \boldsymbol{u}_I \tag{3.173}$$

they have it easier.

To see this we formulate Green's first identity

$$\mathscr{G}(G, u) = \int_\Omega \delta(\boldsymbol{y} - \boldsymbol{x})\, u(\boldsymbol{y})\, d\Omega + \int_{\Gamma_D} \frac{\partial G}{\partial n}\, \bar{u}\, ds_y - a(G, u) = 0 \tag{3.174}$$

where $\bar{u}$ is the prescribed edge displacement on the *Dirichlet-part* Γ_D of the boundary; often $\bar{u} = 0$.

In system II the restraints are relaxed that is the *Dirichlet-part* gets smaller and the *Neumann-part* gets larger, the Green's function G_{II} satisfies

$$-\Delta G_{II} = \delta(\boldsymbol{y} - \boldsymbol{x}) \qquad \frac{\partial G_{II}}{\partial n} = 0 \quad \text{on } \Gamma_N^{II}, \quad G_{II} = 0 \quad \text{on } \Gamma_D^{II} \tag{3.175}$$

and so no additional information is required to evaluate the boundary integral in (3.174) because $\Gamma_D \to \Gamma_D^{II}$ shrinks

$$\mathscr{G}(G_{II}, u_I) = \int_\Omega \delta(\boldsymbol{y} - \boldsymbol{x})\, u_I(\boldsymbol{y})\, d\Omega + \int_{\Gamma_D^{II}} \frac{\partial G_{II}}{\partial n}\, \bar{u}\, ds_y - a(G_{II}, u_I) = 0. \tag{3.176}$$

In structural mechanics this flexibility in the choice of the system II is known as *Method of Reduction*, "Reduktionssatz", [3]. It states that the Dirac delta must not be applied to the original system but that it can be any system that is "contained" in the original system, see Fig. 3.19. Contained means that if the original system is for example statically indeterminate (too many supports) the second system can be a statically determinate system (just the minimum number) which is easier to analyze. In the language of finite elements "contained" means that the two systems must be based on the same set of shape functions but the nodes which are fixed can be different

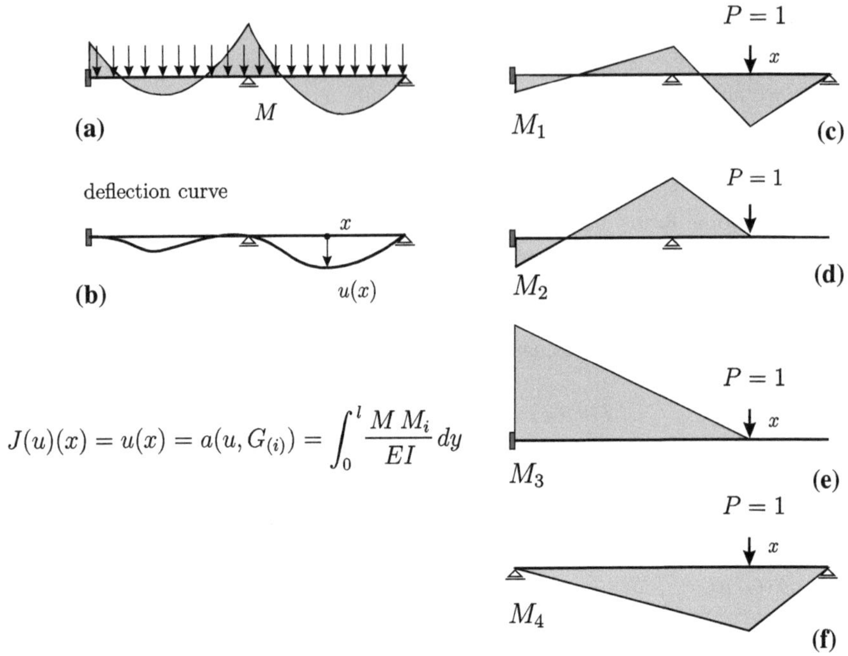

Fig. 3.19 Continuous beam (**a**) and its (**b**) deflection curve and calculating the deflection at the center of the second span with a weak influence function. The unit point load $P = 1$ can be applied to any of the beams on the right. The result is always the same. System **c** is the same as the original system while in systems **d, e, f** the Dirichlet boundary shrinks

or to be more precise: the fixed nodes of the simplified model II must be a subset of the fixed notes of system I. System II can release nodes but it must not keep additional nodes fixed, then Γ_D would grow and we no longer could evaluate the boundary integral in (3.174) because the value of u_I on the added Dirichlet boundary which previously was a part of the Neumann boundary would be unknown: we do not know a-priori how the free edge (Γ_N) of a membrane deflects under pressure.

Thus the stiffness matrix $\boldsymbol{K}$ in (3.173) must belong to the system with the lesser number of fixed nodes, which here we assumed to be the system II. The matrix $\boldsymbol{K}$ could even be the singular system matrix before rows and columns are removed to model fixed supports; in that case the vectors $\boldsymbol{g}_{II}$ and $\boldsymbol{u}_I$ would have to be padded with zeros—corresponding to the position of the fixed supports—to make them full size.

Remark 3.9 The original "Dirichlet-part" in Fig. 3.19 consists of $u(0) = u'(0) = u(x_i) = u(x_e) = 0$. In Fig. 3.19d the end node is released, $u(x_e) \neq 0$, and in Fig. 3.19e also the intermediate support, $u(x_i) \neq 0$, and in Fig. 3.19f the support conditions $u'(0) = 0$ and $u(x_i) = 0$ are released. Each time the new Dirichlet-part is a subset, $\Gamma_D^{II} \subset \Gamma_D^{I}$, of the original part.

3.10 Nodal Form of Influence Functions

The deflection of a taut rope can be written in two ways in integral form

$$u(x) = \int_0^l G(y,x)\, p(y)\, dy = \int_0^l \delta(y-x)\, u(y)\, dy. \tag{3.177}$$

In FE-analysis these two integrals—and of course any other influence function as well—can be evaluated by summing over the nodes.

The Green's function for the vertical displacement of the rope at the point x is the solution of the problem

$$-H \frac{d^2}{dy^2} G(y,x) = \delta(y-x) \tag{3.178}$$

which amounts to the statement that

$$a(G,v) := \int_0^l H\, G'(y,x)\, v'(y)\, dy = \int_0^l \delta(y-x)\, v\, dy \qquad \forall v \in \mathcal{V}. \tag{3.179}$$

Hence the FE-projection onto $\mathcal{V}_h \subset \mathcal{V}$

$$G_h(y,x) = \sum_i^n g_i(x)\, \varphi_i(y) \tag{3.180}$$

leads to the system

$$\boldsymbol{K}\,\boldsymbol{g} = \boldsymbol{j} \tag{3.181}$$

where

$$j_i = \int_0^l \delta(y-x)\varphi_i(y)\, dy = \varphi_i(x) \tag{3.182}$$

$$k_{ij} = a(\varphi_i, \varphi_j) = \int_0^l H\, \varphi_i'\, \varphi_j'\, dx. \tag{3.183}$$

The vector

$$\boldsymbol{g} = \boldsymbol{g}(x) = \{g_1(x), g_2(x), \ldots, g_n(x)\}^T \tag{3.184}$$

is the vector of the nodal values g_i of the Green's function.

Next let

$$u_h(x) = \sum_i^n u_i \, \varphi_i(x) \tag{3.185}$$

the FE-solution of the problem

$$- H \, u''(x) = p(x) \qquad 0 < x < l \qquad u(0) = u(l) = 0. \tag{3.186}$$

That is the vector $\boldsymbol{u}$ of nodal values solves the equation $\boldsymbol{K}\,\boldsymbol{u} = \boldsymbol{f}$ where

$$f_i = \int_0^l p \, \varphi_i \, dx. \tag{3.187}$$

Like the exact solution the FE-solution can be written in two ways in an integral form

$$u_h(x) = \int_0^l G_h(y, x) \, p(y) \, dy = \int_0^l u_h(y) \, \delta(y - x) \, dy \tag{3.188}$$

and we have

$$\begin{aligned} u_h(x) &= \int_0^l G_h(y, x) \, p(y) \, dy = \int_0^l \sum_i g_i(x) \, \varphi_i(y) \, p(y) \, dy = \sum_i g_i(x) \, f_i \\ &= \boldsymbol{g}^T \boldsymbol{f} = \boldsymbol{g}^T \boldsymbol{K} \, \boldsymbol{u} = \boldsymbol{g}^T \boldsymbol{K}^T \boldsymbol{u} = \boldsymbol{j}^T \boldsymbol{u} = \sum_i j_i \, u_i \\ &= \sum_i \varphi_i(x) \, u_i = \int_0^l \sum_i u_i \, \varphi_i(y) \, \delta(y - x) \, dy = \int_0^l u_h(y) \, \delta(y - x) \, dy. \end{aligned} \tag{3.189}$$

That is the vertical displacement $u_h(x)$ at the point x is the scalar product between the vector of nodal values of the Green's function and the equivalent nodal forces of the load case p or, vice versa, between the nodal values of the FE-solution and the equivalent nodal forces j_i of the Green's function

$$u_h(x) = \begin{cases} \displaystyle\sum_i \varphi_i(x) \, u_i = \int_0^l u_h(y) \, \delta(y - x) \, dy = \boldsymbol{j}^T \boldsymbol{u} \\ \displaystyle\int_0^l G_h(y, x) \, p(y) \, dy = \boldsymbol{g}^T \boldsymbol{f}. \end{cases} \tag{3.190}$$

3.10.1 Numerical Effort

What holds true for $u_h(x)$ holds true for any linear functional $J(u_h)$. So if $\boldsymbol{j}$ and $\boldsymbol{g}$ are the vectors of equivalent nodal forces and of nodal displacements respectively of the generalized Green's function then

$$J(u_h) = \boldsymbol{j}^T \boldsymbol{u} = \boldsymbol{j}^T \boldsymbol{K}^{-1} \boldsymbol{f} = \boldsymbol{g}^T \boldsymbol{f} \tag{3.191}$$

or more concisely

$$J(u_h) = \begin{cases} \boldsymbol{j}^T \boldsymbol{u} \\ \boldsymbol{g}^T \boldsymbol{f}. \end{cases} \tag{3.192}$$

Both equations are equivalent. If the focus is only on, say, two or three values of the solution (i.e. calls to two or three functionals) but a long list of load cases is to be inspected then the second formula

$$J(u_h) = \boldsymbol{g}^T \boldsymbol{f} \tag{3.193}$$

is to be preferred because it requires only the calculation of two or three nodal vectors $\boldsymbol{g}(x)$ of the Green's functions. Vice versa if a detailed picture of the solution is required (calls to many different functionals) and the number of load cases is small the first formula

$$J(u_h) = \boldsymbol{j}^T \boldsymbol{u} \tag{3.194}$$

will serve us better.

3.10.2 Sensitivity Plots

The formula $J(u_h) = \boldsymbol{g}^T \boldsymbol{f}$ is the scalar product between the vector $\boldsymbol{g}$, the nodal values of the Green's function, and the vector $\boldsymbol{f}$ of equivalent nodal forces. This scalar product can be written as a sum over the N nodes of the FE-mesh

$$J(u_h) = \sum_{i=1}^{N} \boldsymbol{g}_i^T \boldsymbol{f}_i \qquad i = \text{nodes} \tag{3.195}$$

where the vectors $\boldsymbol{g}_i$ and $\boldsymbol{f}_i$ are the portions of $\boldsymbol{g}$ and $\boldsymbol{f}$ respectively referring to node i

$$\boldsymbol{g} = \{\underbrace{g_1, g_2}_{\boldsymbol{g}_1}, \underbrace{g_3, g_4}_{\boldsymbol{g}_2}, \ldots, g_n\}^T \qquad 2-D \tag{3.196}$$

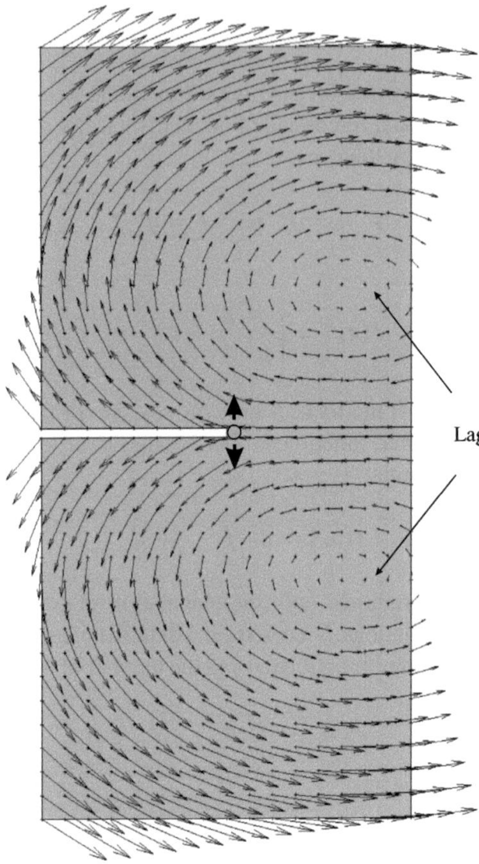

Fig. 3.20 Plot of the nodal vectors $\boldsymbol{g}_i$ of the functional $J(u_h) = \sigma_{yy}$, the vertical stress in the plate close to the slit. The "Lagrange points" are the points where the influence of any force $\boldsymbol{f}_i$ on σ_{yy} is practically zero

So if $\boldsymbol{f}_i$ happens to be orthogonal to $\boldsymbol{g}_i$ at a node then the contribution of the node to $J(u_h)$ is zero. The plot of the vectors $\boldsymbol{g}_i$ therefore represents a sensitivity plot of the functional $J(u_h)$, see Fig. 3.20. Nodal forces $\boldsymbol{f}_i$ which point in the same direction as the $\boldsymbol{g}_i$ are able to maximize their influence on $J(u_h)$.

Note that in Fig. 3.20 there are two calm zones at which the influence of any nodal force on $J(u_h)$ is practically zero. We call these "Lagrange points". In astronomy the Lagrange points are the points at which the graviational influence of the Sun and the Earth balance, they pull with equal but opposite forces on a satellite parked at these points. These Lagrange points can be found in nearly any such plot, see Figs. 3.21 and 3.22. The circular nature of the flow field around these points is a remarkable phenomenon. Continuous fields seem to need such tranquil zones ("fixed points") to reverse directions.

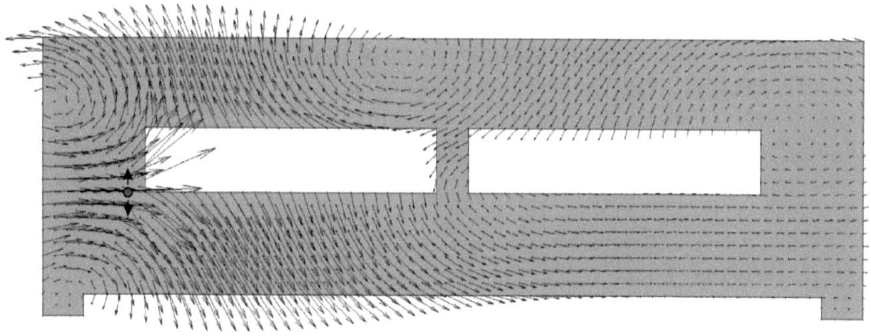

Fig. 3.21 Plot of the nodal vectors $\boldsymbol{g}_i$ of the functional $J(u_h) = \sigma_{yy}$, the vertical stress in the plate close to the corner of the opening

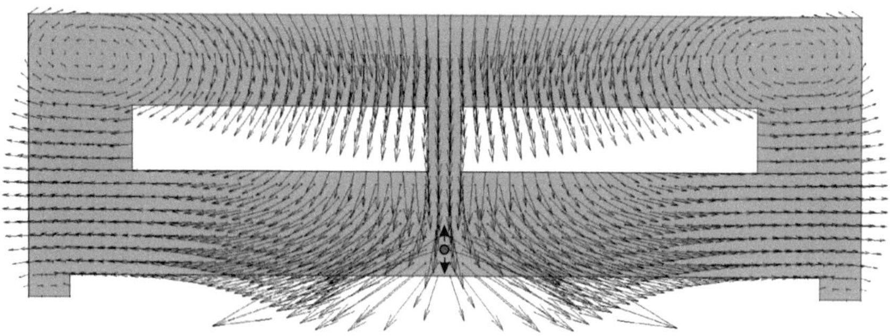

Fig. 3.22 Plot of the nodal vectors $\boldsymbol{g}_i$ of the functional $J(u_h) = \sigma_{yy}$, the vertical stress in the plate near the *bottom* of the plate

Remark 3.10 The Sun, mass M_S at $\boldsymbol{y}_S$, and the Earth, mass M_E at $\boldsymbol{y}_E$, generate at a point $\boldsymbol{x}$ the gravitational potential

$$u(\boldsymbol{x}) = -G\,\frac{M_S}{r_S} - G\,\frac{M_E}{r_E} \qquad r_S = |\boldsymbol{y}_S - \boldsymbol{x}| \quad r_E = |\boldsymbol{y}_E - \boldsymbol{x}| \tag{3.197}$$

where G is the gravitational constant. The gradient of this potential is the force $\boldsymbol{g}$ acting on a mass m located at $\boldsymbol{x}$

$$\boldsymbol{g}(\boldsymbol{x}) = m\,\nabla\,u \tag{3.198}$$

and the Lagrange points are the points where the gradient is zero, where the potential is "flat", where it "takes a break".

The analog to (3.198) is the equation

$$\sigma_{yy} = \boldsymbol{f} \cdot \boldsymbol{G} \stackrel{?}{=} \boldsymbol{f} \cdot \nabla s \tag{3.199}$$

where the load vector $\boldsymbol{f}$ plays the role of the mass m and the role of the gradient is played by the displacement field $\boldsymbol{G}$ generated by the dislocation at the bottom of the slit, see Fig. 3.20. The question is: does there exist a scalar field $s(\boldsymbol{x})$ which has $\boldsymbol{G} = \{G_1, G_2\}^T$ as its gradient? Note that because of $s_{,12} = s_{,21}$ a necessary condition for this to hold is that $G_{1,2} = G_{2,1}$.

3.10.3 What is $\boldsymbol{j}^T \boldsymbol{u}$?

The expression

$$J(\boldsymbol{u}) = \boldsymbol{j}^T \boldsymbol{u} \tag{3.200}$$

is the engineers way of evaluating functionals. He applies J to the single shape functions φ_i, this gives $j_i = J(\varphi_i)$, multiplies it with the nodal displacements u_i and sums over all degrees of freedom.

The deflection of a rope in between the two nodes x_3 and x_4 is

$$u(x) = u_3\, \varphi_3(x) + u_4\, \varphi_4(x) = u_3\, j_3 + u_4\, j_4 \tag{3.201}$$

and the derivative is

$$u'(x) = u_3\, \varphi_3'(x) + u_4\, \varphi_4'(x) = u_3\, \underline{j}_3 + u_4\, \underline{j}_4 \tag{3.202}$$

where the notation $\underline{j}_i$ is for not to confuse the two sets of j_i.

3.11 Nodal Values of Green's Functions

An FE-Green's function is an expansion in terms of the nodal shape functions

$$G_h(y, x) = \sum_i g_i(x)\, \varphi_i(y) \tag{3.203}$$

where $\boldsymbol{g} = \boldsymbol{K}^{-1}\, \boldsymbol{j}$ and $j_i = J(\varphi_i)$. So $\boldsymbol{g}$ is a linear combination of the columns $\boldsymbol{c}_i$ of the inverse

$$\boldsymbol{K}^{-1} = [\boldsymbol{c}_1, \boldsymbol{c}_2, \ldots, \boldsymbol{c}_n]. \tag{3.204}$$

If $J(u)$ is the nodal-value functional $J(u) = u(\boldsymbol{x}_k)$ for a node $\boldsymbol{x}_k$ then $\boldsymbol{j} = \boldsymbol{e}_k$ and $\boldsymbol{g}$ is identical with column $\boldsymbol{c}_k$ of $\boldsymbol{K}^{-1}$.

Any functional applied to the columns $\boldsymbol{c}_i$ gives

$$J(\boldsymbol{c}_i) = \boldsymbol{j}^T\, \boldsymbol{c}_i = \boldsymbol{j}^T\, \boldsymbol{K}^{-1}\, \boldsymbol{e}_i = \boldsymbol{g}^T\, \boldsymbol{e}_i = g_i. \tag{3.205}$$

Applying J to a column $\boldsymbol{c}_i$ (a nodal vector) means applying J to the (scalar-valued) function $c_i(\boldsymbol{x})$ generated by the entries c_{ji} of the column

$$c_i(\boldsymbol{x}) = \sum_j c_{ji}\, \varphi_j(\boldsymbol{x}). \tag{3.206}$$

so that $J(c_i) = \boldsymbol{j}^T\, \boldsymbol{c}_i =: J(\boldsymbol{c}_i)$.

Functionals which take derivatives have vectors $\boldsymbol{g}$ which are difference quotients of the columns $\boldsymbol{c}_i$. For example the influence function for the normal force $N = EA\, u'$ in a truss element has the vector

$$\boldsymbol{g} = \frac{EA}{l_e}(\boldsymbol{c}_i - \boldsymbol{c}_j)\,, \tag{3.207}$$

where the two columns $\boldsymbol{c}_i$ and $\boldsymbol{c}_j$ are associated with the nodal displacements u_i and u_j of the end points of the element of length l_e.

3.11.1 Finite Differences and Finite Elements

Talking about nodal influence functions means to talk about finite differences. The close relationship between finite differences and finite elements has been evident from the start [4]. The stiffness matrix is a difference matrix and its inverse is a fully populated summation matrix—a discrete integral operator.

Both methods extract all information from the nodal values, $J(u) = \boldsymbol{j}^T \boldsymbol{u}$, and so it is no surprise that the nodal forces j_i which generate the Green's function of a functional $J(u)$ are identical with the weights in a finite-difference scheme.

An example in point is the (backward) finite difference scheme for the first derivative at a node x_i

$$u'(x_i) = \frac{u_i - u_{i-1}}{h} = \frac{1}{h} \cdot u_i - \frac{1}{h} \cdot u_{i-1} = j_i \cdot u_i + j_{i-1} \cdot u_{i-1}. \tag{3.208}$$

The weights $\pm 1/h$ are identical with the equivalent nodal forces j_i which generate the FE-Green's function for the first derivative $u_h'(x)$, see Fig. 3.23.

3.11.2 Influence Function for $p(x)$

Given the differential equation

$$-\,u''(x) = p(x) \tag{3.209}$$

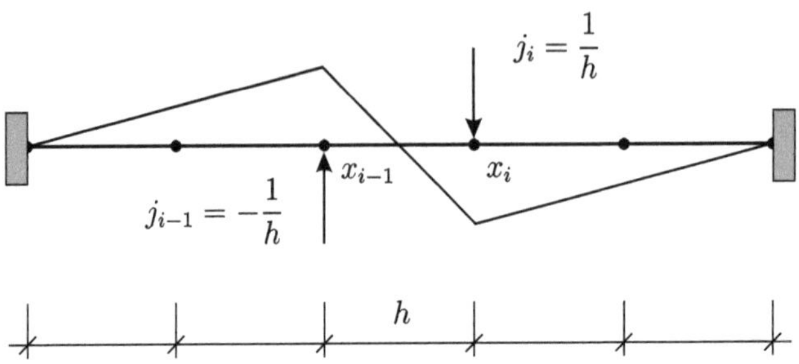

Fig. 3.23 These nodal forces generate the Green's function for $u_h'(x), x \in [x_{i-1}, x_i]$

what is the Green's function for the right-hand side $p(x)$ of the differential equation? That is, which function maps p onto $p(x)$? Of course this is just the Dirac delta

$$p(x) = \int_0^l \delta(y-x)\, p(y)\, dy. \tag{3.210}$$

But does it make sense to calculate the FE-influence function for $p(x)$ if p is given? The answer is that the approximate kernel $\delta_h(y,x)$ allows to map p directly onto $p_h(x)$. This approximate kernel

$$\delta_h(y,x) = \sum_i g_i(x)\, \varphi_i(y) \qquad \boldsymbol{g} = \boldsymbol{K}^{-1}\boldsymbol{j} \tag{3.211}$$

is generated by the equivalent nodal forces

$$j_i = J(\varphi_i) = -\varphi_i''(x) \tag{3.212}$$

and this kernel, see Fig. 3.24c for an example, constitutes the functional

$$J_h(u) = \int_0^l \delta_h(y,x)\, p(y)\, dy. \tag{3.213}$$

Because of $J_h(u) = J_h(u_h)$

$$J_h(u) = \int_0^l \delta_h(y,x)\, p(y)\, dy = J_h(u_h) = p_h(x) \tag{3.214}$$

the approximate kernel $\delta_h(y,x)$ maps p onto $p_h(x)$ that is we have an influence function for $p_h(x)$.

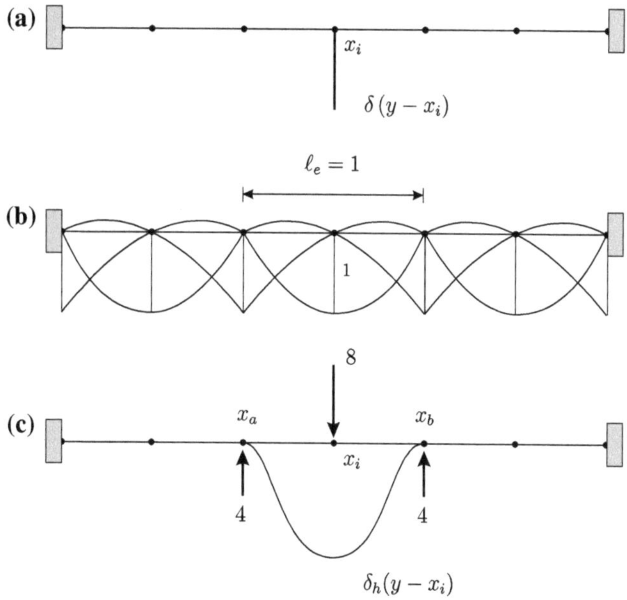

Fig. 3.24 The influence function for $p(x) = -u''(x)$ at x_i is **a** a Dirac delta, **b** quadratic elements, **c** equivalent nodal forces $j_i = -\varphi_i''(x_i)$ and approximate Dirac delta generated by these forces. Because $\varphi_i''(x)$ is constant this is the Dirac delta for any point $x_a \leq x \leq x_b$ on the element

When we use linear elements all the $j_i = -\varphi_i''(x)$ are zero (let x be an internal point) and so also the approximate kernel $\delta_h(y, x)$ is zero and therefore also $p_h(x)$

$$p_h(x) = \int_0^l \delta_h(y, x)\, p(y)\, dy = \int_0^l 0 \cdot p(y)\, dy = 0. \tag{3.215}$$

Which is no surprise.

When quadratic elements are used instead, the shape functions, $l_e = 1$,

$$\varphi_{i-1}(x) = 1 - 3x + 2x^2 \qquad \varphi_i(x) = 4(x - x^2) \qquad \varphi_{i+1}(x) = 2x^2 - x \tag{3.216}$$

have constant second derivatives and so the set of nodal forces

$$j_{i-1} = -4 \qquad j_i = 8 \qquad j_{i+1} = -4 \qquad (l_e = 1) \tag{3.217}$$

which generates the approximate Dirac delta for p_h at an internal point x is the same for each point x of an element. Consequently the function $\delta_h(y, x) = \delta_h(y)$ does not depend on x and this means that p_h is constant on each element

$$p_h = \int_0^l \delta_h(y)\, p(y)\, dy. \tag{3.218}$$

The kernel $\delta_h(y)$ has a tulip-like shape and it is zero outside the element itself, see Fig. 3.24.

To the distributed load p_h inside each element must be added nodal forces

$$P_i = u_h'(x_i^+) - u_h'(x_i^-) \tag{3.219}$$

which result from the possible discontinuity in the first derivative at the nodes. These forces together with the distributed load p_h inside each element constitute the *ersatz load case* which is solved by the FE-program.

Now it is evident that with $l_e \to 0$ the approximate Dirac delta will more and more resemble the true delta and the P_i will tend to zero and so we have a convincing argument that p_h will converge to p and therefore also (any doubts?) u_h to u.

3.11.3 The Foot Print of p_h

This brings us to our next point. Imagine the load p is applied only on a small part Ω_p of the domain Ω. Say a single car is parked on the Golden Gate bridge. All the other lanes are empty. To calculate the displacement u of the bridge we only need to integrate over the part Ω_p of the bridge deck beneath the car (which has morphed into a distributed load p)

$$u(\boldsymbol{x}) = \int_{\Omega_p} G(\boldsymbol{x}, \boldsymbol{y})\, p(\boldsymbol{y})\, d\Omega_{\boldsymbol{y}}. \tag{3.220}$$

In the FE-analysis the car p gets replaced by a work-equivalent load p_h which, unlike the parked car p, spreads over all of Ω and so it seems that it is necessary to integrate over the whole bridge

$$u_h(\boldsymbol{x}) = \int_{\Omega} G_h(\boldsymbol{x}, \boldsymbol{y})\, p_h(\boldsymbol{y})\, d\Omega_{\boldsymbol{y}} \tag{3.221}$$

to generate the FE-solution $u_h(\boldsymbol{x})$ from p_h. But this is not true. The integral stops beyond Ω_p^{+1} which is Ω_p plus a ring of elements of depth 1 which surrounds Ω_p.

Each shape function φ_i whose support[3] has no point in common with Ω_p is orthogonal to p_h and therefore it does not contribute to $u_h(\boldsymbol{x})$. The equivalent nodal forces

[3] support = the points where φ_i is not zero.

$$f_i = \int_\Omega p(\boldsymbol{x})\varphi_i(\boldsymbol{x})\,d\Omega \tag{3.222}$$

of these shape functions are zero and therefore also the equivalent nodal forces in the load case p_h

$$f_i^h = \int_\Omega p_h(\boldsymbol{x})\varphi_i(\boldsymbol{x})\,d\Omega = f_i \tag{3.223}$$

because this identity, $f_i = f_i^h$, is what the Galerkin orthogonality

$$a(u - u_h, \varphi_i) = \int_\Omega (p(\boldsymbol{x}) - p_h(\boldsymbol{x}))\,\varphi_i(\boldsymbol{x})\,d\Omega = f_i - f_i^h = 0 \tag{3.224}$$

is all about.

But the Green's function in (3.221)

$$G_h(\boldsymbol{x}, \boldsymbol{y}) = u_1(\boldsymbol{x})\,\varphi_1(\boldsymbol{y}) + u_2(\boldsymbol{x})\,\varphi_2(\boldsymbol{y}) + \ldots + u_n(\boldsymbol{x})\,\varphi_n(\boldsymbol{y}) \tag{3.225}$$

is an expansion in terms of the φ_i and so only the φ_i which "sense" p will contribute to $u_h(\boldsymbol{x})$ and therefore it is only necessary to integrate over Ω_p^{+1}

$$u_h(\boldsymbol{x}) = \int_{\Omega_p^{+1}} G_h(\boldsymbol{x}, \boldsymbol{y})\,p_h(\boldsymbol{y})\,d\Omega_{\boldsymbol{y}}. \tag{3.226}$$

So in FE-analysis the foot print of a load is nearly preserved. Its work equivalent substitute p_h may extend over all of Ω but effective—with regard to the functionals $J_h(.)$—is only the part of p_h near p.

The real-life functionals $J(.)$ though sense also contributions from outside Ω_p^{+1}. *They do not ignore what lies farther off*. This is an important point because when an engineer designs a structure then he uses the functionals $J(.)$ and not the functionals $J_h(.)$ which is the principal dilemma of finite elements: *The functionals and the solution do not match or else: the engineer, when he uses an FE-program, effectively designs the bridge for the load case p_h and not p.*

3.12 The Inverse Stiffness Matrix

The approximate Green's function for the displacement $u(x)$ at a node x_k has the form

$$G_h(y, x_k) = \sum_i g_i(x_k)\,\varphi_i(y). \tag{3.227}$$

The vector $\boldsymbol{g} = \{g_1, g_2, \ldots, g_n\}^T$ is the solution of the $n \times n$ system

$$\boldsymbol{K}\, \boldsymbol{g} = \boldsymbol{e}_k \qquad \text{(unit vector } \boldsymbol{e}_k), \tag{3.228}$$

which means that the columns $\boldsymbol{c}_k$ of the inverse stiffness matrix $\boldsymbol{K}^{-1}$

$$\boldsymbol{g} = \boldsymbol{K}^{-1} \boldsymbol{e}_k = \boldsymbol{c}_k \tag{3.229}$$

are the nodal displacements which belong to the n Green's functions $G_h(y, x_k)$ of the n nodes x_k

$$G_h(y, x_k) = \sum_i c_{ki}\, \varphi_i(y) = \boldsymbol{c}_k^T\, \varphi_i(y). \tag{3.230}$$

This explains why the inverse of a tridiagonal matrix is fully populated. Even if only one node x_k carries a point load $P = 1$ the whole system takes note. Stated otherwise: the inverse of a difference matrix as $\boldsymbol{K}$ is a summation matrix.

If x is a point in between two nodes then the vector $\boldsymbol{g}$ of the Green's function is the solution of

$$\boldsymbol{g} = \boldsymbol{K}^{-1} \boldsymbol{\varphi}(x) \tag{3.231}$$

where the vector

$$\boldsymbol{\varphi}(x) = \{\varphi_1(x), \varphi_2(x), \varphi_3(x), \ldots, \varphi_n(x)\}^T \tag{3.232}$$

is simply the list of the values of the shape functions at that particular point x. The following theorem summarizes these ideas.

Theorem 3.11 (General form of an FE-Green's function) *Let $\boldsymbol{K}$ be the stiffness matrix.*
(i) The FE-Green's function for $u_h(x)$, the value of the FE-solution at a point x, is

$$G_h(y, x) = \boldsymbol{\varphi}(y)^T\, \boldsymbol{K}^{-1}\, \boldsymbol{\varphi}(x) \tag{3.233}$$

where the vector $\boldsymbol{\varphi}(x)$ is defined in (3.232) and $\boldsymbol{\varphi}(y)$ is the same vector with y substituted for x.
(ii) The generalized Green's function for a linear point functional $J(u)$ is

$$G_h(y, x) = \boldsymbol{\varphi}(y)^T\, \boldsymbol{K}^{-1}\, \boldsymbol{j}(x) \tag{3.234}$$

where

$$\boldsymbol{j}(x) = \{J(\varphi_1), J(\varphi_2), J(\varphi_3), \ldots, J(\varphi_n)\}^T. \tag{3.235}$$

The x in $\boldsymbol{j}(x)$ is the point x in the definition of the point functional, as in $J(u) = u(x)$.

The shape a rope assumes under load is the envelope of infinitely many Green's functions each of which represents the contribution du of an infinitesimal portion $p(y)\,dy$ of the load to the total deflection $u(x)$

$$u(x) = \int_0^l G(y, x)\, p(y)\, dy. \tag{3.236}$$

The FE-solution $u_h(x) = \boldsymbol{u}^T\, \boldsymbol{\phi}(x)$ instead is the sum of a *finite number* of Green's functions weighted with the equivalent nodal forces f_i

$$u_h(x) = f_1\, G_h(x_1, x) + f_2\, G_h(x_2, x) + \ldots + f_n\, G_h(x_n, x)\,, \tag{3.237}$$

because the vector $\boldsymbol{u}$ of the nodal values is

$$\begin{aligned} \boldsymbol{u} &= \boldsymbol{K}^{-1}\boldsymbol{f} = \boldsymbol{K}^{-1}(f_1\,\boldsymbol{e}_1 + f_2\,\boldsymbol{e}_2 + \ldots + f_n\,\boldsymbol{e}_n) \\ &= f_1\,\boldsymbol{c}_1 + f_2\,\boldsymbol{c}_2 + \ldots + f_n\,\boldsymbol{c}_n \end{aligned} \tag{3.238}$$

and column $\boldsymbol{c}_i$ of $\boldsymbol{K}^{-1}$ corresponds to $G_h(x_i, x)$.

What holds true for the FE-solution u_h holds true for the derivatives and stresses as well. The influence function for the transverse force $V_h(x) = H\, u_h'$ in a rope modeled with a series of linear elements has the nodal values

$$\boldsymbol{g}(x) = \boldsymbol{K}^{-1}\left(\frac{H}{h}\,\boldsymbol{e}_k - \frac{H}{h}\,\boldsymbol{e}_{k-1}\right) = \frac{H}{h}(\boldsymbol{c}_k - \boldsymbol{c}_{k-1}) \qquad x_{k-1} < x < x_k. \tag{3.239}$$

This vector is the difference quotient of the columns $\boldsymbol{c}_k$ and $\boldsymbol{c}_{k-1}$ of the inverse matrix $\boldsymbol{K}^{-1}$.

3.12.1 Examples

The bar in Fig. 3.25a is fixed at the left end and consists of five linear elements. The stiffness matrix $\boldsymbol{K}$

$$\boldsymbol{K} = \frac{EA}{l}\begin{bmatrix} 2 & -1 & 0 & 0 & 0 \\ -1 & 2 & -1 & 0 & 0 \\ 0 & -1 & 2 & -1 & 0 \\ 0 & 0 & -1 & 2 & -1 \\ 0 & 0 & 0 & -1 & 1 \end{bmatrix}, \tag{3.240}$$

is a tridiagonal matrix while the inverse

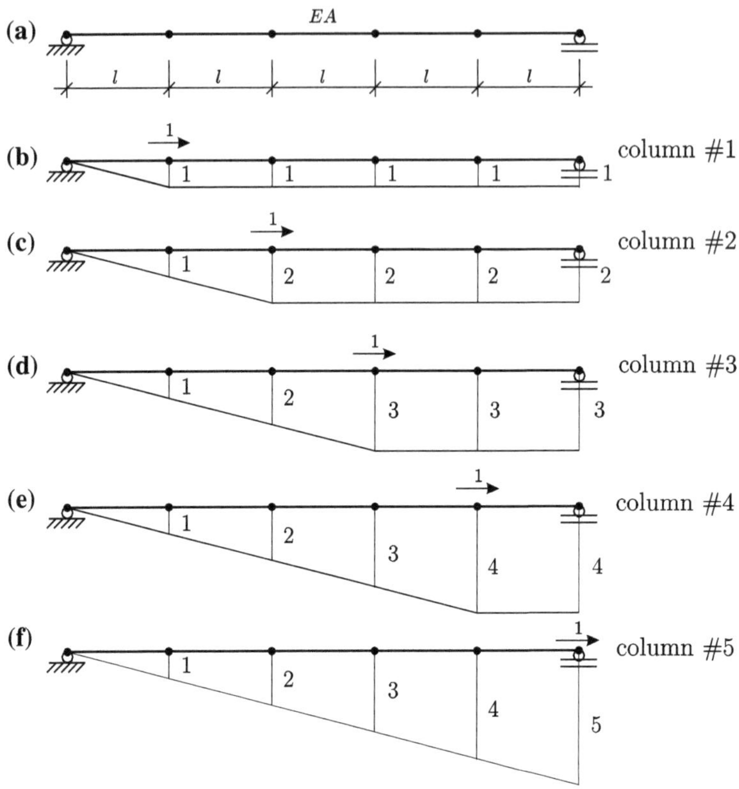

Fig. 3.25 **a** Elastic bar subdivided into five linear elements **b–f** the displacements resulting from the nodal unit loads are the columns of the inverse stiffness matrix (all values times l/EA)

$$\boldsymbol{K}^{-1} = \frac{l}{EA} \begin{bmatrix} 1 & 1 & 1 & 1 & 1 \\ 1 & 2 & 2 & 2 & 2 \\ 1 & 2 & 3 & 3 & 3 \\ 1 & 2 & 3 & 4 & 4 \\ 1 & 2 & 3 & 4 & 5 \end{bmatrix} \tag{3.241}$$

is fully populated. The column $\boldsymbol{c}_k$ of the inverse matrix, see Fig. 3.25b–f, lists the nodal displacements when the bar is subjected to a nodal unit load $P = 1$ at node x_k.

If both ends of the bar are fixed, see Fig. 3.26a, then $\boldsymbol{K}$ is a (4×4) matrix with nearly the same entries as before—only the last entry on the main diagonal changes from 1 to 2—but the inverse is very different, see Fig. 3.26b

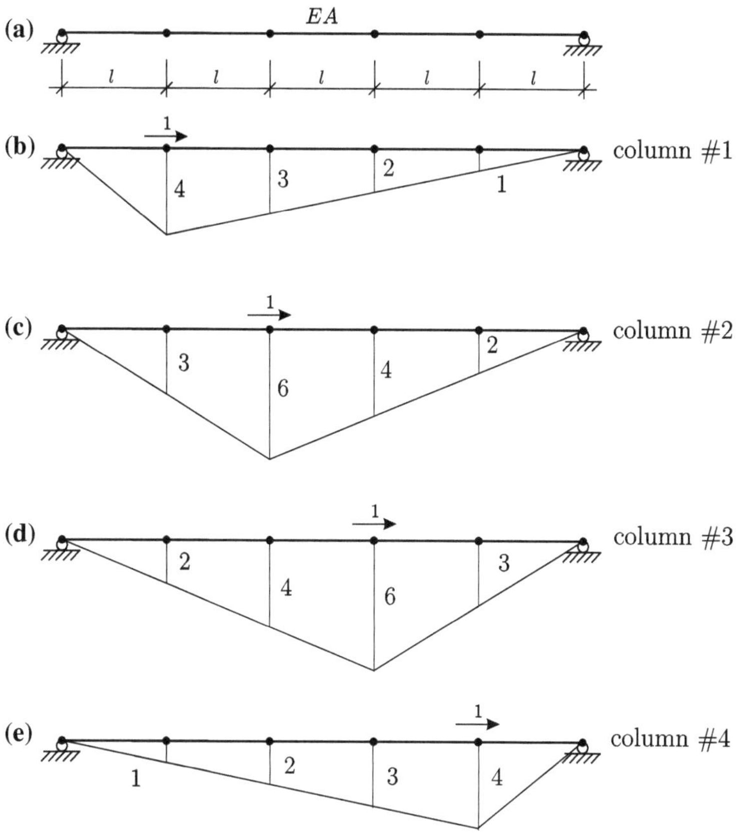

Fig. 3.26 **a** Elastic bar subdivided into five linear elements **b–e** the displacements are the columns of the inverse stiffness matrix (all values times $l/(5\,EA)$)

$$\boldsymbol{K} = \frac{EA}{l}\begin{bmatrix} 2 & -1 & 0 & 0 \\ -1 & 2 & -1 & 0 \\ 0 & -1 & 2 & -1 \\ 0 & 0 & -1 & 2 \end{bmatrix}, \quad \Rightarrow \quad \boldsymbol{K}^{-1} = \frac{l}{5\,EA}\begin{bmatrix} 4 & 3 & 2 & 1 \\ 3 & 6 & 4 & 2 \\ 2 & 4 & 6 & 3 \\ 1 & 2 & 3 & 4 \end{bmatrix}.$$

3.13 Condition of a Stiffness Matrix

If the vector $\boldsymbol{u}$ is the exact solution of the system $\boldsymbol{K}\,\boldsymbol{u} = \boldsymbol{f}$ and the vector $\hat{\boldsymbol{u}}$ an approximate solution then the relative error

$$\frac{|\boldsymbol{u} - \hat{\boldsymbol{u}}|}{|\boldsymbol{u}|} \leq \text{cond}\,(\boldsymbol{K})\frac{|\boldsymbol{f} - \boldsymbol{K}\,\hat{\boldsymbol{u}}|}{|\boldsymbol{f}|} \qquad (3.242)$$

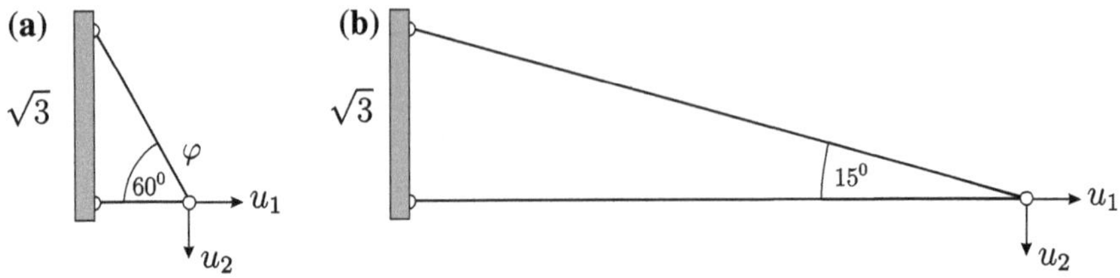

$$\text{cond}(\boldsymbol{K}) = \frac{1.18}{0.317} = 3.72 \qquad \text{cond}(\boldsymbol{K}) = \frac{0.299}{0.0052} = 57.5$$

Fig. 3.27 Two trusses and their condition numbers

depends on the condition of the stiffness matrix. With one degree of freedom, $k\,u = f$, the condition is simply the slope k of the straight line $k\,u$. One measure for the condition of a stiffness matrix $\boldsymbol{K}$ is the ratio of the largest eigenvalue to the smallest eigenvalue

$$\text{cond}(\boldsymbol{K}) = \frac{\lambda_{max}}{\lambda_{min}}. \tag{3.243}$$

A tiny value $\lambda_{min} \sim 0$ means that the forces f_i necessary to squeeze a structure into the shape of the associated eigenvector $\boldsymbol{u}$ are very small

$$\boldsymbol{K}\,\boldsymbol{u} = \lambda_{min}\,\boldsymbol{u} = \boldsymbol{f}. \tag{3.244}$$

The two trusses in Fig. 3.27, the first with normal proportions, the other with a rather elongated shape, have very different condition numbers. We let $EA = 1$ so that the stiffness of an element of length l is $k = EA/l = 1/l$. The stiffness matrix of the first truss ($\varphi = 60^0$) and its inverse are

$$\boldsymbol{K} = \begin{bmatrix} 1.125 & -0.216 \\ -0.216 & 0.375 \end{bmatrix} \qquad \boldsymbol{K}^{-1} = \begin{bmatrix} 1.0 & 0.577 \\ 0.577 & 3.0 \end{bmatrix} \tag{3.245}$$

and the condition number of $\boldsymbol{K}$ is

$$\text{cond}(\boldsymbol{K}) = \frac{\lambda_{max}}{\lambda_{min}} = \frac{1.18}{0.317} = 3.72. \tag{3.246}$$

In the case of the second truss the same matrices are

$$\boldsymbol{K} = \begin{bmatrix} 0.294 & -0.037 \\ -0.037 & 0.01 \end{bmatrix} \qquad \boldsymbol{K}^{-1} = \begin{bmatrix} 6.46 & 24.12 \\ 24.12 & 189.94 \end{bmatrix} \tag{3.247}$$

and the condition number of $\boldsymbol{K}$ is about fifteen times the value of the first matrix

$$\text{cond}(\boldsymbol{K}) = \frac{0.299}{0.0552} = 57.5. \tag{3.248}$$

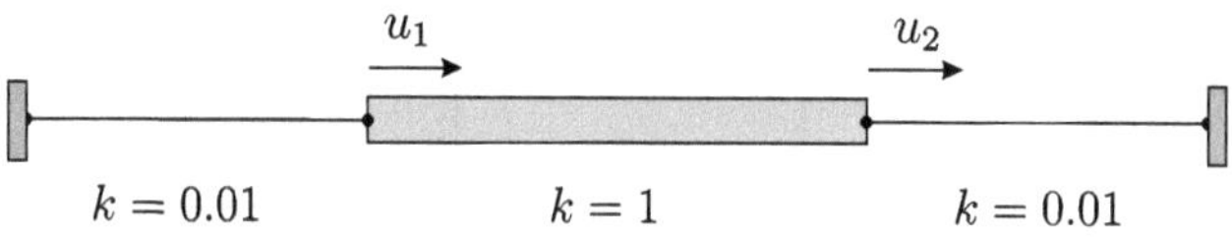

$$\text{cond}(\boldsymbol{K}) = \frac{2.01}{0.01} = 201$$

Fig. 3.28 Three very different rods form a structure

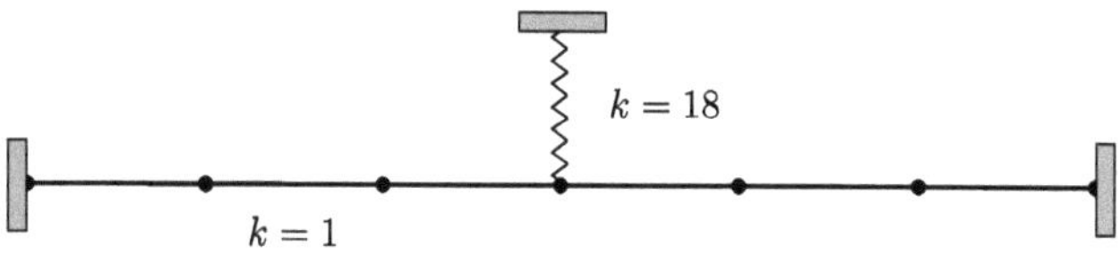

Fig. 3.29 Rope attached to a very stiff spring

Note the large entries in column #2 of $\boldsymbol{K}^{-1}$ which is the (discrete) Green's function for the nodal displacement u_2. A force $f_2 = 1$ would produce a vertical displacement of nearly 190 units! This observation can be made a rule.

Rule of thumb: *A small value λ_{min} implies that small forces $f_i = \lambda_{min}\, v_i$ suffice to push the structure in the direction of the corresponding eigenvector $\boldsymbol{v}$ so some or all columns in $\boldsymbol{K}^{-1}$ must have large entries.*

In such a situation normal service loads would produce very large displacements—probably not a very sound design.

But in other situations it is not the size of the entries which counts but the differences between these entries. The Green's function for the derivative in between two nodes is the difference between the column $\boldsymbol{c}_i$, node x_i, and the column $\boldsymbol{c}_{i+1}$ of the next node x_{i+1}.

The stiffness matrix of the rod in Fig. 3.28 is certainly ill-conditioned

$$\boldsymbol{K} = \begin{bmatrix} 1.01 & -1 \\ -1 & 1.01 \end{bmatrix} \qquad \boldsymbol{K}^{-1} = \begin{bmatrix} 50.2 & 49.8 \\ 49.8 & 50.2 \end{bmatrix} \tag{3.249}$$

and the inverse has large entries but the (discrete) influence function for the slope between the two nodes is small in comparison

$$\boldsymbol{g} = \begin{bmatrix} 50.2 \\ 49.8 \end{bmatrix} - \begin{bmatrix} 49.8 \\ 50.2 \end{bmatrix} = \begin{bmatrix} 0.4 \\ -0.4 \end{bmatrix}. \tag{3.250}$$

On the other hand a single large entry on the diagonal, $k_{ii} \gg 1$, as in Fig. 3.29 will produce large differences between the corresponding columns of $\boldsymbol{K}^{-1}$ so that the Green's function for the slope will be negatively affected by such large jumps in the stiffness

$$
\boldsymbol{K} = \begin{bmatrix} 2 & -1 & 0 & 0 & 0 \\ -1 & 2 & -1 & 0 & 0 \\ 0 & -1 & 20 & -1 & 0 \\ 0 & 0 & -1 & 2 & -1 \\ 0 & 0 & 0 & -1 & 2 \end{bmatrix} \qquad \boldsymbol{K}^{-1} = \frac{1}{20} \begin{bmatrix} 13.5 & 6.9 & 0.4 & 0.2 & 0.1 \\ 6.9 & 13.8 & 0.7 & 0.5 & 0.2 \\ 0.4 & 0.7 & 1.1 & 0.7 & 0.4 \\ 0.2 & 0.5 & 0.7 & 13.8 & 6.9 \\ 0.1 & 0.2 & 0.4 & 6.9 & 13.5 \end{bmatrix}.
$$

Remark 3.11 These effects are independent of the length h of the elements. The entries k_{ij} are of the order $O(1/h)$ and the entries k_{ij}^{-1} are of the order $O(h)$ and the entries $j_i = \varphi_i'(x)$ are of the order $O(1/h)$ so that the $1/h$-effect cancels when the vector $\boldsymbol{g} = \boldsymbol{K}^{-1}\, \boldsymbol{j}$ of the Green's function is calculated.

3.13.1 The Triple Product

Many problems in physics lead to the triple product

$$
\boldsymbol{K}_{n\times n}\, \boldsymbol{u}_{n\times 1} = \boldsymbol{A}_{n\times m}^T\, \boldsymbol{C}_{m\times m}\, \boldsymbol{A}_{m\times n}\, \boldsymbol{u}_{n\times 1} = \boldsymbol{f}_{n\times 1} \tag{3.251}
$$

where $\boldsymbol{A}$ is a rectangular matrix and $\boldsymbol{C}$ a square matrix which depends on the parameters of the problem [5].

In a truss, consisting of m elastic bars, the components of the vector $\boldsymbol{A}\,\boldsymbol{u}$ are the strains $\varepsilon_i = u_i'$ of the m truss elements and the matrix $\boldsymbol{C}$ is a diagonal matrix with entries $c_{ii} = EA_i$, one for each bar, and the vector

$$
\boldsymbol{C}\boldsymbol{A}\,\boldsymbol{u} = \boldsymbol{n} = \{N_1, N_2, \ldots N_m\}^T \tag{3.252}
$$

is the list of the normal forces N_i in the truss elements and

$$
\boldsymbol{A}^T\, \boldsymbol{n} = \boldsymbol{f} \tag{3.253}
$$

formulates the equilibrium conditions at the nodes. If the matrix $\boldsymbol{A}$ is square as in a statically determinate truss, the equilibrium conditions alone suffice to determine the element forces

$$
\boldsymbol{n} = (\boldsymbol{A}^T)^{-1}\, \boldsymbol{f} \tag{3.254}
$$

and these in turn determine the nodal displacements

$$
\boldsymbol{u} = (\boldsymbol{C}\,\boldsymbol{A})^{-1}\, \boldsymbol{n}. \tag{3.255}
$$

Because the matrix $\boldsymbol{A}$ depends only on the geometry of the truss, which is to say the length l_i and angles α_i of the truss elements, it follows that in a statically determinate truss the Green's functions for the internal actions, which are the columns of $(\boldsymbol{A}^T)^{-1}$, only depend on the shape of the truss but not on the coefficients EA_i of the single

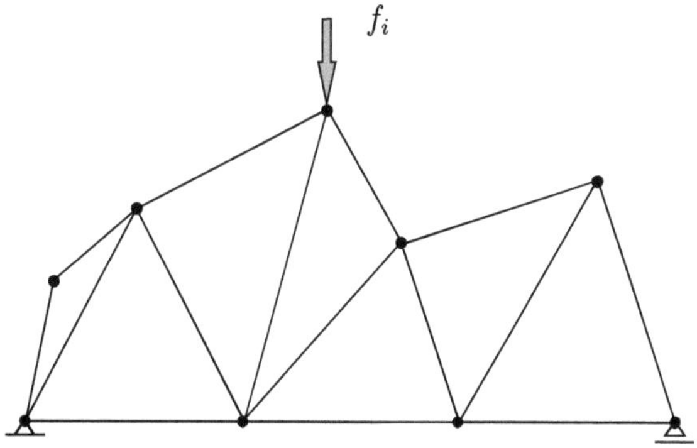

Fig. 3.30 Statically determinate truss: the element forces only depend on the position of the nodes and not on the stiffness of the single elements

truss elements, see Fig. 3.30. With regard to the nodal displacements the situation is different. Their influence functions depend on the matrix $\boldsymbol{C}$.

3.14 Interpolation

In general the FE-solution does not interpolate the exact solution at the nodes x_i. If that were true the vector $\boldsymbol{u}_I$ of the exact nodal displacements

$$\boldsymbol{u}_I = \{u(x_1), u(x_2), \ldots, u(x_n)\}^T \tag{3.256}$$

would be the solution of the system $\boldsymbol{K}\,\boldsymbol{u} = \boldsymbol{f}$ and the interpolant

$$u_I(x) = \sum_{i=1}^{n} u(x_i)\,\varphi_i(x) \tag{3.257}$$

would be identical with the FE-solution. The fact that this is not true is an indication that the interpolant u_I is an inferior approximation in terms of the strain energy metric. The FE-solution beats the interpolant, it has a smaller distance to the exact solution in this metric

$$a(u - u_h, u - u_h) < a(u - u_I, u - u_I). \tag{3.258}$$

Only if the Green's functions of the nodal values $u(x_i)$ lie in $\mathcal{V}_h$ or if the error of the nodal Green's functions is orthogonal to the right hand side p

$$u(x_i) - u_h(x_i) = \int_0^l (G(y, x_i) - G_h(y, x_i))\, p(y)\, dy \tag{3.259}$$

is $u_h(x_i) = u(x_i)$.

A necessary condition for the Green's function of the nodes to lie in $\mathcal{V}_h$ is that the element shape functions φ_i^e are homogeneous solutions of the governing equation. In elementary 1-D problems with only one constant coefficient such as $EA\,u''(x) = p$ or $EI\,u^{IV} = p$ this is the case.[4] The linear shape functions and the Hermite polynomials respectively allow an exact representation of the nodal Green's functions and so $u_h(x_i) = u(x_i)$. But in the case of extended equations (two or more coefficients) such as

$$EA\,u''(x) + c\,u(x) = p \qquad EI\,u^{IV} + c\,u(x) = p \tag{3.260}$$

this is no longer true because the homogeneous solutions of these equations

$$u(x) = c_1\,e^{x\sqrt{c/EA}} + c_2\,e^{-x\sqrt{c/EA}} \tag{3.261}$$

and

$$w(x) = e^{\beta x}(c_1\,\cos\beta\,x + c_2\,\sin\beta\,x) + e^{-\beta x}(c_3\,\cos\beta\,x + c_4\,\sin\beta\,x) \tag{3.262}$$

$$\beta = \sqrt[4]{\frac{c}{EI}} \tag{3.263}$$

respectively are not the typical polynomial shape functions one associates with a FE-trial space $\mathcal{V}_h$. So in these cases the FE-solution does *not* interpolate the exact solution at the nodes.

It can also happen that $\mathcal{V}_h$ is too "smooth". For to generate a Green's function of an equation as $-u'' = p$, we must allow for a discontinuous derivative at the nodes because only then can we model a point force (= unit jump in u') with finite elements. The bubble function at the center node of a quadratic element does not have this property and so $u_h(x_i) \neq u(x_i)$ in general at such nodes, see Fig. 3.31.

This is a hint that FE-analysis must find a balance between the regularity that is required by the strain energy product $a(\varphi_i, \varphi_j)$ and the non-regularity that is necessary in order to come close to the Green's functions.

The fact that quadratic elements do not interpolate the exact solution at internal nodes is not so dramatic as it may sound. In most applications the right-hand side p is constant, then u is quadratic and lies in $\mathcal{V}_h$ and so the error (3.259) is zero at all points x and also at the mid-nodes. Only if u is a third-order (or higher) polynomial will this deficiency become apparent.

Remark 3.12 If the shape functions allow the exact representation of the nodal Green's functions, which implies that they are homogeneous solutions of the governing equation in between the nodes, as is the case in the elementary problems, $-H\,u'' = p$, $-EA\,u'' = p$ and $EI\,u^{IV} = p$, then the error of the Green's functions is zero outside the element which contains the source point [6]. You see this in

[4] Also shear deformations $-GAu'' = p$ belong in this category. Their influence functions are piecewise linear as in a rope.

Fig. 3.31 Quadratic elements do not interpolate the exact solution at the mid-nodes, only at the end nodes of the elements, because the bubble function of the center-node is too smooth, **a** shape functions, **b** point load at mid-node and FE-solution, **c** point load at end node and FE-solution which in this case is exact

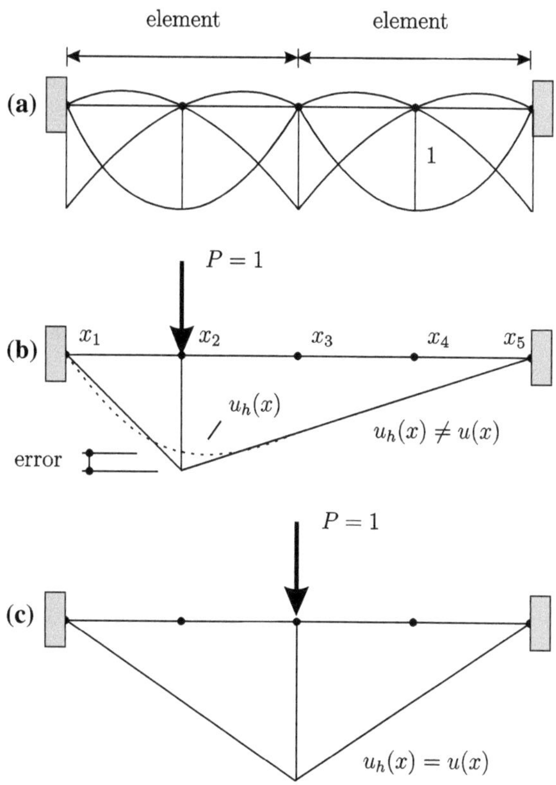

Fig. 1.2. The approximate Green's function misses the peak under the point load but outside the element it is exact.

3.14.1 The Nodal Vector $\boldsymbol{u}_I$

To establish the system of equations for the nodal vector $\boldsymbol{u}_I$ of the interpolant the solution u of the boundary value problem

$$-\Delta u = p \quad \text{on } \Omega \qquad u = 0 \quad \text{on } \Gamma \tag{3.264}$$

is split into the interpolant and a "remainder" term

$$u(x) = u_I(x) + u_p(x) \tag{3.265}$$

which is zero at the nodes, $u_p(x_i) = 0$. Green's first identity

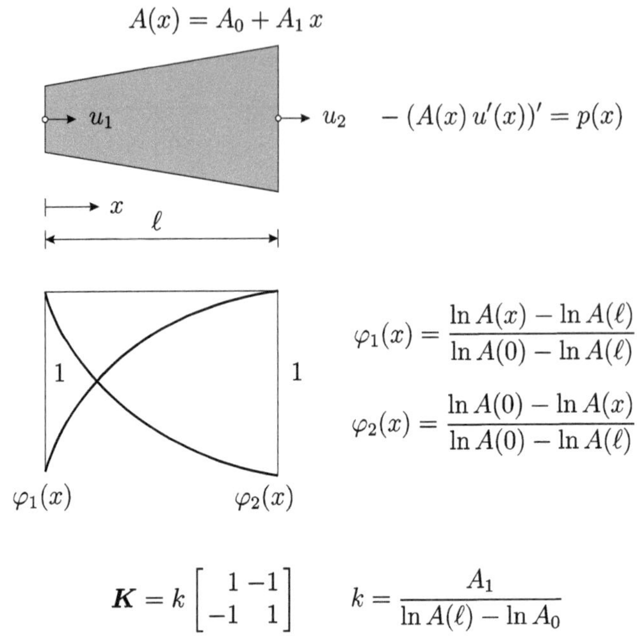

Fig. 3.32 Conic bar element, shape functions and stiffness matrix

$$\mathscr{G}(u, \varphi_i) = \int_\Omega -\Delta\, u\, \varphi_i \, d\Omega - \int_\Gamma \frac{\partial u}{\partial n}\, \varphi_i \, ds - a(u, \varphi_i)$$

$$= \int_\Omega p\, \varphi_i \, d\Omega - a(u_I, \varphi_i) - a(u_p, \varphi_i) = 0 \tag{3.266}$$

implies that the nodal vector $\boldsymbol{u}_I$ solves the system

$$\boldsymbol{K}\, \boldsymbol{u}_I = \boldsymbol{f} + \boldsymbol{k} \tag{3.267}$$

where the vector $\boldsymbol{k}$ has the elements

$$k_i = a(u_p, \varphi_i). \tag{3.268}$$

If $\boldsymbol{k}$ is the zero vector then the FE-solution interpolates u at the nodes.

The equation $\boldsymbol{k} = \boldsymbol{0}$ is equivalent to

$$u(\boldsymbol{x}_i) - u_h(\boldsymbol{x}_i) = \int_\Omega (G(\boldsymbol{y}, \boldsymbol{x}_i) - G_h(\boldsymbol{y}, \boldsymbol{x}_i)\, p(\boldsymbol{y})\, d\Omega_y = 0 \tag{3.269}$$

that is if $k_i = 0$ then the error in the Green's function for $u(\boldsymbol{x})$ at the node $\boldsymbol{x}_i$ is orthogonal to p and vice versa. To see this note that the split $u = u_I + u_p$ implies a corresponding split $p = p_I + p_p = -\Delta u_I - \Delta u_p$ on the right hand side.[5] Because the interpolant u_I lies in $\mathcal{V}_h$ it holds

$$u_I(\boldsymbol{x}_i) = \int_\Omega G_h(\boldsymbol{y}, \boldsymbol{x}_i)\, p_I \, d\Omega_y = \int_\Omega G(\boldsymbol{y}, \boldsymbol{x}_i)\, p_I \, d\Omega_y \tag{3.270}$$

and so it has only to be verified that also the second part p_p of $p = p_I + p_p$ is orthogonal to the error in the Green's function

$$\int_\Omega (G(\boldsymbol{y}, \boldsymbol{x}_i) - G_h(\boldsymbol{y}, \boldsymbol{x}_i))\, p_p \, d\Omega_y = 0 \qquad i = 1, 2, \ldots n. \tag{3.271}$$

But the nodal values of u_p are zero and so the exact nodal Green's functions return zero

$$\int_\Omega G(\boldsymbol{y}, \boldsymbol{x}_i)\, p_p \, d\Omega_y = 0. \tag{3.272}$$

Finally Green's first identity implies

$$\mathcal{G}(u_p, \varphi_i) = \int_\Omega p_p \, \varphi_i \, d\Omega - a(u_p, \varphi_i) = \int_\Omega p_p \, \varphi_i \, d\Omega - 0 = 0 \tag{3.273}$$

for each φ_i and so also the second part

$$\int_\Omega G_h(\boldsymbol{y}, \boldsymbol{x}_i)\, p_p \, d\Omega_y = \sum_i g_i(\boldsymbol{x}_i) \int_\Omega \varphi_i \, p_p \, d\Omega = 0 \tag{3.274}$$

is zero as well because $G_h(y, x_i)$ is just an expansion in terms of the shape functions φ_i. So (3.271) is true.

Example 3.2 The saw blade in Fig. 3.33 is subjected to a unit horizontal point load at its center. The longitudinal displacement of the blade in between two spikes is the solution to the differential equation

$$-\,(EA(x)\, u'(x))' = 0 \tag{3.275}$$

---

[5] p_I and p_p actually are the distributions (= element loads + jump terms) belonging to u_I and u_p respectively. Symbolically this can be expressed in this way.

where

$$A(x) = 1 + 20\,x \tag{3.276}$$

is the longitudinal stiffness; we let $E = 1$.

The homogeneous solution of (3.275)

$$u_0 = c_2 + c_1\,\frac{\ln(1 + 20\,x)}{20} \tag{3.277}$$

comes with two free constants c_1 and c_2 and so the shape functions of a bar element $[0, l]$ of length $l = 1$ are the two functions, see Fig. 3.32,

$$\varphi_1(x) = -\frac{\ln(1 + 20\,x)}{\ln 21} + 1\,, \qquad \varphi_2(x) = \frac{\ln(1 + 20\,x)}{\ln 21}\,. \tag{3.278}$$

Upon substituting these functions into the strain energy product

$$k_{ij} = \int_0^l EA(x)\,\varphi_i'\varphi_j'\,dx\,, \tag{3.279}$$

the entries of the stiffness matrix of a trapezoidal bar element of length $l = 1$ are obtained

$$\boldsymbol{K} = 6.569\begin{bmatrix} 1 & -1 \\ -1 & 1 \end{bmatrix} \qquad 6.569 = \frac{20}{\ln 21}\,. \tag{3.280}$$

The global shape functions $\varphi_i(x)$ of the nodes, which are patched together from the element shape functions, form the basis of the trial and solution space $\mathcal{V}_h$, see Fig. 3.33. Because the φ_i are piecewise homogeneous solutions of the governing equation the exact solution lies in $\mathcal{V}_h$, that is the FE-solution u_h is identical with the exact solution u. So the nodal displacement of the center node can be predicted precisely on this mesh and space $\mathcal{V}_h$. Note that the solution $u(x)$ is the Green's function for the nodal displacement of the center node.

If the unit nodal displacements would be modeled with standard linear shape functions instead, see Fig. 3.33, then there would be a relative large gap between the FE-solution u_h and the exact solution and the normal force of the FE-solution would come out as a zig-zag function—the echo of the zig-zag stiffness distribution because u_h' is constant. The error in the normal force would be even more pronounced if we would substitute for $u(x)$ its interpolant $u_I(x)$ on $\mathcal{V}_h$, see Fig. 3.33f. This is in agreement with the aforementioned fact that the interpolant u_I is less fitting in terms of the strain energy metric than the FE-solution.

The strain energy of the saw blade is the expression

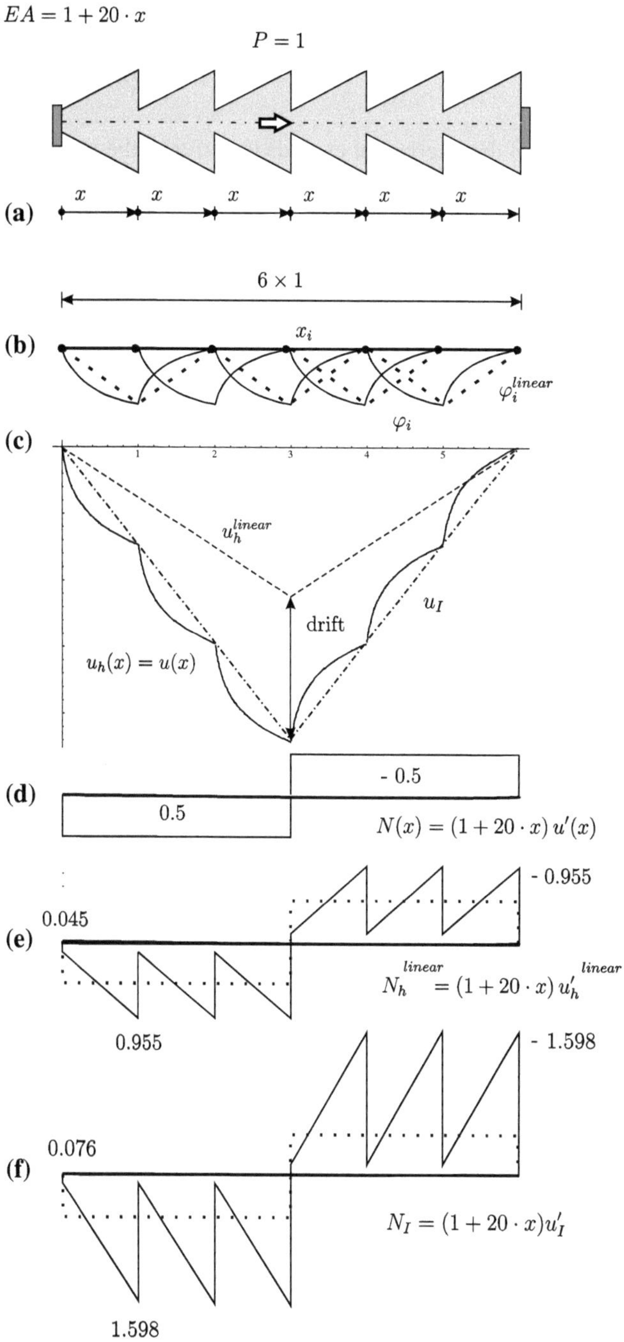

Fig. 3.33 Saw blade

$$a(u,u) = \int_0^l EA(x)\, u'(x)^2\, dx = \int_0^l N(x)\, u'(x)\, dx = \int_0^l \frac{N(x)^2}{EA(x)} dx \tag{3.281}$$

and clearly the error of the interpolant exceeds the error of the linear FE-solution

$$\begin{aligned} a(u-u_h, u-u_h) &= \int_0^l \frac{(N-N_h)^2}{EA(x)} dx \\ &< \int_0^l \frac{(N-N_I)^2}{EA(x)} dx = a(u-u_I, u-u_I) \end{aligned} \tag{3.282}$$

or in pure numbers

$$a(u-u_h, u-u_h) = 0.09 < 0.15 = a(u-u_I, u-u_I). \tag{3.283}$$

So when we operate with the strain energy metric we must sacrifice the "naive" belief that next to the exact solution the interpolant is second in rank.

3.15 Infinite Stresses

Singular points are points where the stresses become infinite. Such hot spots typically lie on the edge, at corner points or at points where the boundary conditions change, see Fig. 3.34.

If we believe that influence functions can predict also these stresses—not directly at the hot spot itself, but close by—then we have a problem: How does a dislocation (= influence function for the stress σ_{yy} at the crack tip) make that the lower and upper edge of the plate, see Fig. 3.35, disappear from the screen and move in opposite directions, to $\pm\infty$? This is the only solution possible if we believe that the stress becomes singular at the crack tip

$$\sigma(\boldsymbol{x}) = \int_\Gamma G(\boldsymbol{y}, \boldsymbol{x})\, p(\boldsymbol{y}) ds_{\boldsymbol{y}} = \infty. \tag{3.284}$$

How does this happen? How can a dislocation produce an infinite displacement field?

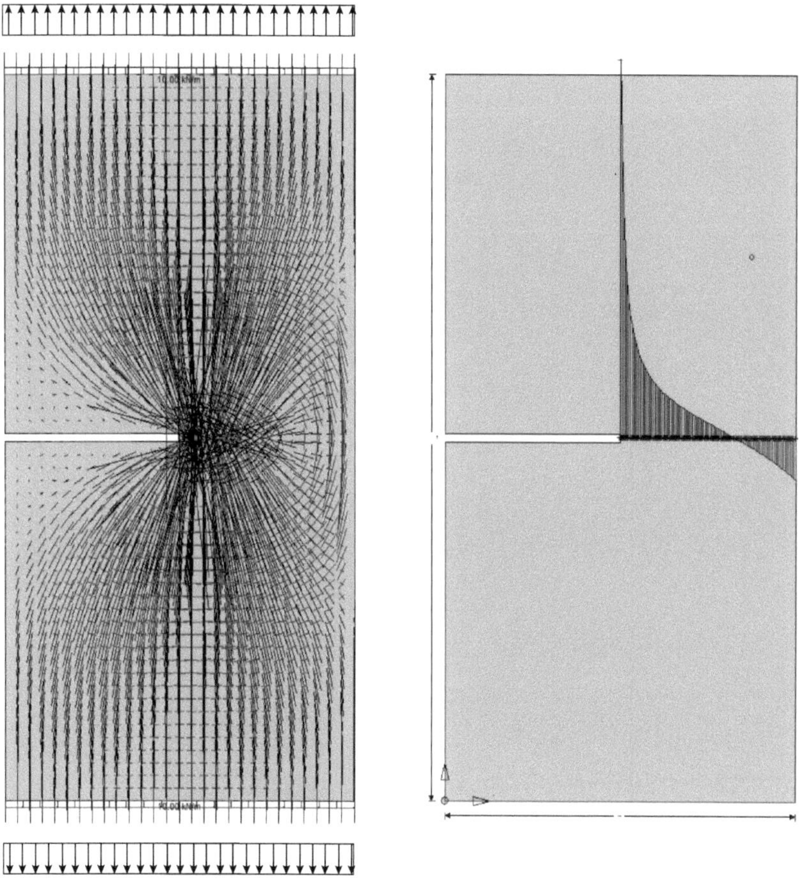

Fig. 3.34 Stresses in a plate with a slit

3.15.1 Archimedes' Lever

Archimedes knew how this could happen and he promptly claimed that he knew how to lift the Earth, s. Fig. 3.36a.

The force exerted by his hand times distance from his hand to fulcrum equals mass of Earth times distance from Earth to fulcrum.

Where Archimedes would place the Earth we place a support and we ask for the reaction force R_A which is necessary to balance the action of a single force P (Archimedes' hand) at the end of the cantilever beam.

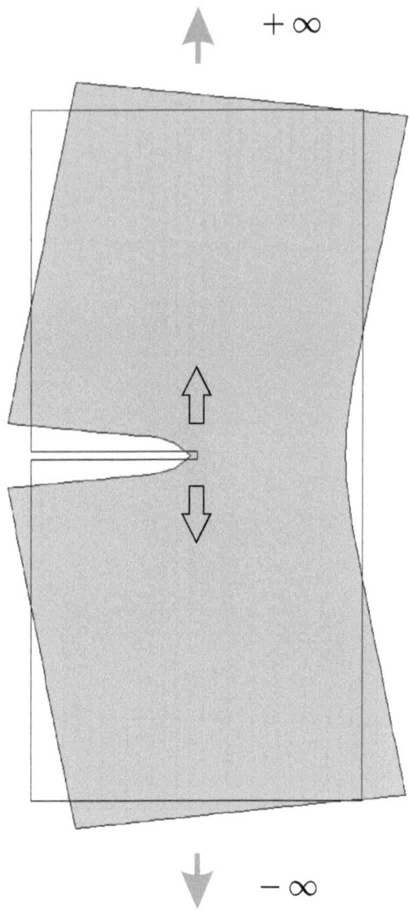

Fig. 3.35 A dislocation at the crack tip must produce infinite displacements

The influence function for the support reaction R_A is generated by pushing the left end of the beam down by one unit length and as a reaction to this the beam rotates about the fulcrum.

In linear mechanics rotations are pseudo-rotations: the distance x of a point from the fulcrum and its vertical lift y form a right angle triangle

$$\tan\varphi = \frac{y}{x} \tag{3.285}$$

where φ is the rotation angle. In the case of the support the lift is $y = 1$ and so the lever ultimately will perform a 90^0 rotation when the distance h between the two supports tends to zero

$$\tan\varphi = \lim_{h\to 0}\frac{1}{h} = \infty \tag{3.286}$$

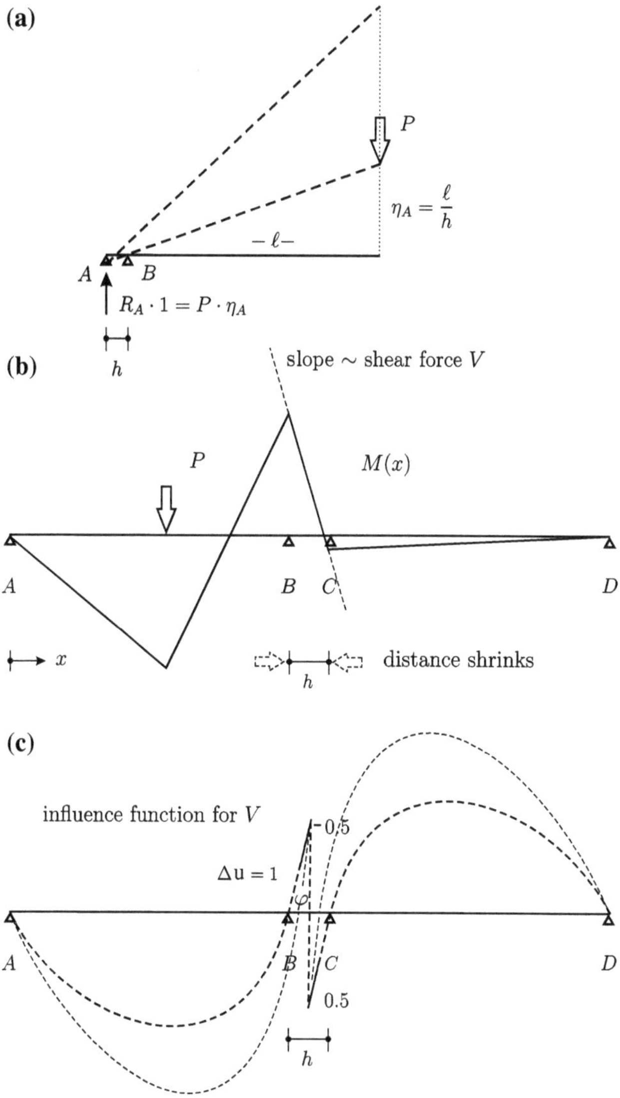

Fig. 3.36 How influence functions become infinite: **a** Archimedes' lever, **b** moment $M(x)$ and **c** influence function for the shear force

and this means that an infinite force

$$R_A = \lim_{h \to 0} \frac{1}{h} \, l \, P \tag{3.287}$$

is necessary to balance even the tiniest force P at the other end of the cantilever.

Actually Archimedes applied a "one-sided" dislocation. (i) He inserted a shear-hinge (support reaction = shear force V) next to the support and (ii) he spread this hinge by one unit in vertical direction. But the left side of the hinge cannot move, it is attached to the support, so the right side of the hinge must go all the way, that is it moves one unit down and so it rotates the beam by 90^0 if the distance h shrinks to zero.

We suspect that in most such cases the $\tan \varphi = 1/h$ effect is the culprit which produces the additional singularity in the Green's function. A unit dislocation of a point infinitely close to a fixed point kicks the displacement field into overdrive and sends it to the "stars". That is we presume that doubly singular influence functions have the form

$$\frac{1}{|\boldsymbol{\xi} - \boldsymbol{x}|} r^{-1} f(\varphi) \tag{3.288}$$

where $\boldsymbol{\xi}$ is the singular point on the boundary (the point to which the source point will drift in the end), $\boldsymbol{x}$ is the current position of the source point inside Ω, $r = |\boldsymbol{x} - \boldsymbol{y}|$ is the distance from $\boldsymbol{x}$ to the points $\boldsymbol{y}$ on Ω and φ is the angle of the radius vector from $\boldsymbol{x}$ to $\boldsymbol{y}$. So there is a "double singularity", the natural r^{-1}, and the $1/|\boldsymbol{\xi} - \boldsymbol{x}|$ singularity.

3.15.2 Continuous Beam

A similar effect can be observed in the case of the continuous beam in Fig. 3.36b where the point load produces a zig-zag bending moment in the beam. Because the shear force is proportional to the slope of the bending moment $M(x)$

$$V(x) = -EI\, M'(x) \tag{3.289}$$

the shear force will become infinite if the distance h between the two interior supports shrinks to zero because the bending moment M has in the limit an infinite slope. Consequently the influence function for $V(x)$

$$V(x) = G(y_P, x) \cdot P \tag{3.290}$$

must develop in the limit, $h \to 0$, an infinite peak at the foot of the point load P. Is this true?

The influence function for $V(x)$ is the response of the continuous beam to a unit dislocation at the source point x, the mid-point between the two supports. Because of the symmetry of the system such a dislocation $[[G]] = 1$ will manifest itself as a shear deformation $G = \pm 0.5$ of the source point and this means that the two end points of the central span rotate counter-clockwise by the same angle

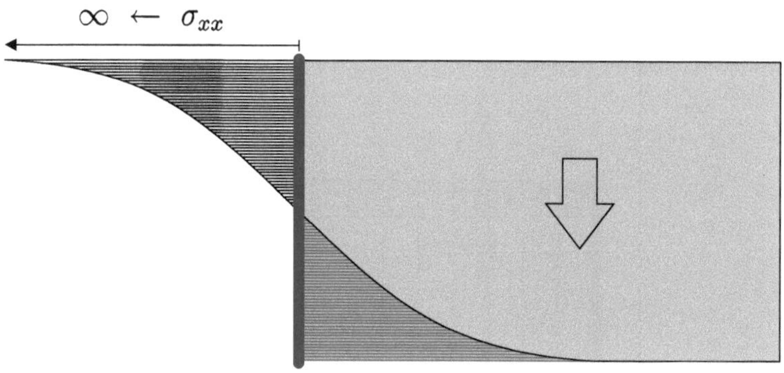

Fig. 3.37 Gravity load in a cantilever plate, the stresses in the extreme fibers tend to $\pm\infty$

$$\tan\varphi = \frac{0.5}{0.5\,h} = \frac{1}{h} \tag{3.291}$$

and in the limit, $h \to 0$, will this rotation cause a lift of infinite magnitude, $G(y_P, x) = \infty$, at the foot of P.

3.15.3 Cantilever Plate

But even in standard situations we observe infinite stresses as in the case of the cantilever plate in Fig. 3.37 which carries its own weight and can do so only by producing infinite stresses in the extreme fibers. We may assume that this would also happen if the weight were replaced by one single force P acting at an, more or less, arbitrary single point $\boldsymbol{y}_P$.

If this is true then the influence function for the edge stress σ_{xx} of the upper fiber must have an infinite value at almost any point of the plate

$$\sigma_{xx}(\boldsymbol{x}) = G(\boldsymbol{y}_P, \boldsymbol{x}) \cdot P = \infty \cdot P. \tag{3.292}$$

Now the influence function is generated by a horizontal unit dislocation of the corner point. The nodal forces which produce the FE-version of the influence function are the stresses σ_{xx} of the nodal shape functions at the corner point and this means that only the element to which the corner point belongs carries nodal forces. These forces are proportional to E/h, where E is the modulus of elasticity ($E = 2.1\ 10^5$ N/mm^2 (steel)) and h is the element length.

In a numerical test, see Fig. 3.38, on an adaptively refined mesh of bilinear finite elements the vertical displacement due to the dislocation grew indeed exponentially with $h \to 0$.

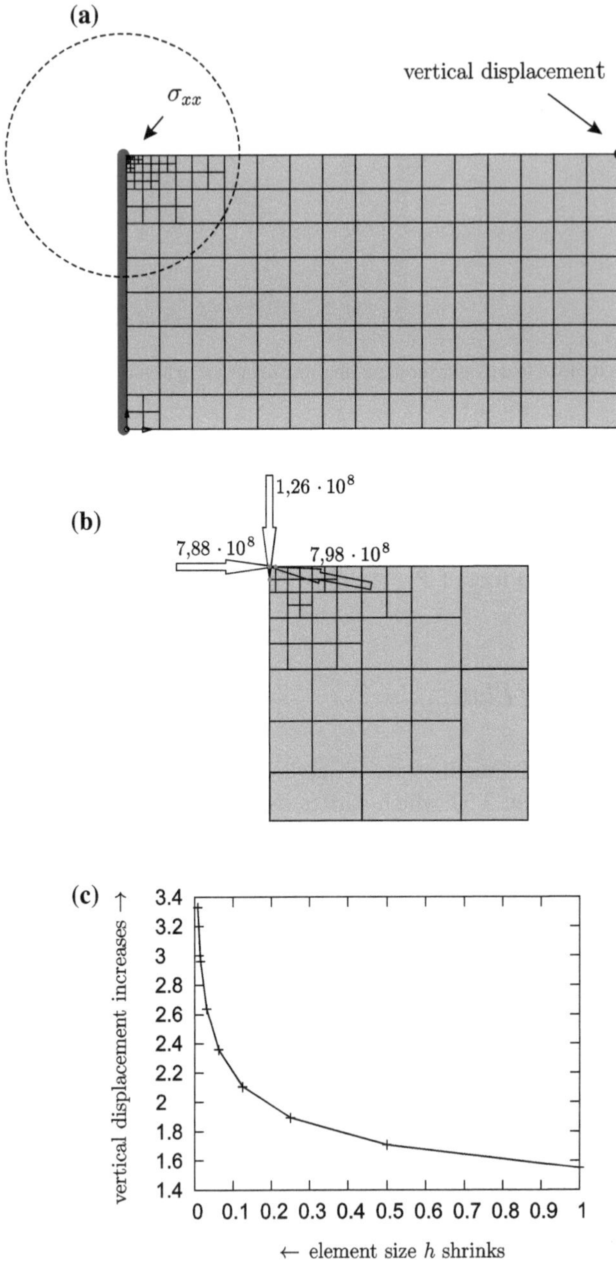

Fig. 3.38 Generating the influence function for the stress σ_{xx} at the corner point, **a** mesh, **b** equivalent nodal forces that generate the influence function, **c** vertical displacement of the *upper right corner point* as a function of the element length h

So to infinite stresses must correspond infinite displacements of the influence functions. No other solution is possible if we trust that the concept of influence functions is also valid in the neighborhood of singular points.

Remark 3.13 Why the stresses become infinite will be discussed in Sect. 3.19 and also why rounding off the corner points will eliminate the singular stresses.

3.15.4 Summary

There are three kinds of infinity

1. The influence function is infinite at the source point.
2. The integral of the influence function is unbounded, is infinite.
3. The influence function is pointwise infinite almost everywhere.

In 2-D and 3-D problems most influence functions are infinite at the source point as for example the fundamental solution of the Laplace operator

$$g(\boldsymbol{y}, \boldsymbol{x}) = \begin{cases} \dfrac{1}{2\pi} \ln r & \text{(2-D)} \\ \dfrac{1}{4\pi} r & \text{(3-D)} \end{cases} . \tag{3.293}$$

An example of the second type is the influence function for the vertical stress σ_{yy} in the soil underneath the edge of a rigid punch, see Fig. 2.9, p. 76. The value of σ_{yy} is infinite because the integral

$$\sigma_{yy}(\boldsymbol{x}) = \bar{u} \int_{\Gamma_D} t(\boldsymbol{y}, \boldsymbol{x})\, ds_{\boldsymbol{y}} \tag{3.294}$$

is unbounded. The kernel $t(\boldsymbol{y}, \boldsymbol{x})$ is the soil pressure on the underside of the rigid punch if a unit dislocation (inf. func. for σ_{yy}) occurs directly under the edge of the rigid punch. The kernel $t(\boldsymbol{y}, \boldsymbol{x})$ has a singularity of type $1/r^2$ and so its area cannot be measured, it is infinite. The slightest downward movement $\bar{u}$ of the rigid punch produces an infinite stress σ_{yy} underneath the tip of the punch.

3.16 Why Do Singularities Matter?

Or to be more precise: Why do poorly resolved singularities matter? After all the FE-influence function

$$u_h(\boldsymbol{x}) = \int_\Omega G_h(\boldsymbol{y}, \boldsymbol{x})\, p(\boldsymbol{y})\, d\Omega_y \tag{3.295}$$

always looks the same. Where do the singularities—if any—hide? They hide in the kernel $G_h(\boldsymbol{y}, \boldsymbol{x})$. To see this recall that solutions of the Poisson equation, $-\Delta u = p$, (and similarly of other equations as well) can also be written in the form

$$u(\boldsymbol{x}) = \int_\Gamma \left[g(\boldsymbol{y}, \boldsymbol{x}) \frac{\partial u}{\partial n}(\boldsymbol{y}) - \frac{\partial g(\boldsymbol{y}, \boldsymbol{x})}{\partial n} u(\boldsymbol{y}) \right] ds_y + \int_\Omega g(\boldsymbol{y}, \boldsymbol{x})\, p(\boldsymbol{y})\, d\Omega_y \tag{3.296}$$

where $g(\boldsymbol{y}, \boldsymbol{x}) = 1/(2\,\pi) \ln r$ is the fundamental solution.

This formula can also be applied to the Green's function $G_h(\boldsymbol{y}, \boldsymbol{x})$ itself

$$\begin{aligned} G_h(\boldsymbol{y}, \boldsymbol{x}) = &\int_\Gamma \left[g(\boldsymbol{\xi}, \boldsymbol{y}) \frac{\partial G_h}{\partial n}(\boldsymbol{\xi}, \boldsymbol{x}) - \frac{\partial g(\boldsymbol{\xi}, \boldsymbol{y})}{\partial n} G_h(\boldsymbol{\xi}, \boldsymbol{x}) \right] ds_\xi \\ &+ \underbrace{\int_\Omega g(\boldsymbol{\xi}, \boldsymbol{y})\, \delta(\boldsymbol{\xi} - \boldsymbol{x})\, d\Omega_\xi}_{= g(\boldsymbol{y}, \boldsymbol{x})} \end{aligned} \tag{3.297}$$

and here one sees how $G_h(\boldsymbol{y}, \boldsymbol{x})$ depends on its slope $\partial G_h / \partial n$ on the boundary. If the true slope becomes singular at a corner point then the FE-slope probably will only be a poor approximation and then this defect will have a negative influence on $G_h(\boldsymbol{y}, \boldsymbol{x})$ and therewith finally also on u_h.

3.17 Nature Makes No Jumps: Finite Elements Do

On crossing the interface between two neighboring elements the stresses jump. This means that also the influence functions must jump. At any given point on Ω the influence functions for the stresses at two infinitely close points—separated only by a mesh line—differ by a distinct margin. Why is this?

The answer can be seen in Fig. 3.39. The equivalent nodal forces which generate the influence function for σ_{xx} at the point $\boldsymbol{x}_1$ are the stresses of the shape functions of the neighboring nodes at that point. Because only shape functions whose support contains the point $\boldsymbol{x}_1$ contribute to the action only the four nodes of the element that contains the point $\boldsymbol{x}_1$ will carry nodal forces. The moment the point wanders into the next element, $\boldsymbol{x}_1 \rightarrow \boldsymbol{x}_2$, the previously loaded nodes will be unloaded and the forces will appear at the four corner nodes of the next element.

This sudden switch in the load carrying nodes is responsible for the jump in the influence function and therewith for the jump in the stresses.

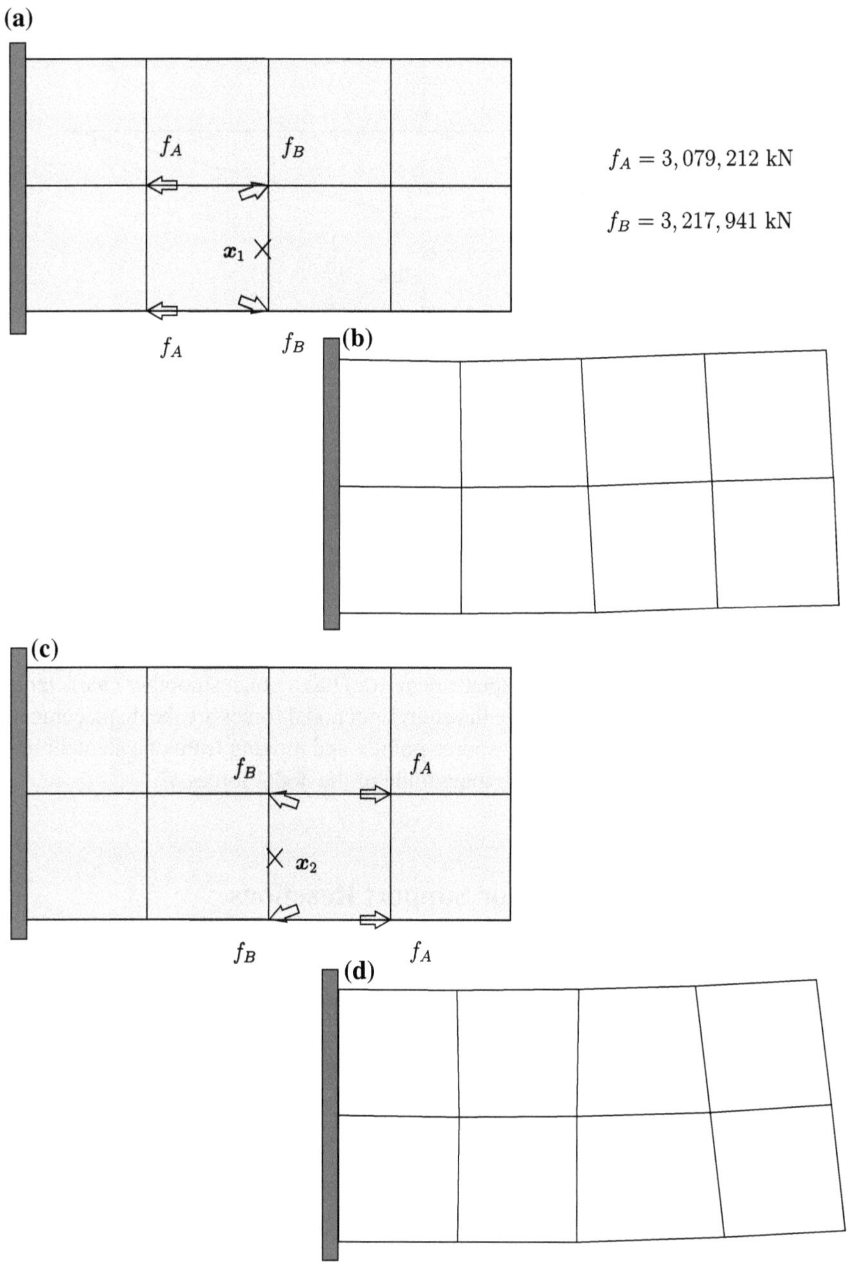

Fig. 3.39 FE-influence functions for σ_{xx} at two neighboring points, **a** equivalent nodal forces which generate the influence function for σ_{xx} at $\boldsymbol{x}_1$, **b** influence function, **c** equivalent nodal forces for the influence function for σ_{xx} at $\boldsymbol{x}_2$, **d** influence function

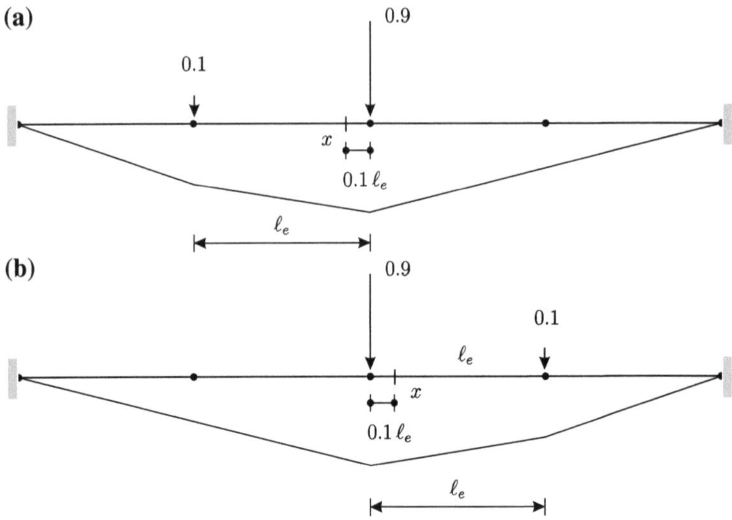

Fig. 3.40 FE-influence functions for the displacement of a rope: different locations (**a**) and (**b**) of the source point x

The influence function for a displacement $u(\boldsymbol{x})$ has a much smoother character as can be seen in Fig. 3.40. In this case the equivalent nodal forces are the displacements $\varphi_i(x)$ of the shape functions at the source point x and moving from one element into the next element hardly affects the magnitude of the nodal forces f_i.

3.18 Influence Functions for Support Reactions

The influence function G for the support reaction R of the continuous beam, see Fig. 3.41a, b, is the deflection of the beam when the support settles by one unit.

The two parts G_1 and G_2, in the first and in the second span respectively, are homogeneous solutions and they are joined together at the interface, $G_1 = 1 = G_2$ and $G_1' = G_2'$, $M_1 = M_2$ so that indeed

$$\mathscr{B}(G_1, w)_{\Omega_1} + \mathscr{B}(G_2, w)_{\Omega_2} = V^- - V^+ + \int_0^l G(y, x)\, p(y)\, dy = 0 \qquad (3.298)$$

where the *jump* in the shear force equals the support reaction $R = V^+ - V^-$.

To obtain an FE-approximation of the influence function the support reactions of the shape functions φ_i are applied as nodal forces

$$j_i = J(\varphi_i) = R(\varphi_i) = a(\varphi_X, \varphi_i). \qquad (3.299)$$

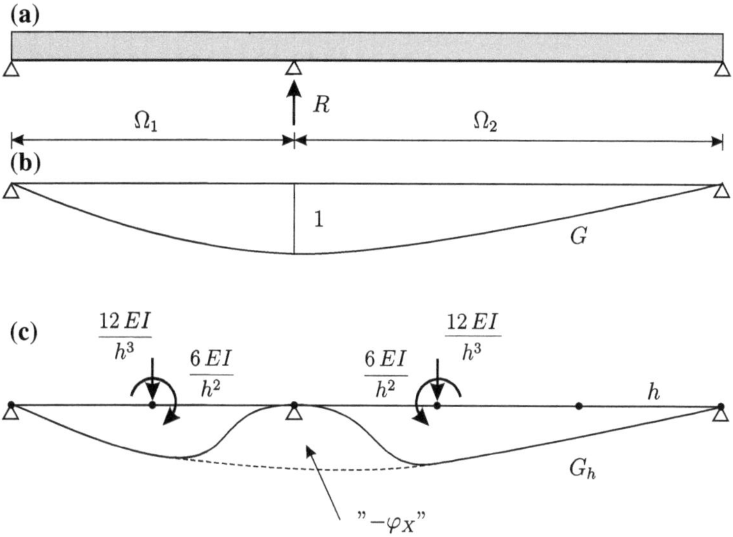

Fig. 3.41 Beam, **a** support reaction R, **b** influence function for support reaction R, **c** FE-approximation of the influence function on $\mathcal{V}_h$; the nodal forces f_i are the support reactions R of the shape functions

But the result is not what is to be expected, see Fig. 3.41c. What is wrong?

In some sense nothing is wrong. Technically one could argue that because the test and trial space $\mathcal{V}_h$ does not contain the nodal shape function of the support G cannot lie in $\mathcal{V}_h$.

An engineer knows that the exact solution $G = G_0 + 1 \cdot \varphi_X$ can be split into a homogeneous solution G_0 and the shape function φ_X of the support so that he would choose for G_h the *ansatz*

$$G_h(x) = G_0^h(x) + 1 \cdot \varphi_X(x) = \sum_i g_i\, \varphi_i(x) + 1 \cdot \varphi_X(x). \tag{3.300}$$

Because G_0 is a homogeneous solution ($p = 0$) the variational problem takes the form

$$a(G_h, \varphi_i) = 0 \qquad \forall\, \varphi_i \in \mathcal{V}_h \tag{3.301}$$

or

$$a(G_0^h, \varphi_i) = -a(\varphi_X, \varphi_i) \qquad \forall\, \varphi_i \in \mathcal{V}_h \tag{3.302}$$

which leads to

$$\boldsymbol{K}\, \boldsymbol{g} = -\boldsymbol{f} \tag{3.303}$$

where

$$f_i = a(\varphi_X, \varphi_i) = R(\varphi_i) \tag{3.304}$$

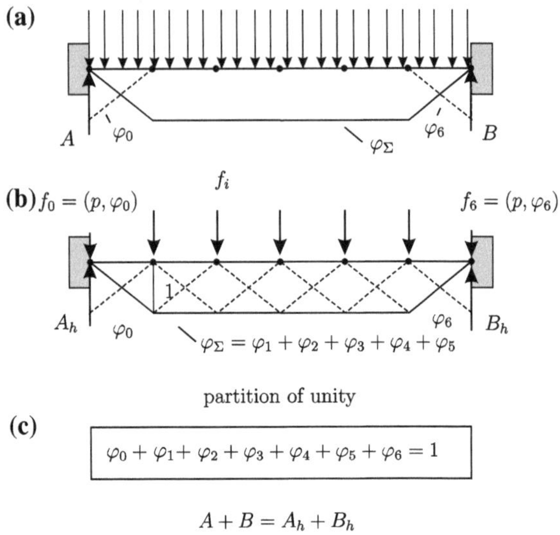

Fig. 3.42 Taut rope, **a** virtual displacement φ_Σ, **b** equivalent nodal forces, **c** the shape functions form a partition of unity

is the support reaction which belongs to the shape function φ_i. This way the engineer obtains the exact solution, $G_h = G_0^h + \varphi_X = G$, while the FE-method internally operates with a "wrong" influence function, the one in Fig. 3.41c.

3.18.1 Global Equilibrium

But why does an FE-solution satisfy the global equilibrium condition if it cannot generate the exact Green's function for the support reaction? The answer is that part of the influence function which is missing in Fig. 3.41c, namely φ_X, is exactly the part which renders the nodal force f_i which is directly applied at the support.

$$f_i = \int_0^l \varphi_X \, p \, dy \tag{3.305}$$

This is why global equilibrium is established.

So nothing is lost, only the load is distributed differently. Global equilibrium means that the sum of the FE-support reactions balances the applied load, as in the case of the rope see Fig. 3.42, that is

$$A + B = \int_0^l p \, dx = A_h + B_h . \tag{3.306}$$

The sum $A_h + B_h$ equals $A + B$ because the shape functions φ_i form a *partition of unity*

$$\sum_{i=0}^{6} \varphi_i = \varphi_0(x) + \varphi_1(x) + \ldots + \varphi_6(x) = 1 \qquad 0 < x < l. \tag{3.307}$$

Because the FE-load case p_h, here the nodal forces f_i, is equivalent to p with respect to all test functions $v_h \in \mathcal{V}_h$ we have

$$\int_0^l p\,\varphi_\Sigma\,dx = \int_0^l p_h\,\varphi_\Sigma\,dx = \sum_{i=1}^{5} f_i\,\varphi_\Sigma(x_i). \tag{3.308}$$

To the equivalent nodal forces f_i in the interior the engineer adds the nodal forces f_0 and f_6 at the supports

$$f_0 = \int_0^l p\,\varphi_0\,dx \qquad f_6 = \int_0^l p\,\varphi_6\,dx \tag{3.309}$$

so that the total sum

$$A_h + B_h = \sum_{i=0}^{6} f_i = \int_0^l p \sum_{i=0}^{6} \varphi_i(x)\,dx = \int_0^l p \cdot 1\,dx = A + B \tag{3.310}$$

equals the applied load.

This may be formulated as a rule: If the unrestrained space $\mathcal{V}_h^+$, that is simply the set of all trial functions φ_i of a mesh (before any boundary conditions are taken into account) contains the rigid body modes, translations and (engineering) rotations, then global equilibrium is established, that is the resultant $\boldsymbol{R}_h$ of the FE-forces coincides in length, direction and orientation with the resultant $\boldsymbol{R}$ of the applied forces.

Remark 3.14 But a handicap remains. The total sum of the forces which flow from inside a region Ω to the edge cannot exceed in magnitude the resultant internal load R_{int}^h summed up by the function φ_Σ (the handicapped 1 with the ramps at the end)

$$R_{int}^h = \int_\Omega \varphi_\Sigma\,p\,d\Omega < \int_\Omega 1 \cdot p\,d\Omega = R \tag{3.311}$$

so that if p is a uniform load then the resultant shear force V, the integral along the edge, will be less than the exact value.

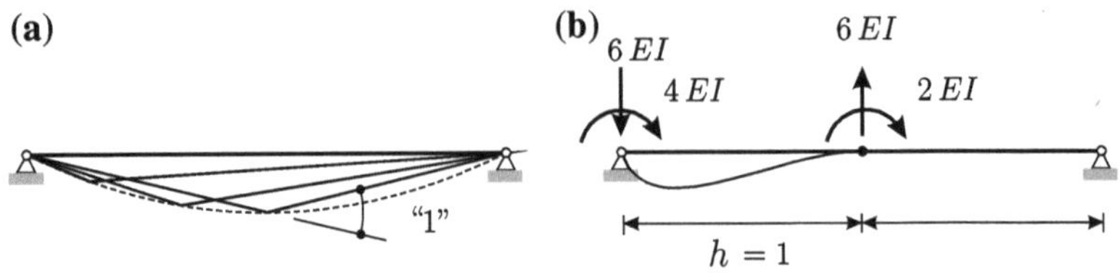

Fig. 3.43 Influence functions for $M(x)$ **a** the more the source point nears the hinged support the smaller the influence function gets, **b** the FE-influence function for $M(0)$ ignores this tendency; it only gets smaller if $h \to 0$

3.18.2 A Paradox?

The more the source point nears the hinged support the more the influence function for $M(x)$ nears zero because at a hinged support M is zero, see Fig. 3.43a. But the FE-approximation of the Green's function for $M(0)$ is not zero and this is no contradiction because a distributed load p will generate a nodal moment $f = (p, \varphi_i)$ at the support and the deflection in Fig. 3.43b is the influence function for $M_h(0) = f$. Only if the element size h tends to zero will also the FE-influence function tend to zero. Note that the influence function is exact, has the value zero, outside the element itself.

3.19 The Path the Load Takes

The stress distribution in a plate, see Fig. 3.44, resembles the flow of a fluid. At prominent points the flow is strong and rapid while at other points it is more relaxed; sometimes it even seems to loose drive and direction. We can clearly discern a principal route which the streamlines follow.

On this route lie the most energetic streamlines. Now energy is measured in terms of strains and stresses. For the stresses (~strains) at a point to be large dislocations at the point must produce large displacements in the loaded zone. This distinguishes the main path from the "also runs". The impression of a path is due to the fact that dislocations at neighboring points and in similar directions yield approximately the same displacements in the region of the load.

In real structures the streamlines are continuous functions while in FE-analysis they jump and change directions on crossing interelement boundaries.

The pattern that a specific load generates is an intrinsic property of the structure. The 1-D example of a truss, see Fig. 3.45, may illustrate this. The size of the normal force N_i in a truss element l_i is proportional to the displacement Δ_i the load experiences when the truss element is spread by one unit length, see Fig. 3.45b, c.

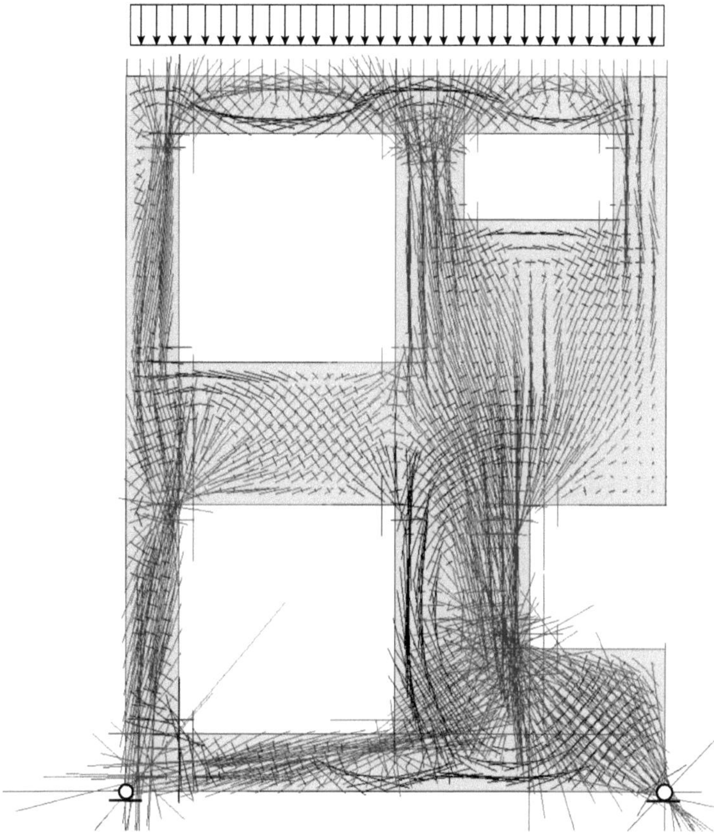

Fig. 3.44 Principal stresses in a plate

The element which carries the largest portion of the load is the element whose dislocation effects the largest displacement Δ_i of the load. The energy in a single truss element is

$$\frac{N_i^2}{EA}\, l_i = \frac{P^2\, \Delta_i^2}{EA}\, l_i \tag{3.312}$$

and so the energy balance $W_i = W_e$ requires that

$$\frac{1}{2} \sum_i \frac{P^2\, \Delta_i^2}{EA}\, l_i = \frac{1}{2}\, P\, u = \frac{1}{2}\, P \cdot (P\, u_1) \tag{3.313}$$

where u_1 is the displacement of the load if $P = 1$ or

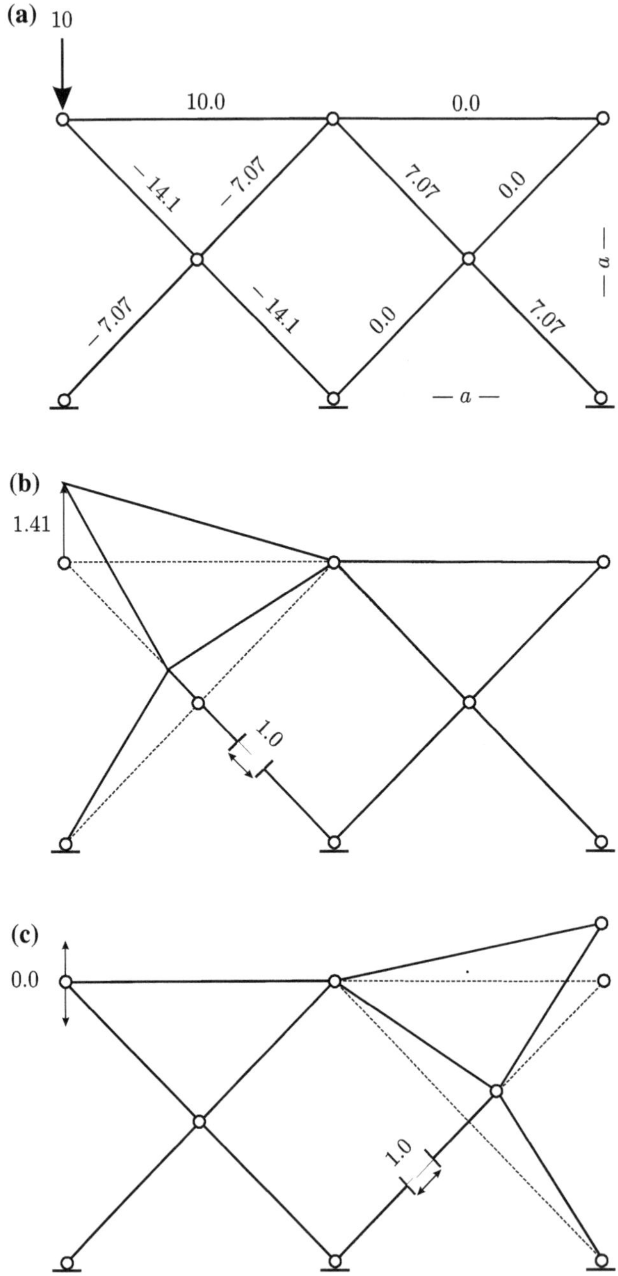

Fig. 3.45 Truss, **a** normal forces, **b** and **c** influence functions for normal forces. In linear mechanics the rule is: if a point rotates about a pole to the *right* or to the *left* of it the point moves straight *up* or *down*

$$\frac{1}{2}\sum_i \frac{\Delta_i^2}{EA} l_i = \frac{1}{2} u_1. \tag{3.314}$$

In other situations, when the shear forces dominate, the streamlines form two nearly equally strong orthogonal directions, see Fig. 3.46. When we represent all the pairs of mutually orthogonal streamlines in a cross section by just two arrows, see Fig. 3.46b, then their vector sum must be equal to the applied force P. In terms of work this means: two dislocations in the direction of the two inclined streamlines lift the foot of the point load by one unit of displacement.

It is now also clear why the stresses become singular at the fixed edge. The more the streamlines approach the fixed edge the more they flatten (because the edge is fixed in vertical direction, $u_y = 0$) and the more the streamlines must stretch so that their ever smaller and smaller vertical components manage to balance the load.

Hang a lantern on a rope! Before you can pull the rope really tight the rope will snap. This is the same situation.

When the corners are rounded off then the streamlines can rotate and so they have it easier to balance the applied load, see Figs. 3.47 and 3.48—no need to develop infinite stresses.

3.20 The Path the Influence Function takes

Influence functions are like fluids which spread from the source point all over the problem domain, and in so doing they have the tendency to ebb away and to ramify. One and the same point can be reached on different paths and so it is important that all parts of a problem domain are modeled with the correct stiffness. If the flow runs against artificial obstacles—because of an error in the modeling—or if the stiffness of a single zone through which the ripples propagate is not modeled correctly then the signal that arrives at the point will be disturbed and will no longer represent an accurate, reliable value.

In the design of planar reinforced concrete structures it has become fashionable to divide structures such as wall plates into beam-like zones and zones where the stress distribution is more complex and requires a true 2-D model. Originally this zone model was introduced to characterize the stress distribution in a structure. Today engineers often assume that the stress distribution in beam-like zones is "exact", because of the 1-D pattern, and that only the 2-D zones require special attention. But also internal actions in 1-D elements, as the normal force $N = \sigma_{yy} \cdot A$ in the cross section of the beam-like column in Fig. 3.49, can be wrong if the zones through which the influence function for σ_{yy} propagates are not modeled careful enough.

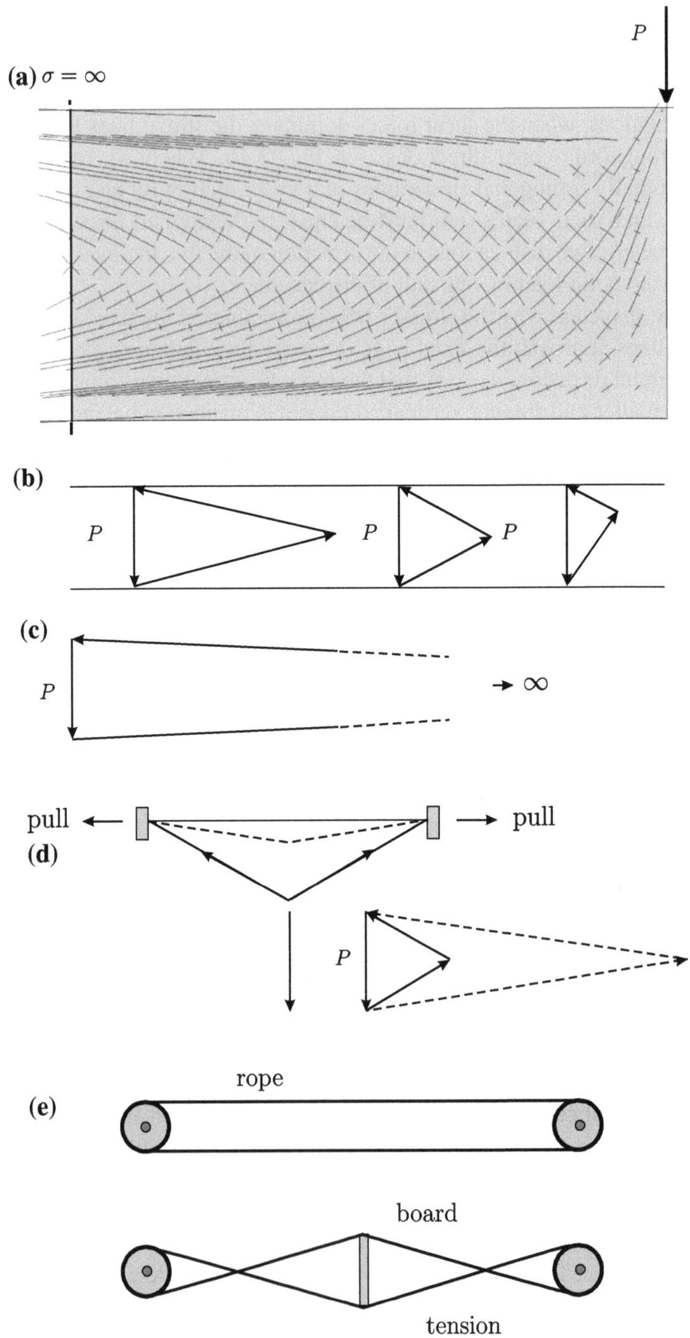

Fig. 3.46 Cantilever plate, **a** principal stresses ("streamlines"), **b** force triangles in different cross sections, **c** at the fixed edge the apex of the force triangle nears the point ∞, **d** street lantern—the same principle **e** a board placed between two ropes exploits the same effect

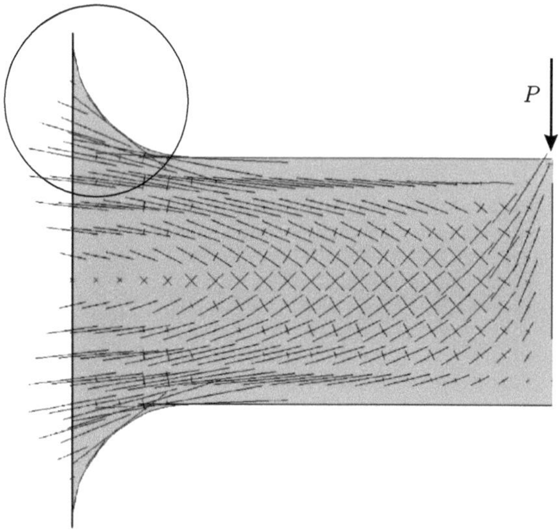

Fig. 3.47 The benefit of rounding off the corners: the streamlines can rotate and so they have it easier to balance the load

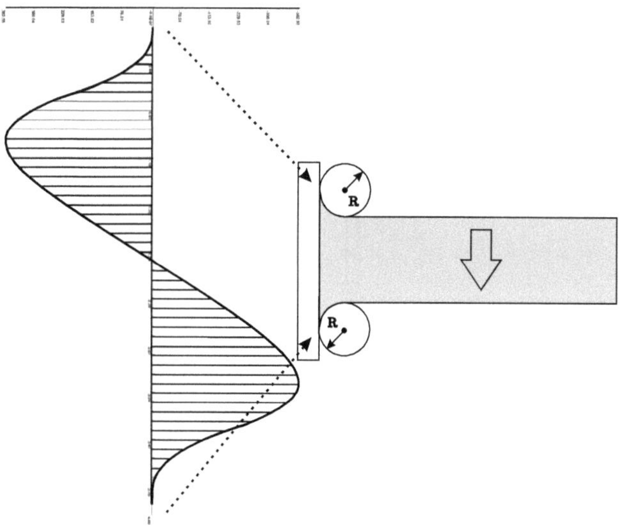

Fig. 3.48 Stress distribution (σ_{xx}) along the edge of the plate when the corners are rounded off

3.21 Mixed Problems

The logic of Green's functions directly carries over to mixed problems: also FE-solutions of mixed problems are based on an approximation of the pertinent Green's functions.

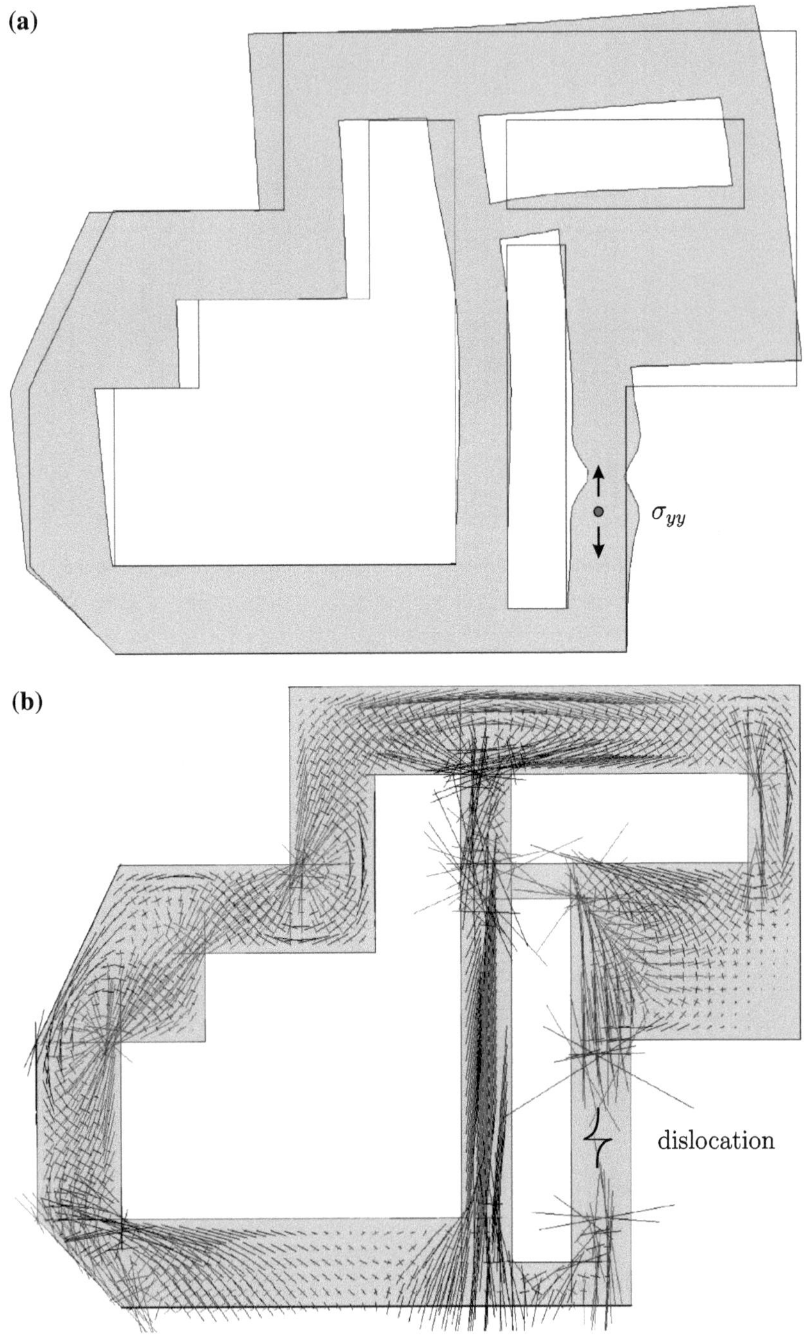

Fig. 3.49 Wall plate, **a** influence function for the stress σ_{yy}, **b** stress field generated by the dislocation at the source point

The boundary value problem

$$-\Delta u = p \qquad \text{on } \Omega \qquad u = 0 \qquad \text{on } \Gamma \tag{3.315}$$

can be split into a coupled system

$$\nabla w - \boldsymbol{\sigma} = \mathbf{0}_{(2)} \tag{3.316}$$

$$-\text{div}\,\boldsymbol{\sigma} = p_{(1)} \tag{3.317}$$

for the two functions w and $\boldsymbol{\sigma}$ or $\boldsymbol{v} = \{w, \boldsymbol{\sigma}\}^T$.

To this system belongs the identity

$$\mathcal{G}(\boldsymbol{v}, \hat{\boldsymbol{v}}) = \int_\Omega \left[(\nabla w - \boldsymbol{\sigma}) \cdot \hat{\boldsymbol{\sigma}} - \text{div}\,\boldsymbol{\sigma}\,\hat{w}\right] d\Omega + \int_\Gamma \boldsymbol{\sigma} \cdot \boldsymbol{n}\,\hat{w}\,ds$$
$$\underbrace{- \int_\Omega (\nabla w \cdot \hat{\boldsymbol{\sigma}} + \nabla \hat{w} \cdot \boldsymbol{\sigma} - \boldsymbol{\sigma} \cdot \hat{\boldsymbol{\sigma}})\,d\Omega}_{a(\boldsymbol{v}, \hat{\boldsymbol{v}})} = 0 \tag{3.318}$$

and so the boundary value problem is equivalent to the variational problem

$$\boldsymbol{v} \in \mathcal{V} \qquad a(\boldsymbol{v}, \hat{\boldsymbol{v}}) = \int_\Omega p\,\hat{w}\,d\Omega \qquad \forall\,\hat{\boldsymbol{v}} \in \mathcal{V} \tag{3.319}$$

where $\mathcal{V}$ is the usual space of trial and test functions with zero displacements, $w = 0$, on the edge Γ.

In FE-analysis we would construct with piecewise polynomial shape functions φ_i (scalars) and $\boldsymbol{\psi}_j$ (vector-valued) a subset $\mathcal{V}_h \subset \mathcal{V}$ so that

$$w = \sum_{i=1}^{d} u_i\,\varphi_i(\boldsymbol{x}) \qquad \boldsymbol{\sigma} = \sum_{j=1}^{s} s_j\,\boldsymbol{\psi}_j(\boldsymbol{x}). \tag{3.320}$$

The *Babuška–Brezzi condition* requires that $d < s$. Vector fields ψ_j with odd indices, $j = 1, 3, 5, \ldots$, have only horizontal components and vector fields $j = 2, 4, 6, \ldots$ only vertical components

$$\boldsymbol{\psi}_{j=odd}(\boldsymbol{x}) = \begin{bmatrix} \cdot \\ 0 \end{bmatrix} \qquad \boldsymbol{\psi}_{j=even}(\boldsymbol{x}) = \begin{bmatrix} 0 \\ \cdot \end{bmatrix}. \tag{3.321}$$

For the FE-solution (3.320) to be the best fit on $\mathcal{V}_h$ the error in the strain energy product must be orthogonal to all $d + s$ test functions

$$\boldsymbol{v}_i := \begin{bmatrix} \varphi_i(\boldsymbol{x}) \\ \boldsymbol{0} \end{bmatrix}, i = 1 \ldots d, \qquad \boldsymbol{v}_i := \begin{bmatrix} 0 \\ \psi_i(\boldsymbol{x}) \end{bmatrix}, i = d+1 \ldots s \tag{3.322}$$

that is it must hold

$$a(\boldsymbol{v}, \hat{\boldsymbol{v}}_i) = \int_\Omega p\, \hat{w}_i \, d\Omega \qquad i = 1 \ldots d+s \tag{3.323}$$

or

$$\begin{bmatrix} \boldsymbol{0}_{(d\times d)} & \boldsymbol{A}_{(d\times s)} \\ \boldsymbol{A}^T_{(s\times d)} & \boldsymbol{B}_{(s\times s)} \end{bmatrix} \begin{bmatrix} \boldsymbol{u} \\ \boldsymbol{s} \end{bmatrix} = \begin{bmatrix} \boldsymbol{p} \\ \boldsymbol{0} \end{bmatrix} \qquad \begin{matrix} d \text{ equations} \\ s \text{ equations} \end{matrix} \tag{3.324}$$

where p_i is the right-hand side of (3.323) for indices in the range $1 \leq i \leq d$; beyond that end point the $\hat{w}_i$ are zero.

Writing $\boldsymbol{K}\boldsymbol{v} = \boldsymbol{f}$ for this system the same algebra as before can be applied.

Let $\boldsymbol{J}(\boldsymbol{v})$ some linear functional on $\mathcal{V}$. On $\mathcal{V}_h$ it can be written as a scalar product

$$J(\boldsymbol{v}) = \boldsymbol{j}_u^T \boldsymbol{u} + \boldsymbol{j}_s^T \boldsymbol{s} =: \boldsymbol{j}^T \boldsymbol{v} \tag{3.325}$$

where

$$(j_u)_i = J(\boldsymbol{v}_i) \qquad (j_\sigma)_i = J(\psi_i) \tag{3.326}$$

and so

$$J(\boldsymbol{v}) = \boldsymbol{j}^T \boldsymbol{v} = \boldsymbol{j}^T \boldsymbol{K}^{-1} \boldsymbol{f} = \boldsymbol{g}^T \boldsymbol{f} \tag{3.327}$$

if $\boldsymbol{g}$ is the nodal vector of the Green's function $\boldsymbol{K}\,\boldsymbol{g} = \boldsymbol{j}$.

Example 3.3 The equation

$$-u'' = p \qquad u(0) = u(\ell) = 0 \tag{3.328}$$

can be split into the system

$$u' - \sigma = 0 \tag{3.329}$$
$$\sigma' = p \tag{3.330}$$

and the variational form of this boundary value problem for $\boldsymbol{v} = \{u, \sigma\}^T$ is

$$a(\boldsymbol{v}, \hat{\boldsymbol{v}}) := \int_0^l (u'\,\hat{\sigma} + \hat{u}'\,\sigma - \sigma\,\hat{\sigma})\, dx = \int_0^l p\,\hat{u}\, dx. \tag{3.331}$$

The rod in Fig. 3.50a is subdivided into 5 elements of length $l_e = 1$ and the longitudinal displacement u is approximated with piecewise linear shape functions and the stresses within the five elements with piecewise constant functions σ_j

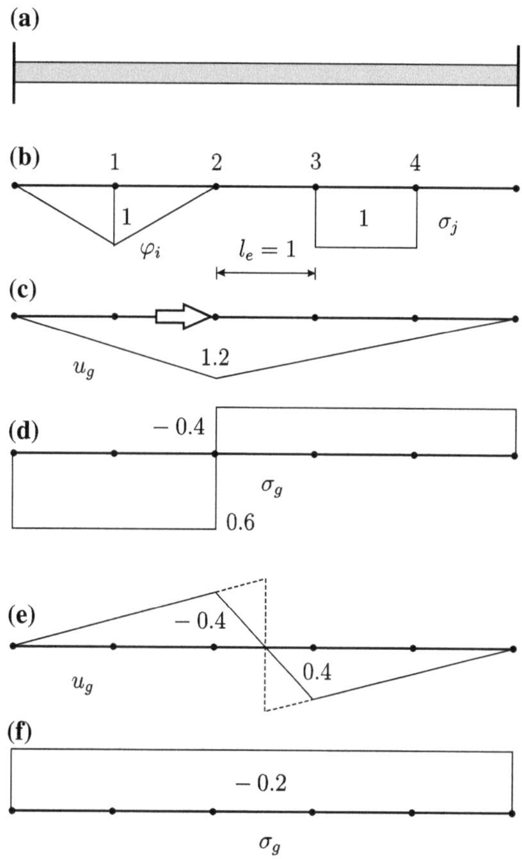

Fig. 3.50 Influence functions for a rod, mixed formulation, **a** rod, **b** elements and shape functions, **c** influence function for u_2 (displacements), **d** stresses, **e** influence function for the stress σ in element Ω_3 (displacements), **f** stresses

$$u = \sum_{i=1}^{4} u_i\,\varphi_i(x) \qquad \sigma = \sum_{j=1}^{5} s_j\,\sigma_j. \tag{3.332}$$

This results in the system

$$\begin{bmatrix} \mathbf{0}_{(4\times4)} & \boldsymbol{A}_{(4\times5)} \\ \boldsymbol{A}^T_{(5\times4)} & \boldsymbol{B}_{(5\times5)} \end{bmatrix} \begin{bmatrix} \boldsymbol{u} \\ \boldsymbol{s} \end{bmatrix} = \begin{bmatrix} \boldsymbol{p} \\ \mathbf{0} \end{bmatrix} \tag{3.333}$$

where

$$\boldsymbol{A} = \begin{bmatrix} 1 & -1 & 0 & 0 & 0 \\ 0 & 1 & -1 & 0 & 0 \\ 0 & 0 & 1 & -1 & 0 \\ 0 & 0 & 0 & 1 & -1 \end{bmatrix} \qquad a_{ij} = \int_0^l \varphi_i'\,\sigma_j\,dx \tag{3.334}$$

and $\boldsymbol{B} = -l_e \cdot \boldsymbol{I}$ because $b_{ij} = -(\sigma_j, \sigma_j) = -l_e \cdot \delta_{ij}$.

Let the functional $J(\boldsymbol{v}) = u(x_2)$ be the value of the solution at the node x_2 then the vector of the equivalent nodal forces which generates the Green's function consists of the elements

$$j_i = \varphi_i(x_2) \qquad i = 1, 2, 3, 4\,, \qquad j_i = 0 \qquad i = 5, 6, \ldots 9 \tag{3.335}$$

that is

$$\boldsymbol{j} = [0, 1, 0, 0, 0, 0, 0, 0, 0]^T \tag{3.336}$$

and the nodal values $\boldsymbol{g} = \boldsymbol{K}^{-1}\boldsymbol{j}$ of the Green's function are

$$\boldsymbol{g} = \{\underbrace{0.6, 1.2, 0.8, 0.4}_{u_i}, \underbrace{0.6, 0.6, -0.4, -0.4, -0.4}_{s_i}\}^T \tag{3.337}$$

and if $J(\boldsymbol{v}) = \sigma_3$, the stress at the midpoint $\bar{x}_3$ in element 3, then the vector $\boldsymbol{f}$ has the elements

$$f_i = 0\,, \quad i = 1, 2, 3, 4\,, \qquad f_i = \sigma_i(\bar{x}_3)\,, \quad i = 5, 6, \ldots 9 \tag{3.338}$$

that is

$$\boldsymbol{f} = [0, 0, 0, 0, 0, 0, 1, 0, 0]^T \tag{3.339}$$

because only σ_3 (= dof # 7) has a non-zero value $\sigma_3 = 1$ at $\bar{x}_3$, and the nodal values $\boldsymbol{g} = \boldsymbol{K}^{-1}\boldsymbol{f}$ of the Green's function are

$$\boldsymbol{g} = \{\underbrace{-0.2, -0.4, 0.4, 0.2}_{u_i}, \underbrace{-0.2, -0.2, -0.2, -0.2, -0.2}_{s_i}\}^T. \tag{3.340}$$

3.21.1 Tottenham's Equation for Mixed Problems

It is evident that Tottenham's equation also holds true for mixed problems, that is if

$$u(\boldsymbol{x}) = \int_\Omega (G(\boldsymbol{y}, \boldsymbol{x})\, p(\boldsymbol{y}) + \boldsymbol{\sigma}_G(\boldsymbol{y}, \boldsymbol{x}) \cdot \boldsymbol{\sigma}_0(\boldsymbol{y}))\, d\Omega_{\boldsymbol{y}} \tag{3.341}$$

is the integral representation of the exact solution then

$$u_h(\boldsymbol{x}) = \int_\Omega (G_h(\boldsymbol{y}, \boldsymbol{x})\, p(\boldsymbol{y}) + \boldsymbol{\sigma}_{G_h}(\boldsymbol{y}, \boldsymbol{x}) \cdot \boldsymbol{\sigma}_0(\boldsymbol{y}))\, d\Omega_{\boldsymbol{y}} \tag{3.342}$$

is the integral representation of the FE-solution where G_h and $\boldsymbol{\sigma}_{G_h}$ are the FE-approximations of the kernel functions.

3.22 Condensation of a Stiffness Matrix

Green's functions also play a prominent role in the condensation of a stiffness matrix.

Let us assume that the domain Ω is split into a domain Ω_1 and a domain Ω_2, see Fig. 3.51. The displacements on Ω_1 are represented by shape functions φ_i and the associated nodal values $\boldsymbol{u}_1$, and the displacements on the domain Ω_2 by shape functions ψ_i and nodal values $\boldsymbol{u}_2$. The functions φ_i and ψ_i form a partition of unity of Ω.

Consequently the stiffness matrix $\boldsymbol{K}$ is partitioned as follows

$$\begin{bmatrix} \boldsymbol{A} & \boldsymbol{B} \\ \boldsymbol{B}^T & \boldsymbol{C} \end{bmatrix} \begin{bmatrix} \boldsymbol{u}_1 \\ \boldsymbol{u}_2 \end{bmatrix} = \begin{bmatrix} \boldsymbol{f}_1 \\ \boldsymbol{f}_2 \end{bmatrix}, \qquad (3.343)$$

and

$$\begin{bmatrix} a(\varphi_i, \varphi_i) & a(\psi_j, \varphi_i) \\ a(\varphi_j, \psi_i) & a(\psi_i, \psi_i) \end{bmatrix} \begin{bmatrix} u_1 \\ u_2 \end{bmatrix} = \begin{bmatrix} (p, \varphi_i) \\ (p, \psi_i) \end{bmatrix} \qquad (3.344)$$

respectively. Green's first identity

$$\mathscr{G}(u, \hat{u}) = \int_\Omega -\Delta u\, \hat{u}\, d\Omega + \int_\Gamma \frac{\partial u}{\partial n}\, \hat{u}\, ds - a(u, \hat{u}) = 0 \qquad (3.345)$$

allows to replace the strain energy product δW_i by exterior virtual work δW_e

$$\delta W_i = a(\varphi_i, \varphi_j) = (\varphi_i, p(\varphi_j)) = \delta W_e\,. \qquad (3.346)$$

If the system (3.343) is solved for $\boldsymbol{u}_2$ it follows

$$\boldsymbol{u}_2 = -\boldsymbol{C}^{-1}\boldsymbol{B}^T\boldsymbol{u}_1 + \boldsymbol{C}^{-1}\boldsymbol{f}_2, \qquad (3.347)$$

which allows to write the system in terms of $\boldsymbol{u}_1$ and the load vectors $\boldsymbol{f}_1$ and $\boldsymbol{f}_2$,

$$\boldsymbol{A}\boldsymbol{u}_1 - \boldsymbol{B}\boldsymbol{C}^{-1}\boldsymbol{B}^T\boldsymbol{u}_1 = \boldsymbol{f}_1 - \boldsymbol{B}\boldsymbol{C}^{-1}\boldsymbol{f}_2\,. \qquad (3.348)$$

Let us study the part

$$(\boldsymbol{B}\boldsymbol{C}^{-1}\boldsymbol{B}^T)_{ij} = B_{il}C^{-1}_{lk}B_{kj} \qquad (3.349)$$

of the system. Based of the previous analysis it is

$$B_{il} = a(\varphi_i, \psi_l) = (p(\varphi_i), \psi_l) \quad \text{and} \quad B_{kj} = a(\psi_k, \varphi_j) = (\psi_k, p(\varphi_j)) \qquad (3.350)$$

and so (3.349) can be written as

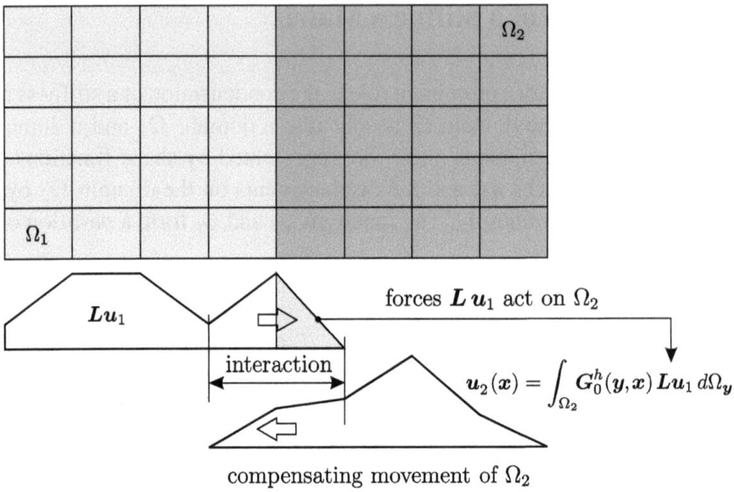

Fig. 3.51 Schematic representation of the interaction between domains Ω_1 and Ω_2

$$
\begin{aligned}
(\boldsymbol{B}\boldsymbol{C}^{-1}\boldsymbol{B}^T)_{ij} &= (p(\varphi_i), \psi_l)\, C_{lk}^{-1}\, (\psi_k, p(\varphi_j)) \\
&= \int_\Omega \int_\Omega p(\varphi_i)(\boldsymbol{x})\, G_h^{(2)}(\boldsymbol{y}, \boldsymbol{x})\, p(\varphi_j)(\boldsymbol{y})\, d\Omega_{\boldsymbol{y}}\, d\Omega_{\boldsymbol{x}}\,,
\end{aligned} \tag{3.351}
$$

where $G_h^{(2)}$ is the Green's function on the domain Ω_2,

$$
G_h^{(2)}(\boldsymbol{y}, \boldsymbol{x}) = \boldsymbol{\psi}(\boldsymbol{x})^T \boldsymbol{C}^{-1} \boldsymbol{\psi}(\boldsymbol{y}) = \psi_i(\boldsymbol{x})\, C^{-1}\, \psi_j(\boldsymbol{y})\,. \tag{3.352}
$$

It can also be expressed in this way: the displacement field $u(\boldsymbol{x}) = \varphi(\boldsymbol{x})^T\, \boldsymbol{u}_1$ on Ω_1 generates forces p in Ω and therewith also on Ω_2—at least in the short overlap between Ω_1 and Ω_2. The shape function $\varphi_i \in \Omega_1$ which are associated with the nodes at the interface extend into Ω_2 for a short range. The forces acting on Ω_2 induce a compensating movement

$$
\tilde{\boldsymbol{u}}(\boldsymbol{x}) = \int_\Omega G_h^{(2)}(\boldsymbol{y}, \boldsymbol{x}) p(\varphi_j)(\boldsymbol{y})\, d\Omega_{\boldsymbol{y}} \tag{3.353}
$$

in the short overlap, see Fig. 3.51. The work done by the forces p which were generated by the test functions φ_i on acting through $\boldsymbol{u}_2$ are just the contributions in $\boldsymbol{B}\boldsymbol{C}^{-1}\boldsymbol{B}^T$. If the element size h tends to zero the overlap also tends to zero and the forces $p(\varphi_j)$ concentrate on the interface Γ_I, $p(\varphi_j) \to \boldsymbol{t}(\varphi_j)$, that is then the expression above tends to the boundary integral

$$\tilde{\boldsymbol{u}}(\boldsymbol{x}) = \int_{\Gamma_I} \boldsymbol{G}(\boldsymbol{y}, \boldsymbol{x})\, \boldsymbol{t}(\varphi_j)(\boldsymbol{y})\, d\Omega_{\boldsymbol{y}}\,, \tag{3.354}$$

which for points $\boldsymbol{x}$ on the interface, $\boldsymbol{u}_1(\boldsymbol{x}) = \boldsymbol{u}_2(\boldsymbol{x})$ is the operator which is the inverse to the *Poincaré-Steklov-Operator*

$$\boldsymbol{t}_1(\boldsymbol{x}) = \int_{\Gamma_I} \boldsymbol{t}_y \boldsymbol{t}_x \boldsymbol{G}_0(\boldsymbol{y}, \boldsymbol{x})\, \boldsymbol{u}_1(\boldsymbol{y})\, ds_{\boldsymbol{y}}\,, \tag{3.355}$$

which establishes the link between the edge forces $\boldsymbol{t}_1$ and the edge displacements $\boldsymbol{u}_1$.

Imagine all internal degrees of freedom are eliminated from a stiffness matrix $\boldsymbol{K}$ by condensation—only the u_i on the boundary are left—then the modified equation $\bar{\boldsymbol{K}}\bar{\boldsymbol{u}} = \bar{\boldsymbol{f}}$ (all terms live on the boundary) resembles a discrete *Poincaré-Steklov-operator*. You press on a rubber ball (nodal forces f_i) and the deformed shape of the ball comes out as $\bar{\boldsymbol{u}} = \bar{\boldsymbol{K}}^{-1}\bar{\boldsymbol{f}}$.

3.23 *p*-Method

To facilitate the formulations in the following we write the boundary value problem in an "abstract" form

$$L\,u = p \qquad \text{on } \Omega \qquad u = 0 \qquad \text{on } \Gamma \tag{3.356}$$

and we use the short-hand notation (u, v) for domain integrals so that Green's first identity on $\mathcal{V} \times \mathcal{V}$ reads

$$\mathscr{G}(u, v) = (L\,u, v) - a(u, v) = 0. \tag{3.357}$$

This identity allows to recast the original boundary value problem in a weak form

$$a(u, v) = (p, v) \qquad \forall\, v \in \mathcal{V} \tag{3.358}$$

where $\mathcal{V}$ is the solution and trial space

$$\mathcal{V} := \{v \in H^1(\Omega)\,|v = 0\}. \tag{3.359}$$

In the p-method the trial space is enriched with higher degree polynomials, that is *bubble-functions* are added to each element so that the FE-solution has the form (we drop the usual subscript h on the FE-solution u_h) $u = u_\ell + u_p$. Here u_ℓ is the original low degree field and u_p are the added polynomials. If this split is repeated on the side of the test functions as well, $\hat{u} = \hat{u}_\ell + \hat{u}_p$, then the variational formulation becomes

$$a(u_\ell + u_p, \hat{u}_\ell + \hat{u}_p) = (p, \hat{u}_\ell + \hat{u}_p) \quad \forall\ (\hat{u}_\ell + \hat{u}_p). \tag{3.360}$$

Because the test functions are independent of each other these are actually two sets of equations

$$a(u_\ell, \hat{u}_\ell) + a(u_p, \hat{u}_\ell) = (p, \hat{u}_\ell) \quad \forall\, \hat{u}_\ell \tag{3.361}$$
$$a(u_\ell, \hat{u}_p) + a(u_p, \hat{u}_p) = (p, \hat{u}_p) \quad \forall\, \hat{u}_p \tag{3.362}$$

or after integration by parts (Green's first identity)

$$a(u_\ell, \hat{u}_\ell) + (u_p, L\,\hat{u}_\ell) = (p, \hat{u}_\ell) \quad \forall\, \hat{u}_\ell \tag{3.363}$$
$$(L\,u_\ell, \hat{u}_p) + (L\,u_p, \hat{u}_p) = (p, \hat{u}_p) \quad \forall\, \hat{u}_p. \tag{3.364}$$

The last equation is identical with

$$(L\,u_p, \hat{u}_p) = -(L\,u_\ell - p, \hat{u}_p) \quad \forall\, \hat{u}_p\,. \tag{3.365}$$

The ansatz u_p is the solution to the differential equation $L\,u_p = p - (L\,u_\ell)$ and therefore can be represented by the associated Green's function

$$u_p = \int_\Omega G_p(y, x)\,(p - (L\,u_\ell))\; d\Omega_y := L_p^{-1}(p - L\,u_\ell)\,. \tag{3.366}$$

The lower index p at G_p is to indicate that the kernel G_p is the Green's function on the trial space $\mathcal{V}_p$, which is "the other half" of the space $\mathcal{V}_\ell$, that is $\mathcal{V} = \mathcal{V}_\ell + \mathcal{V}_p$.

Substituting this expression into (3.363) it follows

$$a(u_\ell, \hat{u}_\ell) - (L_p^{-1} L\,u_\ell, L\,\hat{u}_\ell) = (p, \hat{u}_\ell) - (L_p^{-1} p, L\,\hat{u}_\ell) \quad \forall\, \hat{u}_\ell. \tag{3.367}$$

To the Eqs. (3.361) and (3.362) corresponds in the FE-method the system

$$\begin{bmatrix} \boldsymbol{A} & \boldsymbol{B} \\ \boldsymbol{B}^T & \boldsymbol{C} \end{bmatrix} \begin{bmatrix} \boldsymbol{u}_\ell \\ \boldsymbol{u}_p \end{bmatrix} = \begin{bmatrix} \boldsymbol{f}_\ell \\ \boldsymbol{f}_p \end{bmatrix} \tag{3.368}$$

and the inverse $\boldsymbol{C}^{-1}$ is the FE-approximation of the Green's function in (3.366), so that after the usual transformation

$$(\boldsymbol{A} - \boldsymbol{B}\,\boldsymbol{C}^{-1}\boldsymbol{B}^T)\,\boldsymbol{u}_\ell = \boldsymbol{f}_\ell - \boldsymbol{B}\boldsymbol{C}^{-1}\,\boldsymbol{f}_p \tag{3.369}$$

we recover exactly (3.367).

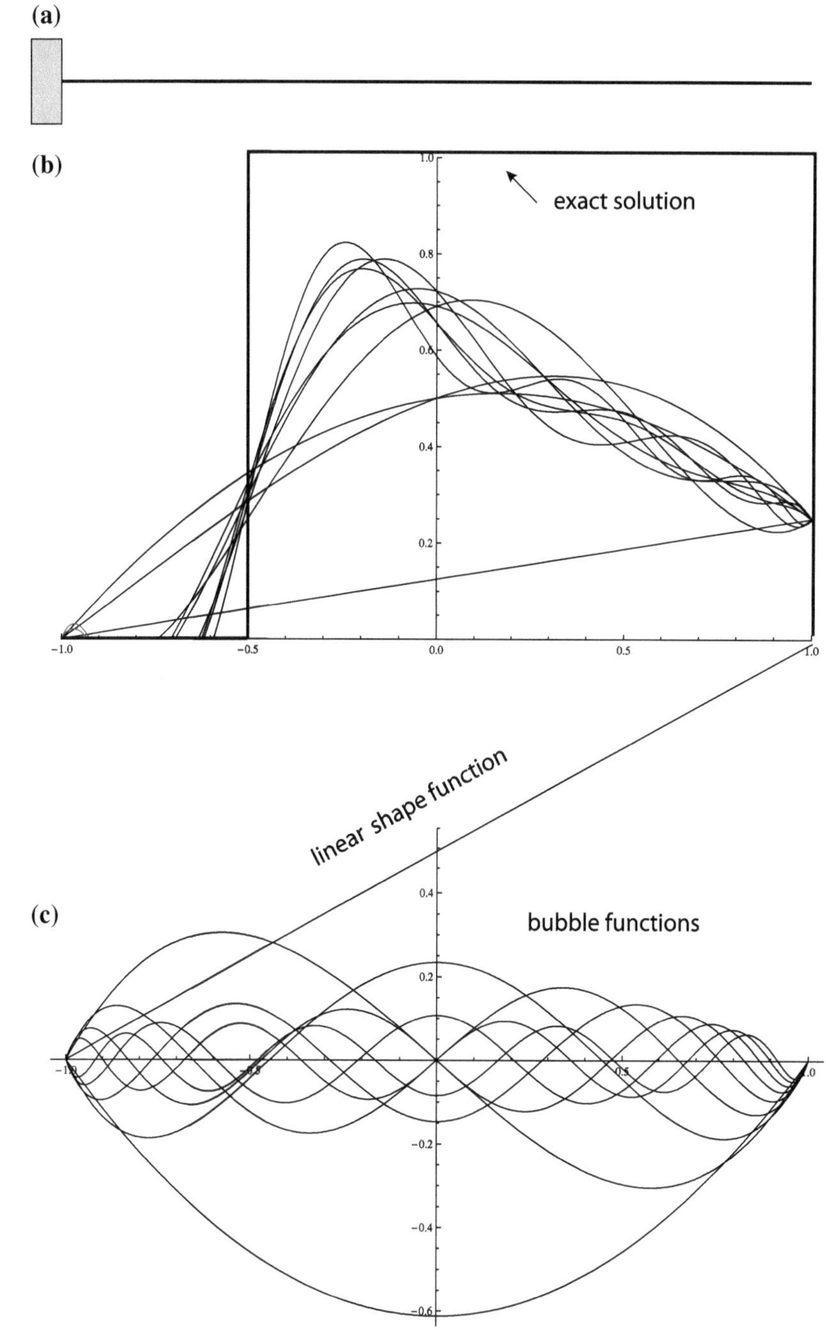

Fig. 3.52 The influence function for the normal force $N(-0,5)$ is a dislocation, a unit step function. **a** bar, **b** FE-solution for different polynomial degrees, **c** shape functions

Equation (3.366) implies, that it is the task of the refinement u_p to correct the error $p - L\,u_\ell$ of the original coarse solution u_ℓ. But we must be careful. In the p-method the approximate kernel G_p^h is an expansion in terms of smooth functions u_p. But such a smooth kernel is not very well qualified to model discontinuities of higher derivatives of the solution. The p-method achieves good results where the solution is smooth. Then the kernel G_p^h has all the advantages. But not if in the middle of a plate element the load drops abruptly to zero (= discontinuity in the second derivative) or line loads cross the element (= discontinuity in the first derivative. In these cases a local enrichment with smooth functions will not help much. The reason is that the Schur-complement $(\boldsymbol{A} - \boldsymbol{B}^T\,\boldsymbol{C}^{-1}\boldsymbol{B})\,\boldsymbol{u}_1 = \boldsymbol{f}_1 - \boldsymbol{B}^T\,\boldsymbol{C}^{-1}\,\boldsymbol{f}_2$ depends on the Green's function.

To cite an example we consider a bar $[-1, +1]$ which is fixed on its left side and modeled by just one linear element. To the original linear shape function are added 10 additional functions (P_i = Lagrange polynomial), see Fig. 3.52d

$$\varphi_i(x) = \frac{P_{i+1}(x) - P_{i-1}(x)}{\sqrt{2}\sqrt{2(i+1)-1}} \qquad i = 1, 2, 3, \ldots p. \tag{3.370}$$

At count $p = 3$ the system is

$$\boldsymbol{K} = \frac{EA}{2}\begin{bmatrix} 1\,0\,0\,0 \\ 0\,2\,0\,0 \\ 0\,0\,2\,0 \\ 0\,0\,0\,2 \end{bmatrix}\begin{bmatrix} u_2 \\ u_3 \\ u_4 \\ u_5 \end{bmatrix} = \begin{bmatrix} f_2 \\ f_3 \\ f_4 \\ f_5 \end{bmatrix} \qquad f_i = N(\varphi_i)(-0.5)\,.$$

so that the stiffness matrix consists of the matrix $\boldsymbol{A}$ (just 1 element) and the 3×3 matrix $\boldsymbol{C}$. The matrices $\boldsymbol{B}$ and $\boldsymbol{B}^T$ are zero.

For a test the influence function for the normal force N at the quarter point $x = -0.5$ was calculated. The influence function is a unit step function, a simple translation of all parts to the right of the point by one unit. Obviously does the p-method have difficulties to approximate the true solution, see Fig. 3.52.

References

1. Tottenham H (1970) Basic principles. In: Tottenham H, Brebbia C (eds) Finite element techniques in structural mechanics. Southampton University Press, Southampton
2. Jakobsen B, Rasendahl F (1994) The Sleipner platform accident. Struct Eng Int 3:190–194
3. Hartmann F (1985) The mathematical foundation of structural mechanics. Springer, Berlin
4. Strang G, Fix GJ (1973) An analysis of the finite element method. Prentice Hall, Englewood Cliffs
5. Strang G (2007) Computational science and engineering. Wellesley-Cambridge Press, Wellesley
6. Hartmann F, Katz C (2007) Structural analysis with finite elements, 2nd edn. Springer, Berlin

Chapter 4
The Discretization Error

In this chapter we discuss the connection between Green's functions and the error of FE-solutions, how the error of an FE-solution can be traced back to the error in the approximation of the Green's function. We begin with classical asymptotic error analysis where tools of functional analysis allow to express the energy error in powers of the mesh-width h. In goal-oriented refinement where the focus is on minimizing the error in certain functionals the adaptive refinement is steered by two errors, the error in the primal, the original, problem and the error in the dual problem, the approximation of the Green's function with finite elements.

The technique of goal-oriented refinement can also be applied to nonlinear problems where the dual problem is the approximation of the Green's function at the current linearization point. In the case of nonlinear functionals the Green's function is taken as the Green's function of the linearized functionals.

Because FE-solutions, in general, do not interpolate the exact solution at the nodes they exhibit a certain drift, a mismatch at the nodes. This is the prime reason why the output the approximate Green's functions produce is not exact. If the drift is uniform, or nearly so, then it is called pollution: badly resolved singularities or inconsistencies in the discretization make that the solution gets shifted in a certain direction. Often the cause of these shifts lies outside the zone where the shift is observed. Pollution is "silent", it is not accompanied by oscillating stresses and it cannot be discovered solely by a local analysis.

The chapter closes with a remark about the strong singularities in Green's functions. These singularities are normally much higher than the typical stress singularities at singular points on the boundary but surprisingly an FE-program manages to approximate these functions relatively well. This is a kind of a paradox: (weak) singularities on the boundary are hard to approximate while (strong) singularities in the interior can be resolved quite easily.

F. Hartmann, *Green's Functions and Finite Elements*,
DOI: 10.1007/978-3-642-29523-2_4, © Springer-Verlag Berlin Heidelberg 2013

4.1 Asymptotic Error Analysis

To solve the boundary value problem

$$-\Delta u = p \quad \text{on } \Omega \subset \mathbb{R}^2 \quad u = 0 \quad \text{on } \Gamma \tag{4.1}$$

with finite elements the problem is recast in weak form

$$a(u, v) = (p, v) \quad \forall\, v \in \mathcal{V} \tag{4.2}$$

and the best approximation u_h on a given subset $\mathcal{V}_h \subset \mathcal{V}$ is found by solving the system

$$a(u_h, \varphi_i) = (p, \varphi_i) \quad i = 1, 2, \ldots, n. \tag{4.3}$$

The only terms which allow to trace the error of the FE-solution are the error forces, see Fig. 4.1,

$$r_e := p - (-\Delta u_h) \quad \text{on element } \Omega_e\,, \tag{4.4}$$

the difference between the original right-hand side p and the right-hand side p_h of the FE-solution on each element Ω_e, and the jumps l_k in the normal derivative in between the elements, see Fig. 4.2,

$$l_k := \frac{\partial u^+}{\partial n} - \frac{\partial u^-}{\partial n} \quad \text{on } \Gamma_k \tag{4.5}$$

which can be interpreted as line loads l_k which produce the kinks in the fabric of the membrane along the mesh lines Γ_k. For technical reasons these forces l_k are split into two halves and are distributed evenly to the two sides Γ_e and $\Gamma_{e'}$ of the two elements Ω_e and $\Omega_{e'}$ which border on the mesh line Γ_k,

$$j_e := 0.5\, l_k \quad j_{e'} = 0.5\, l_k\,. \tag{4.6}$$

It seems evident that the quality of the FE-solution should be linked to the size of the error forces r_e and j_e. But in which way?

To work this out note first that the error $e = u - u_h$ solves the variational problem

$$a(e, v) = a(u, v) - a(u_h, v) = (p, v) - (p_h, v) \quad \forall\, v \in \mathcal{V} \tag{4.7}$$

or

$$a(e, v) = (p - p_h, v) := \sum_e \left\{ \int_{\Omega_e} r_e\, v\, d\Omega + \int_{\Gamma_e} j_e\, v\, ds \right\}. \tag{4.8}$$

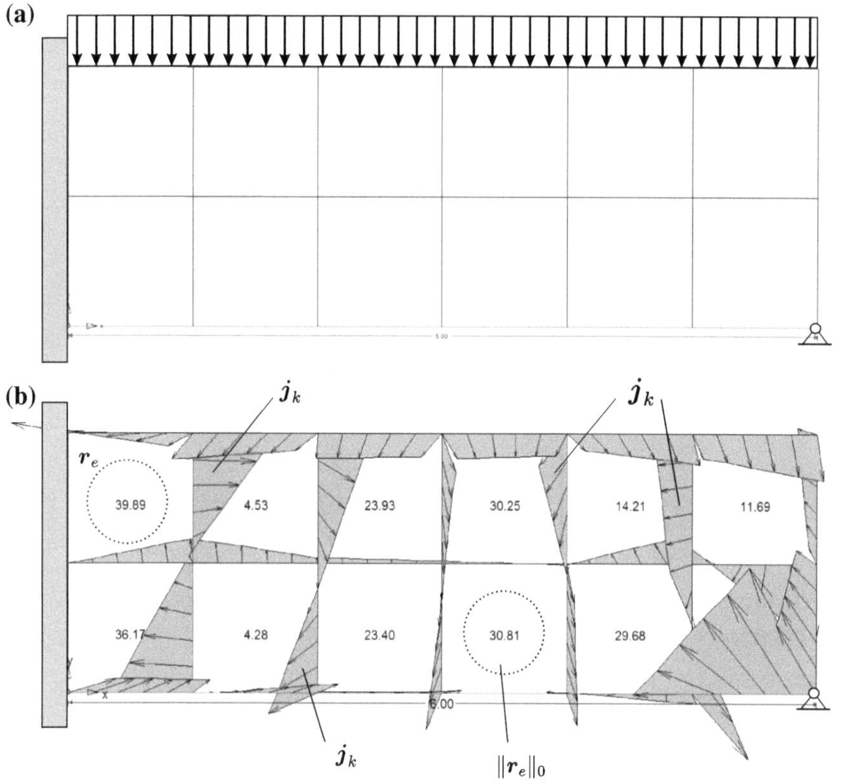

Fig. 4.1 Plate problem, **a** the original load case $\boldsymbol{p}$ and **b** the FE-load case $\boldsymbol{p}_h$. The numbers inside the elements are the L_2-norms $= ((r_x^e, r_x^e) + (r_y^e, r_y^e))^{1/2}$ of the element error forces

Because of the Galerkin orthogonality the residuals are orthogonal to all $v \in \mathcal{V}_h$ and therefore also to the function $u_I \in \mathcal{V}_h$, which interpolates the solution u (if u were given it could be interpolated with the functions in $\mathcal{V}_h$ and this would be the function u_I)

$$0 = \sum_e \left\{ \int_{\Omega_e} r_e\, u_I\, d\Omega + \int_{\Gamma_e} j_e\, u_I\, ds \right\} . \tag{4.9}$$

Hence it follows

$$a(e, v) = \sum_e \left\{ \int_{\Omega_e} r_e\, (v - u_I)\, d\Omega + \int_{\Gamma_e} j (v - u_I)\, ds \right\} \tag{4.10}$$

or if e itself is chosen as virtual displacement

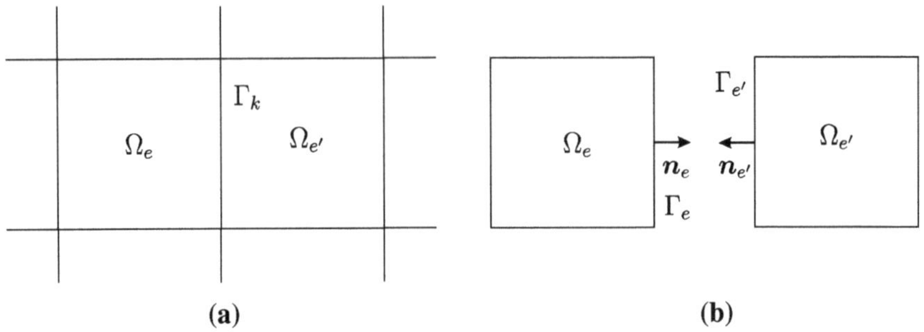

Fig. 4.2 **a** Two elements e and e' which border on a mesh line Γ_k, **b** normal vectors on the two sides Γ_e and $\Gamma_{e'}$

$$a(e,e) = \sum_e \left\{ \int_{\Omega_e} r_e\,(e - u_I)\,d\Omega + \int_{\Gamma_e} j_e\,(e - u_I)\,ds \right\}. \tag{4.11}$$

The load case $p - p_h$, that is the error forces $\{r_e, j_e\}$ of each element $e = 1, 2, \ldots N_e$, is orthogonal to the part u_h of the error $e = u - u_h$ and so

$$a(e,e) = \sum_e \left\{ \int_{\Omega_e} r_e\,(u - u_I)\,d\Omega + \int_{\Gamma_e} j_e\,(u - u_I)\,ds \right\}. \tag{4.12}$$

The next steps are an exercise in approximation theory. If u is a regular function then the interpolation error $u - u_I$ is bounded by a certain power of the element size h times some constant. Combining this with the Cauchy–Schwarz inequality provides an estimate for the square of the error e in the energy norm

$$\|e\|_E^2 \le c_1 \sum_e \left\{ h_e^2 \|r\|^2_{L_2(\Omega_e)} + h_e \|j_e\|^2_{L_2(\Gamma_e)} \right\} \tag{4.13}$$

and after some further steps also for the norm itself

$$\|e\|_E \le c_2 \sum_e \left\{ h_e \|r_e\|_{L_2(\Omega_e)} + h_e^{0.5} \|j_e\|_{L_2(\Gamma_e)} \right\}. \tag{4.14}$$

All terms in this equation, with the exception of the constants c_i, can be calculated. The constants $c_i, i = 1, 2$ depend on how well the functions in $\mathcal{V}_h$ can interpolate functions and on how smooth the function u is (what is not known a priori).

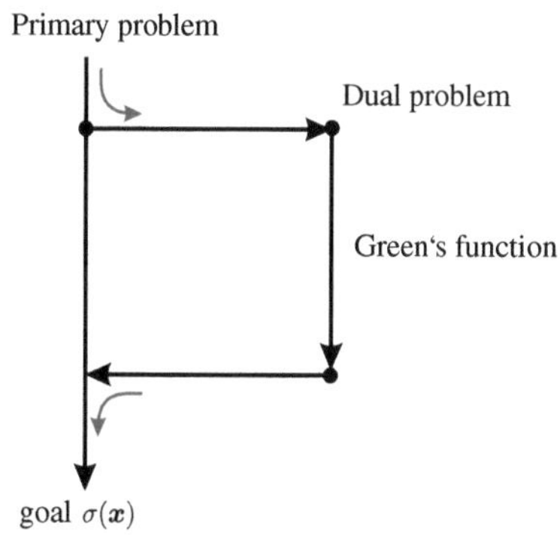

Fig. 4.3 In goal-oriented adaptive refinement two problems are solved: the primal problem (original problem) and the dual problem (Green's function)

So the sum

$$\eta = \sum_e \eta_e := \sum_e \left\{ \|r_e\|_{L_2(\Omega_e)} + h_e^{0.5} \|j_e\|_{L_2(\Gamma_e)} \right\} \tag{4.15}$$

can serve as an error indicator. If H (the Greek capital letter η) is the total energy error then the ratio

$$\eta_e^{rel} = \frac{\eta_e}{H} \tag{4.16}$$

indicates how much an element contributes to the overall error and in adaptive finite element methods those elements where this ratio exceeds a certain threshold value are refined; only the unknown value H must be replaced by some estimate $\tilde{H}$.

4.2 Goal-Oriented Refinement

The same logic is now applied to the error in a functional

$$J(e) = u(\boldsymbol{x}) - u_h(\boldsymbol{x}) = \int_\Omega [G(\boldsymbol{y}, \boldsymbol{x}) - G_h(\boldsymbol{y}, \boldsymbol{x})]\, p(\boldsymbol{y})\, d\Omega_{\boldsymbol{y}}\,. \tag{4.17}$$

Because of the Galerkin orthogonality the FE-load case p_h can be added to this equation without changing the math

$$J(e) = u(\boldsymbol{x}) - u_h(\boldsymbol{x}) = \int_\Omega [G(\boldsymbol{y}, \boldsymbol{x}) - G_h(\boldsymbol{y}, \boldsymbol{x})]\, (p(\boldsymbol{y}) - p_h(\boldsymbol{y}))\, d\Omega_{\boldsymbol{y}}\,, \tag{4.18}$$

and a switch to the internal energy

$$J(e) = u(\boldsymbol{x}) - u_h(\boldsymbol{x}) = a(G[\boldsymbol{x}] - G_h[\boldsymbol{x}], u - u_h) \tag{4.19}$$

allows to apply the Cauchy–Schwarz inequality (ignoring for the moment that the Green's function has infinite energy)

$$\begin{aligned} |J(e)| &= |u(\boldsymbol{x}) - u_h(x)| = |a(G[\boldsymbol{x}] - G_h[\boldsymbol{x}], u - u_h)| \\ &\leq a(G[\boldsymbol{x}] - G_h[\boldsymbol{x}], G[\boldsymbol{x}] - G_h[\boldsymbol{x}]) \cdot a(u - u_h, u - u_h) \\ &\leq \|G[\boldsymbol{x}] - G_h[\boldsymbol{x}]\|_E \cdot \|u - u_h\|_E \,. \end{aligned} \tag{4.20}$$

Hence the upper limit for $|J(e)|$ is the product of two errors: the error in the Green's function and the error of the FE-solution, both errors measured in the energy norm.

If η_p^e and η_G^e are two error estimators on an element Ω_e for the primal and dual error respectively then follows

$$|J(e)| = |u(\boldsymbol{x}) - u_h(\boldsymbol{x})| \leq \sum_e \eta_G^e \cdot \eta_p^e \tag{4.21}$$

which is the basic tool in goal-oriented adaptive refinement, see Fig. 4.3.

Strictly speaking does this estimate only hold true if the energy of the Green's function is bounded, $a(G, G) < \infty$, which would be the case if the point load were spread over a small circular patch with radius $\rho > 0$.

4.3 Comparison

To demonstrate the difference in meshes generated by a standard ("global") adaptive refinement and a goal-oriented refinement the two techniques were applied to the plate in Fig. 4.4 which carries a point load at its upper left corner.

The corner points of the plate are singular points and this is registered by the algorithm which mainly refines the mesh near these points, Fig. 4.4a. When the focus is on a particular value, say the vertical stress $J(\boldsymbol{u}) = \sigma_{yy}$ at the edge of one of the openings, then the goal-oriented refinement which evaluates both error indicators, for $u - u_h$ and $G - G_h$, almost uniquely concentrates on the neighborhood of the source point, Fig. 4.4b. In Fig. 4.5a is plotted the vertical displacement of the Green's function for σ_{yy} and in Fig. 4.5b the horizontal component. Both components are of course in-plane components. Only for illustrative purposes are the function values plotted above (+) and below (−) the mid-surface of the plate.

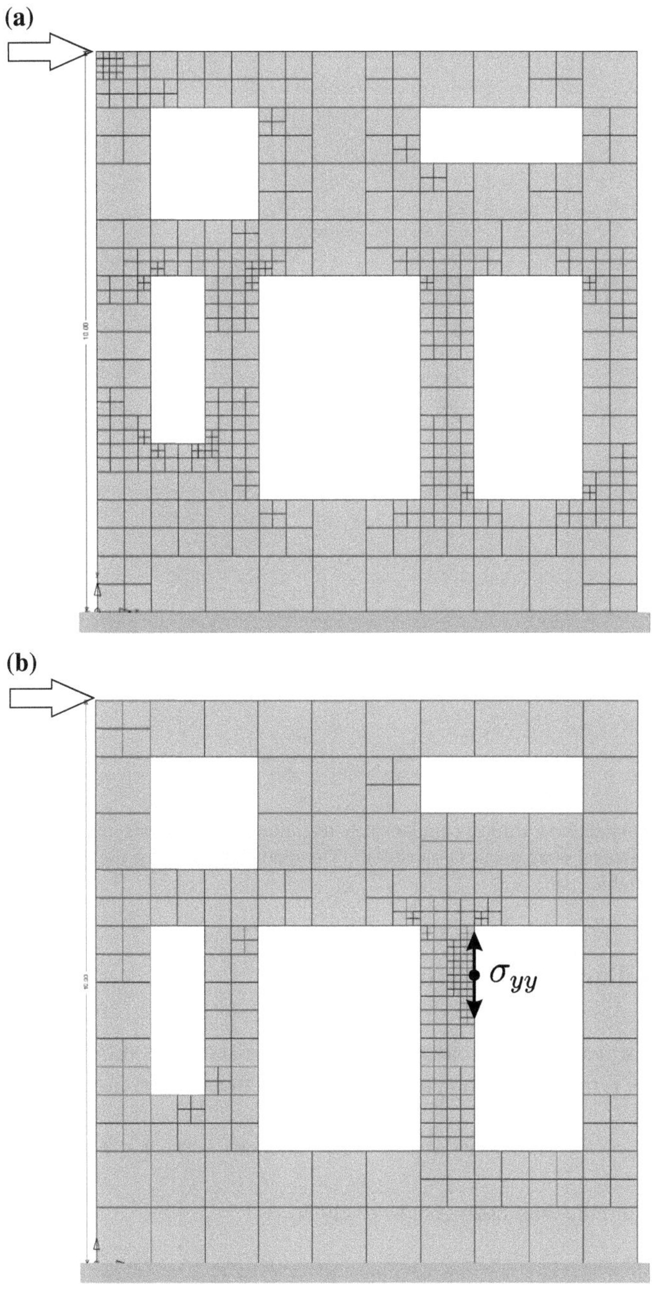

Fig. 4.4 Adaptive refinement, **a** standard refinement, **b** goal-oriented refinement $J(\boldsymbol{u}) = \sigma_{yy}$ at the edge of the opening

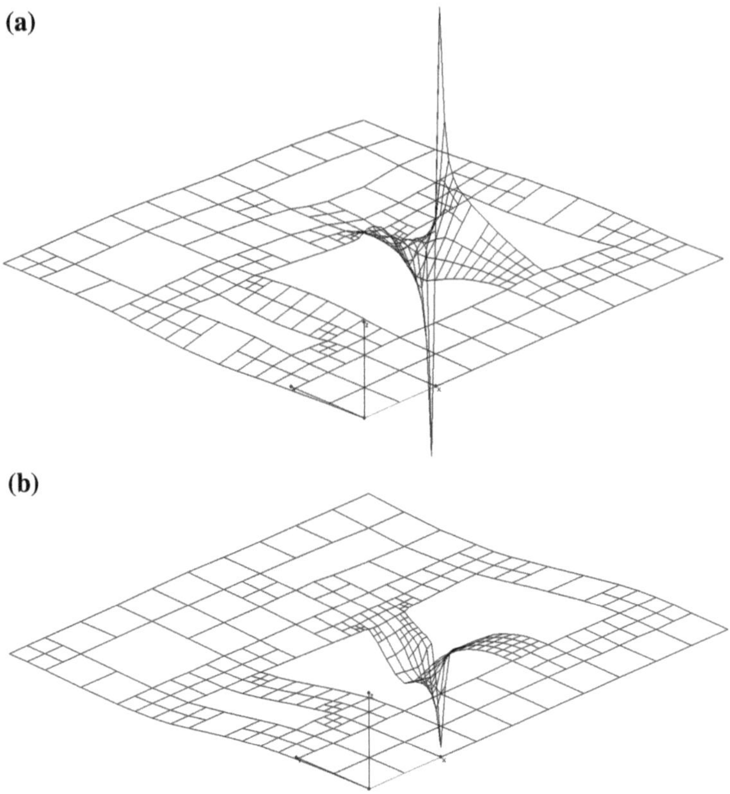

Fig. 4.5 3-D plot of the **a** vertical component (y-direction) of the Green's function for $J(\boldsymbol{u}) = \sigma_{yy}$ and **b** the horizontal component (x-direction). The dipole character of the influence function is easily recognizable

4.4 Primal and Dual Error

The *primal solution* is the solution u of the boundary value problem (4.1) and the *dual solution* is the Green's function for $u(\boldsymbol{x})$; it is the solution to the boundary value problem

$$-\Delta G(\boldsymbol{y}, \boldsymbol{x}) = \delta(\boldsymbol{y} - \boldsymbol{x}) \quad G(\boldsymbol{y}, \boldsymbol{x}) = 0 \quad \boldsymbol{y} \in \Gamma\,. \tag{4.22}$$

The primal and the dual errors respectively are

$$e = u - u_h \quad e^* = G - G_h \tag{4.23}$$

and the functional

$$\rho(u_h)(\cdot) = (p - p_h, \cdot) \tag{4.24}$$

is called the *primal residual* and the functional

$$\rho^*(G_h)(\cdot) = J(\cdot) - J_h(\cdot) \tag{4.25}$$

is called the *dual residual*.

Imagine p represents a stone A and p_h a second stone B. Which is heavier? What you do to resolve the problem is you lift first the stone p by an amount v and then p_h by the same amount v. The difference in the work done

$$\rho(u_h)(v) = (p - p_h, v) \tag{4.26}$$

is the primal residual. It gives you the necessary clue.

In the same way imagine you have two lenses, δ and δ_h, one perfect the other looks suspicious. What you do to compare δ_h with δ is you send two identical beams of light (v) through both lenses and you measure the aberration

$$\rho^*(G_h)(v) = J(v) - J_h(v) = \int_\Omega \delta\, v\, d\Omega - \int_\Omega \delta_h\, v\, d\Omega = v(\boldsymbol{x}) - v_h(\boldsymbol{x}) \tag{4.27}$$

or, as we would say, the difference in the displacement, $v(\boldsymbol{x}) - v_h(\boldsymbol{x})$.

So the primal residual provides a hint how far off the FE-solution u_h is by shaking the membrane and measuring the virtual work done by the *ersatz* load case p_h and comparing it with the work done by the exact load p. Or a beam of light is sent through the "lense" $J_h(v)$ and we look where it lands on the screen. This is what is done in the *patch test*. We fabricate a solution, send it through the FE-code and hope to see a perfect image on the screen.

4.5 An Analysis of the Goal-Oriented Error Estimator

The error in a functional

$$J(e) = u(\boldsymbol{x}) - u_h(\boldsymbol{x}) = \int_{\Omega_p} [G(\boldsymbol{y}, \boldsymbol{x}) - G_h(\boldsymbol{y}, \boldsymbol{x})]\, p(\boldsymbol{y})\, d\Omega_{\boldsymbol{y}} \tag{4.28}$$

is a consequence of the dual error $G - G_h$, the difference between G and G_h in the region Ω_p where the load p is applied. But why does in (4.20) figure also the primal error $\|u - u_h\|_E$?

The reason is that (4.28) is the exact formula for the error while (4.20) is an estimate of this error based on the Cauchy–Schwarz inequality (L_2-scalar product)

$$|J(e)| \leq \|G - G_h\|_0 \cdot \|p\|_0\,. \tag{4.29}$$

Firstly the Galerkin orthogonality allows to add p_h to (4.28), see (4.18),

$$J(e) = \int_{\Omega} [G(\mathbf{y}, \mathbf{x}) - G_h(\mathbf{y}, \mathbf{x})]\,[p(\mathbf{y}) - p_h(\mathbf{y})]\, d\Omega_{\mathbf{y}}\,. \tag{4.30}$$

Secondly it may be assumed that the load $p - p_h$, which is equivalent to $Lu - Lu_h$, (L is the differential operator in $Lu = p$) is a linear and continuous functional on $\mathcal{V}$, that is for each $v \in \mathcal{V}$ the integral

$$(p - p_h, v) = \int_{\Omega} (p - p_h)\, v\, d\Omega \tag{4.31}$$

is bounded

$$|(p - p_h, v)| \leq \|p - p_h\|_* \|v\| \tag{4.32}$$

where $\|\cdot\|$ is the norm on $\mathcal{V}$ and $\|\cdot\|_*$ is the norm on the dual space $\mathcal{V}^*$ (the set of all linear and continuous functionals on $\mathcal{V}$). Let $u - u_h$ be the solution to the boundary value problem with the right hand side $p - p_h$, that is $L(u - u_h) = p - p_h$, then the norm of $u - u_h$ is equal to the norm of $p - p_h$,

$$\|p - p_h\|_* = \|u - u_h\|\,. \tag{4.33}$$

Because the norm $\|\cdot\|$ and the energy norm are equivalent on $\mathcal{V}$ it follows

$$|(p - p_h, v)| \leq \|p - p_h\|_* \|v\| \leq c \cdot \|u - u_h\|_E\, \|v\|_E\,. \tag{4.34}$$

Setting $v = G[\mathbf{x}] - \mathbf{G}_h[\mathbf{x}]$ provides the estimate

$$\begin{aligned} |J(e)| &= |\int_{\Omega} [G(\mathbf{y}, \mathbf{x}) - G_h(\mathbf{y}, \mathbf{x})]\,(p(\mathbf{y}) - p_h(\mathbf{y})]\, d\Omega_{\mathbf{y}}| \\ &\leq c \cdot \|G[\mathbf{x}] - G[\mathbf{x}]\|_E\, \|u - u_h\|_E\,, \end{aligned} \tag{4.35}$$

which is the same result as in (4.20). Hence the primal error is an estimate of the error $|p - p_h|$.

Adaptive methods draw on these error indicators. Where the error forces are large the mesh is refined and by repeating this step it is hoped that the FE-solution improves. But often the refinements are only indicators of hot spots, of points, mostly on the boundary, where the exact solution becomes singular and then an adaptive refinement can only try to restrict the influence of the singularity without actually removing it.

4.6 The Algebra of the Residuals

The primal residual is zero for functions $v_h \in \mathcal{V}_h$

$$\rho(u_h)(v_h) = (p - p_h, v_h) = a(e, v_h) = 0\,. \tag{4.36}$$

This is the Galerkin orthogonality. Substituting for v_h the exact Green's function G produces the error in the corresponding functional $J(e)$

$$(p - p_h, G) = \rho(u_h)(G) = J(e)\,. \tag{4.37}$$

The dual residual exhibits a similar behavior. It is zero for functions $v_h \in \mathcal{V}_h$

$$J(v_h) - J_h(v_h) = 0 \tag{4.38}$$

and it attains at u the value $J(e)$ because

$$\begin{aligned} \rho^*(G_h)(u) &= (p, e^*) = (p, G) - (p, G_h) \\ &= J(u) - J_h(u) = J(u) - J(u_h) = J(e)\,. \end{aligned} \tag{4.39}$$

So there is a certain symmetry in the error functionals

$$(p - p_h, G) = \rho(u_h)(G) = J(e) = \rho^*(G_h)(u) = (G - G_h, p) \tag{4.40}$$

or in words: *error in* $p \times G$ = *error in* $G \times p$. This symmetry is only observed in linear problems.

Remark 4.1 Arbitrary functions $\psi_h \in \mathcal{V}_h$ can be subtracted in (4.37) from G without changing the result

$$(p - p_h, G - \psi_h) = \rho(u_h)(G - \psi_h) = (p, G - \psi_h) - a(u_h, G - \psi_h) = J(e), \tag{4.41}$$

because

$$\begin{aligned} a(u, G - \psi_h) - a(u_h, G - \psi_h) &= a(u - u_h, G) - a(u - u_h, \psi_h) \\ &= a(u - u_h, G) = J(e)\,. \end{aligned}$$

This also holds true in (4.39); for each $u - \varphi_h$ with $\varphi_h \in \mathcal{V}_h$ we have

$$a(u - \varphi_h, G) - a(u - \varphi_h, G_h) = (p, G - G_h) + a(u_h, G - G_h) = (p, G - G_h)\,. \tag{4.42}$$

4.7 Goal-Oriented Refinement for Nonlinear Problems

The method of goal-oriented refinement can also be applied to nonlinear problems. This is done in three steps:

(i) Linear and nonlinear functionals $J(\boldsymbol{u})$ can be written as, see Sect. 2.9.3,

$$J(\boldsymbol{u}) = \boldsymbol{j}_u^T \boldsymbol{u} \tag{4.43}$$

where the component j_{u_i} of the vector $\boldsymbol{j}_u$ is the Gateaux derivative of $J(\boldsymbol{u})$ in the direction of the shape function φ_i. For linear functional $\boldsymbol{j}_u$ is a constant vector while for nonlinear functionals it depends on $\boldsymbol{u}$.

(ii) Evaluating the functional $J(\boldsymbol{u})$ under the side condition $\boldsymbol{k}(\boldsymbol{u}) = \boldsymbol{f}$ leads to the Lagrange functional

$$\mathcal{L}(\boldsymbol{u}, \boldsymbol{\lambda}) = J(\boldsymbol{u}) - \boldsymbol{\lambda}^T(\boldsymbol{k}(\boldsymbol{u}) - \boldsymbol{f}) = \boldsymbol{u}^T \boldsymbol{j}_u - \boldsymbol{\lambda}^T(\boldsymbol{k}(\boldsymbol{u}) - \boldsymbol{f}) \tag{4.44}$$

and the stationary point $\{\boldsymbol{u}, \boldsymbol{\lambda}\}$ is determined by the equations

$$\boldsymbol{k}(\boldsymbol{u}) = \boldsymbol{f} \quad \boldsymbol{K}_T(\boldsymbol{u})\,\boldsymbol{\lambda} = \boldsymbol{j}_u \tag{4.45}$$

where $\boldsymbol{K}_T$ is the tangent stiffness matrix at the point $\boldsymbol{u}$.

(iii) The tangent stiffness matrix appears also when the equation $\boldsymbol{k}(\boldsymbol{u}) = \boldsymbol{f}$ is solved by Newton's method

$$\boldsymbol{K}_T(\boldsymbol{u}_i)\,\boldsymbol{e}_{i+1} \simeq \boldsymbol{f} - \boldsymbol{k}(\boldsymbol{u}_i) \quad \boldsymbol{e}_i = \boldsymbol{u}_{i+1} - \boldsymbol{u}_i\,. \tag{4.46}$$

Given a symmetric matrix such as $\boldsymbol{K}_T(\boldsymbol{u}_i)$ the following identity ($\pi = \pi^T$)

$$\mathscr{B}(\boldsymbol{e}, \boldsymbol{\lambda}) = \boldsymbol{\lambda}^T \boldsymbol{K}_T(\boldsymbol{u}_i)\,\boldsymbol{e} - \boldsymbol{e}^T \boldsymbol{K}_T(\boldsymbol{u}_i)\,\boldsymbol{\lambda} = 0 \tag{4.47}$$

holds true for all vectors $\boldsymbol{e}$ and $\boldsymbol{\lambda}$. Eventually the second matrix $\boldsymbol{K}_T$ has to be replaced by its transpose if $\boldsymbol{K}_T$ is not symmetric.

Conclusion: If the vector $\boldsymbol{\lambda}_i$ is a solution of

$$\boldsymbol{K}_T(\boldsymbol{u}_i)\,\boldsymbol{\lambda}_i = \boldsymbol{j}_u \tag{4.48}$$

and $\boldsymbol{e} = \boldsymbol{e}_{i+1} = \boldsymbol{u}_{i+1} - \boldsymbol{u}_i$ then

$$J(\boldsymbol{e}_{i+1}) = \boldsymbol{\lambda}_i^T\,(\boldsymbol{f} - \boldsymbol{k}(\boldsymbol{u}_i)) + \text{remainder}\,. \tag{4.49}$$

This is not the same result $J(\boldsymbol{u}) = \boldsymbol{\lambda}^T \boldsymbol{f}$ as in linear analysis (the vector $\boldsymbol{\lambda}$ is of course identical with the vector $\boldsymbol{g}$ in the preceding chapters). This equation only allows to make (educated) guesses about the size of the error $J(\boldsymbol{e}_{i+1})$ with the Green's function at the linearization point.

But this is motivation enough to apply goal-oriented adaptive refinement as in linear problems. The primal problem is (4.45_1) and the dual problem is (4.45_2).

4.7.1 Estimates

While the previous results were based mainly on linear algebra we present in the following some background material to show how by an elaborate mathematical analysis estimates for the error $J(u) - J(u_h)$ can be obtained in a nonlinear context [1].

In Sect. 2.9 it was demonstrated that if a linear functional $J(u)$ is to be evaluated under the side condition that u is a solution to the variational problem $a(u, v) = (p, v)\ \forall\, v \in \mathcal{V}$ then this can be formulated in the context of Lagrange multipliers as the problem to find a stationary point of the functional

$$\mathcal{L}(u, G) = J(u) - (a(u, G) - (p, G)) \tag{4.50}$$

because λ is identical with the Green's function G. In linear problems $a(.,.)$ is a bilinear form and $(p, .)$ is a linear form.

We say a functional such as

$$\Pi(u) = \frac{1}{2}\, a(u, u) - (p, u) \tag{4.51}$$

is stationary at u if the Gateaux derivative is zero "in all directions $\delta u \in \mathcal{V}$" that is

$$\Pi'(u)(\delta u) := \frac{d}{d\,\varepsilon}\Pi(u + \varepsilon\delta u)|_{\varepsilon=0} = a(u, \delta u) - (p, \delta u) = 0 \quad \forall\, \delta u \in \mathcal{V} \tag{4.52}$$

so that $\{u, G\}$ is a stationary point of $\mathcal{L}(u, G)$ (two arguments instead of one) if both Gateaux derivatives are zero in all possible directions δu, δG

$$\mathcal{L}'_u(u, G)(\delta u) = \frac{d}{d\,\varepsilon}\mathcal{L}(u + \varepsilon\delta u, G)|_{\varepsilon=0} = 0 \tag{4.53}$$

$$\mathcal{L}'_G(u, G)(\delta G) = \frac{d}{d\,\eta}\mathcal{L}(u, \lambda + \eta\,\delta G)|_{\eta=0} = 0\,. \tag{4.54}$$

In the nonlinear case $a(\cdot\,;\cdot)$ is a semilinear form, that is it is nonlinear in the first argument and linear in the second and the variational formulation that underlies the FE-method has the form

$$A(u; \delta u) := a(u; \delta u) - (p, \delta u) = 0 \quad \forall\, \delta u \in \mathcal{V} \tag{4.55}$$

so that the Lagrange functional (4.50) is

$$\mathcal{L}(u, G) := J(u) - A(u; G)\,. \tag{4.56}$$

The Gateaux-derivatives of the functionals with respect to u are

$$J'_u(u; \delta u) := \frac{d}{d\epsilon}\big(J(u + \epsilon\, \delta u)\big)\Big|_{\epsilon=0} \tag{4.57}$$

and

$$A'_u(u; \delta u, G) := \frac{d}{d\epsilon}\big(A(u + \epsilon\, \delta u; G)\big)\Big|_{\epsilon=0} \tag{4.58}$$

while the derivative with respect to G

$$A'_G(u; G, \delta G) := \frac{d}{d\eta}\big(A(u; G + \eta\, \delta G)\big)\Big|_{\eta=0} = A(u; \delta G)\,, \tag{4.59}$$

is identical with $A(u; \Delta G)$ because $A(\cdot; \cdot)$ is linear in the second argument.

Hence the variational formulation of the primal (1st equ.) and dual problem (2nd equ.) are

$$\mathcal{L}'(u, G)(\delta u, \delta G) := \left\{ \begin{array}{l} -A(u; \delta G) \\ J'_u(u; \delta u) - A'_u(u; \delta u, G) \end{array} \right\} = 0 \quad \forall\, \delta u, \forall\, \delta G \tag{4.60}$$

and the Galerkin approximation $\{u_h, G_h\}$ is the solution of

$$\mathcal{L}'(u_h, G_h)(\delta u_h, \delta G_h) := \left\{ \begin{array}{l} -A(u_h; \delta G_h) \\ J'_u(u_h; \delta u_h) - A'_u(u_h; \delta u_h, G_h) \end{array} \right\} = 0 \quad \forall\, \delta u_h, \forall\, \delta G_h\,. \tag{4.61}$$

After these preliminary steps we can now concentrate on the error $J(u) - J(u_h)$. To this end we let $\boldsymbol{x} = \{u, G\}$ and $\boldsymbol{x}_h = \{u_h, G_h\}$ and we note that

$$\mathcal{L}(\boldsymbol{x}) - \mathcal{L}(\boldsymbol{x}_h) = J(u) - J(u_h)\,. \tag{4.62}$$

By doing a Taylor expansion of the left an estimate for the right-hand side can be formulated. This is the central idea.

Under the assumption that $\mathcal{L}(\boldsymbol{x})$ is a three-times differentiable function and $\boldsymbol{x}$ is a stationary point, $\mathcal{L}'(\boldsymbol{x})(\boldsymbol{\delta x}) = 0\ \forall\, \boldsymbol{\delta x}$, and $\boldsymbol{x}_h$ is the FE-approximation to $\boldsymbol{x}$ holds [1],

$$\mathcal{L}(\boldsymbol{x}) - \mathcal{L}(\boldsymbol{x}_h) = \frac{1}{2}\,\mathcal{L}'(\boldsymbol{x}_h)(\boldsymbol{x} - \boldsymbol{x}_h) + \mathcal{R}_h^{(3)} \tag{4.63}$$

where the remainder term $\mathcal{R}_h^{(3)}$ is cubic in the error $\boldsymbol{e} = \boldsymbol{x} - \boldsymbol{x}_h$.

Let $\boldsymbol{y}_h = \{\varphi_i, \psi_j\}$ be an arbitrary pair of test functions on $\mathcal{V}_h \times \mathcal{V}_h$, that is the φ_i approximate u_h and the ψ_j the function G_h (normally the two sets are the same, $\psi_i \equiv \varphi_i$) then holds

$$\begin{aligned}\mathcal{L}'(\boldsymbol{x}_h)(\boldsymbol{x} - \boldsymbol{y}_h) &= \mathcal{L}'_u(u_h, G_h)(u - \varphi_i) + \mathcal{L}'_G(u_h, \psi_j)(G - \psi_j)\\ &= J'_u(u_h)(u - \varphi_i) - A'(u_h)(u - \varphi_i, \psi_j) - A(u_h)(G - \psi_j)\\ &= \rho^*(u_h, \psi_j)(u - \varphi_i) + \rho(u_h)(G - \psi_j)\end{aligned} \tag{4.64}$$

and this immediately implies that

$$J(u) - J(u_h) = \rho^*(u_h, G_h)(u - u_h) + \rho(u_h)(G - G_h)\,. \tag{4.65}$$

This is just the sum of the dual and the primal residual which in nonlinear problems are no longer the same, though the difference between the two

$$\rho^*(u_h, G_h)(u - u_h) = \rho(u_h)(G - G_h) + \Delta\rho \tag{4.66}$$

is, ($e = u - u_h, e^* = G - G_h$),

$$\Delta\rho = \int_0^1 \{A''(u_h + s\,e)(e, e, G_h + s\,e^*) - J''(u_h + s\,e)(e, e)\} \tag{4.67}$$

of second order so that the simplified error representation is

$$J(u) - J(u_h) = \rho(u_h)(G - G_h) + \mathcal{R}_h^{(2)} \tag{4.68}$$

or explicitly

$$J(u) - J(u_h) = a(u_h; G - G_h) - (p, G - G_h) + \mathcal{R}_h^{(2)}\,. \tag{4.69}$$

Basically what this means is that the error in the functional is—we simplify somewhat and let $a(u_h; G - G_h) \simeq (p_h, G - G_h)$—a sum of two effects: the error in the right-hand side, $p - p_h$, and the error in the Green's function, $G - G_h$

$$J(u) - J(u_h) \sim (p_h, G - G_h) - (p, G - G_h) + \text{remainder} \tag{4.70}$$

The problem with this equation is that the exact Green's function G is unknown and for this estimate to be useful G must be replaced by a higher order approximation [1].

4.7.2 Nonlinear Functionals

In the FE-context the weak formulation of the dual problem leads to the system

$$\boldsymbol{K}_T(\boldsymbol{u}_i)\,\boldsymbol{g}_i = \boldsymbol{j}_u \tag{4.71}$$

where the vector $\boldsymbol{j}_u$ has as its components

$$J'(u_h;\varphi_i) \quad i = 1, 2, \ldots n \tag{4.72}$$

the Gateaux-derivatives of $J(u_h)$ in the direction of φ_i.

We give one example for how to handle such a nonlinear functional. In nonlinear elasticity the stress at a point $\boldsymbol{x}$ is defined as

$$J(\boldsymbol{u}) = \sigma_{ij}(\boldsymbol{u})(\boldsymbol{x})\,. \tag{4.73}$$

The Gateaux-derivative of this nonlinear functional is

$$J'(\boldsymbol{u};\boldsymbol{v}) := \left[\frac{d}{d\epsilon}\,J(\boldsymbol{u}+\epsilon\,\boldsymbol{v})\right]_{\epsilon=0} = \left[\frac{d}{d\epsilon}\,\sigma_{ij}(\boldsymbol{u}+\epsilon\,\boldsymbol{v})\right]_{\epsilon=0} \tag{4.74}$$

or with $\boldsymbol{S} = \boldsymbol{C}[\boldsymbol{E}(\boldsymbol{u})]$,

$$\left[\frac{d}{d\epsilon}\,\boldsymbol{S}(\boldsymbol{u}+\epsilon\,\boldsymbol{v})\right]_{\epsilon=0} = \boldsymbol{C}\left[\frac{d}{d\epsilon}\,\boldsymbol{E}(\boldsymbol{u}+\epsilon\,\boldsymbol{v})\right]_{\epsilon=0} = \boldsymbol{C}[\boldsymbol{E}_{\boldsymbol{u}}(\boldsymbol{v})] =: \hat{\boldsymbol{S}} \tag{4.75}$$

where $\boldsymbol{E}_{\boldsymbol{u}}(\boldsymbol{v})$ is the Gateaux-derivative of the *Green–Lagrangian* strain tensor, see Sect. 6.1. Hence the equivalent nodal forces

$$j_k = J'(\boldsymbol{u}_h;\varphi_k) = \hat{\sigma}_{ij}(\varphi_k)(\boldsymbol{x}) \tag{4.76}$$

are the components $\hat{\sigma}_{ij}$ of the tangent stress tensor $\boldsymbol{C}[\boldsymbol{E}_{\boldsymbol{u}_h}(\varphi_k)]$.

4.7.3 Implementation

The total load $\boldsymbol{f}$ is partitioned into M equal parts so that the load can be applied in single steps

$$\boldsymbol{f}(t) = q(t)\,\boldsymbol{f} = \frac{t}{M}\,\boldsymbol{f} \qquad t = 1, 2, \ldots M \tag{4.77}$$

and at each stage t the mesh is adaptively refined based on the combined error indicator of the primal and the dual problem.

- Generate an initial mesh $\mathcal{T}_t$ and let $t = 0$.

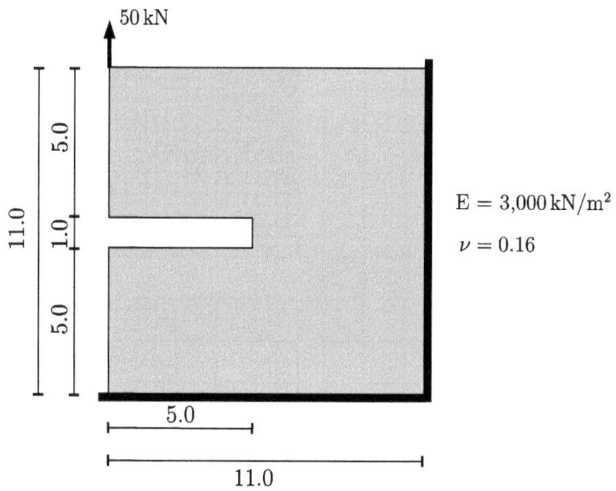

Fig. 4.6 Slit plate

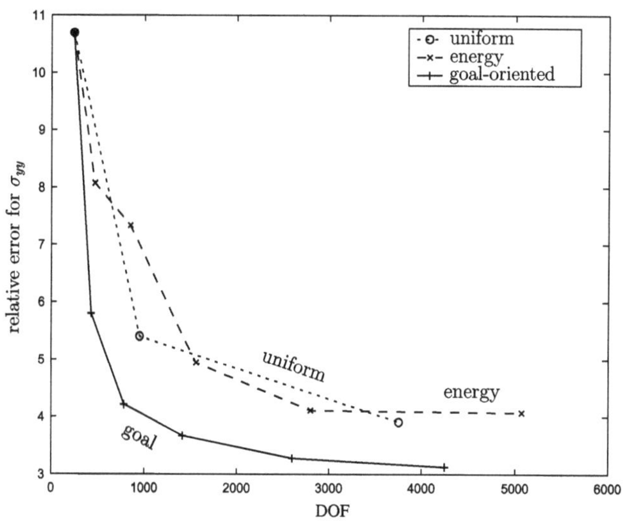

Fig. 4.7 Results for different refinement strategies

- (A) for $t = 1$ up to M (loop over all load steps M)

$$q(t) = \frac{t}{M} \quad \text{actual load parameter}$$

1. Solve the nonlinear equation with Newton-Raphson (iterator i).
   Let $i = 0$, $\boldsymbol{u}_{\Delta_t} = 0$, $\boldsymbol{u}_t^{(0)} = \boldsymbol{u}_{t-1}$ and solve the equations

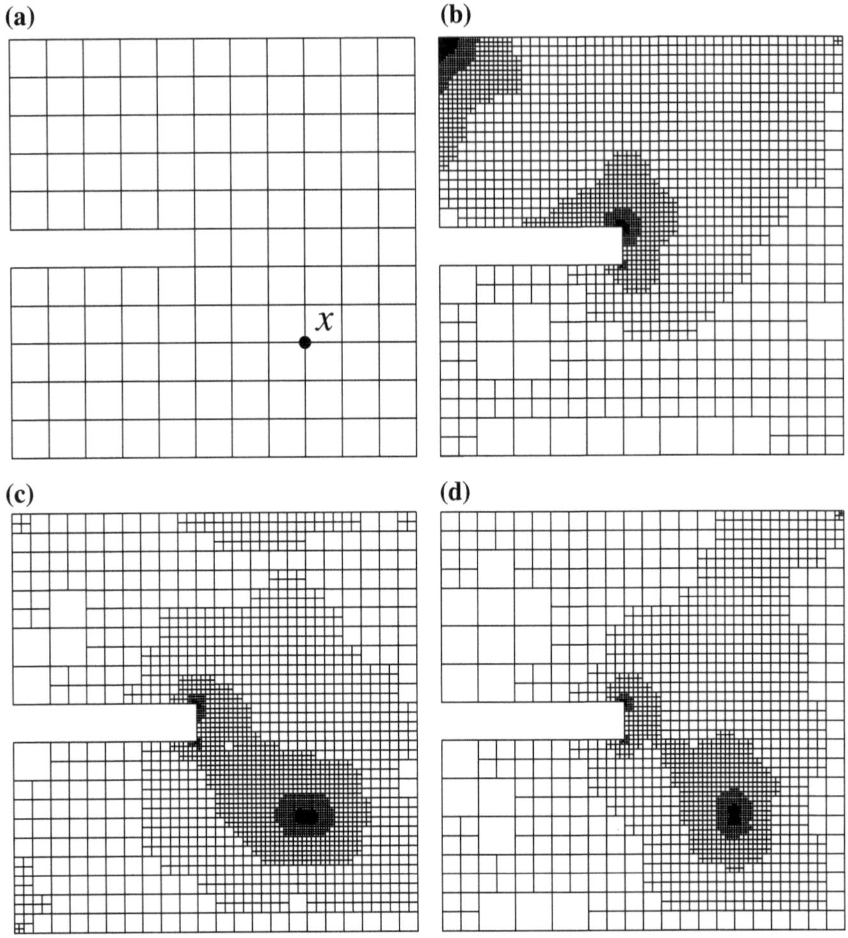

Fig. 4.8 Meshes **a** initial mesh, **b** refinement based on energy norm, **c** goal-oriented refinement for $\sigma_{xx}(\boldsymbol{x})$ and **d** for $\sigma_{yy}(\boldsymbol{x})$

$$\begin{aligned} \boldsymbol{K}_T(\boldsymbol{u}_t^{(i)})\,\boldsymbol{u}_{\Delta_t}^{(i+1)} &= q(t)\cdot \boldsymbol{f} - \boldsymbol{k}(\boldsymbol{u}_t^{(i)}) \\ \boldsymbol{u}_t^{(i+1)} &= \boldsymbol{u}_{\Delta_t}^{(i+1)} + \boldsymbol{u}_t^{(i)} \end{aligned}$$

Let $i = i + 1$ and repeat until convergence is achieved.

2. Calculate the error indicators of the primal problem $\eta_e^{(p)}$.
3. Formulate the dual problem at the current equilibrium position $\boldsymbol{u}_t$:

$$\boldsymbol{K}_T(\boldsymbol{u}_t)\,\boldsymbol{g}_t = \boldsymbol{j}_t \tag{4.78}$$

4. Calculate the error indicator of the dual problem $\eta_e^{(g)}$.

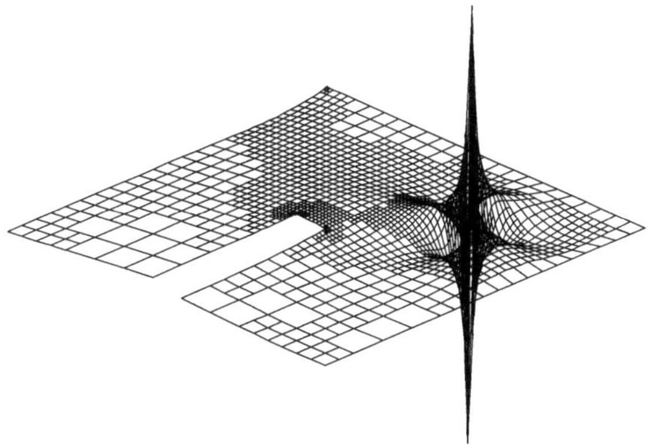

Fig. 4.9 Green's function (at an equilibrium point) for the stress σ_{yy}

5. Refine the mesh:
   - Determine the combined error indicators

$$\eta_e = \eta_e^{(p)} \cdot \eta_e^{(g)} \,. \tag{4.79}$$

   - Calculate the error estimator

$$J(e) \approx \eta = \sum_{\Omega_e} \eta_e \,. \tag{4.80}$$

   - IF $|\eta| \leq$ TOL (global tolerance) $\rightarrow$ $t = t + 1$, GOTO (A).
   - IF $|\eta_e| \leq \mathrm{TOL}_e$ (local element tolerance) $\rightarrow$ refine the element Ω_e.
   - Generate a new mesh $\mathcal{T}_{t+1}$, transfer data, set $t = t + 1$.
   - GOTO 1

The plate in Fig. 4.6 with a slit served as a test example [2]. The plate is subjected to a point load at its upper left corner and the focus was on the horizontal and vertical stress, $J(\boldsymbol{u}) = \sigma_{xx}$ and $J(\boldsymbol{u}) = \sigma_{yy}$ respectively, at one of the four quarter points $\boldsymbol{x}$. The results in Fig. 4.7 confirm that the goal-oriented refinement achieves the best results. In Fig. 4.8 are plotted the different meshes and in Fig. 4.9 is plotted the Green's function for $\sigma_{yy}(\boldsymbol{x})$.

4.8 Drift

If the nodal values $u(\boldsymbol{x}_i)$ of the exact solution were given $u(\boldsymbol{x})$ could be interpolated at the nodes with the nodal shape functions $\varphi_i \in \mathcal{V}_h$ of the mesh

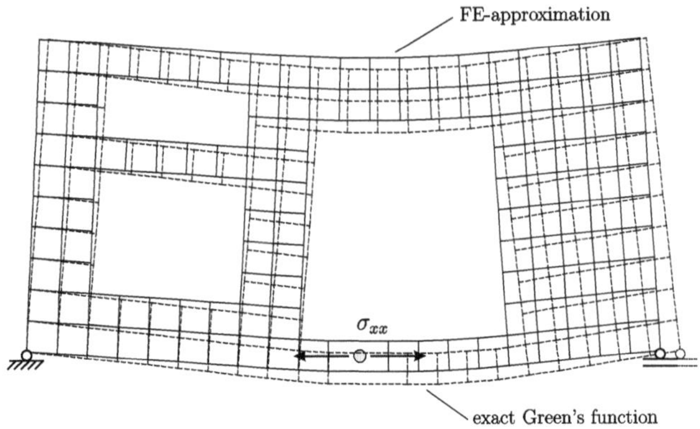

Fig. 4.10 Symbolic representation of the drift of an FE-solution, here the Green's function for the stress σ_{xx}

$$u_I(\boldsymbol{x}) = \sum_{i=1}^{n} u(\boldsymbol{x}_i)\,\varphi_i(\boldsymbol{x})\,. \tag{4.81}$$

This function, the interpolant u_I, has zero drift at the nodes

$$u(\boldsymbol{x}_i) - u_h(\boldsymbol{x}_i) = 0 \tag{4.82}$$

but it deviates from the true solution in between the nodes.

This observation suggests to split the error $e_h(\boldsymbol{x}) = u(\boldsymbol{x}) - u_h(\boldsymbol{x})$ of an FE-solution into a *local* and a *global* error

$$e_h(\boldsymbol{x}) = u(\boldsymbol{x}) - u_I(\boldsymbol{x}) + u_I(\boldsymbol{x}) - u_h(\boldsymbol{x}) = e_h^{loc}(\boldsymbol{x}) + e_h^{glob}(\boldsymbol{x})\,. \tag{4.83}$$

The local error is that part of the solution which is missed by the interpolant and the *drift*—the mismatch between u and u_h at the nodes—represents the global error, s. Fig. 4.10 and 3.33 p. 177,

$$e_h^{loc}(\boldsymbol{x}) = u(\boldsymbol{x}) - u_I(\boldsymbol{x}) \quad e_h^{glob}(\boldsymbol{x}) = u_I(\boldsymbol{x}) - u_h(\boldsymbol{x})\,. \tag{4.84}$$

A slightly different definition of these two errors is obtained if the interpolant is replaced by a function $\tilde{u}_I$ which—as u_I interpolates u at the end nodes—but which additionally provides a local best fit in the sense of the energy metric [3]. But asymptotically these two definitions are equivalent.

That FE-solutions have a drift at the nodes is not a failure in itself because the interpolant is not the function which minimizes the energy error $\|u - u_h\|_E$ and so FE-solutions *must* have a drift. Only in elementary 1-D problems is the FE-solution identical with the interpolant, $u_I(x) = u_h(x)$, see Sect. 3.14.

In FE-analysis interest often focuses on the error on a certain patch Ω_p of the mesh, and then local and global may refer to contributions to the error from sources inside or outside the patch, respectively. In this context the local error is also termed the *near-field error* and the global error is referred to as the *far-field error*.

The local error would be for example the error from a poorly resolved local load while the global error could have its cause in a singularity at the boundary.

4.9 Combination of Modeling and Discretization Error

In any engineering numerical analysis the error has two parts: the discretization error and the modeling error. The modeling error is the error between the physical exact solution, let us call it u_c, and the engineering solution u based on some simplified assumptions so that the problem becomes numerically tractable, can be reduced to a Poisson equation or to a 2-D elasticity problem, etc.

The combination of these two errors, the modeling error, $u_c - u$, and the discretization error, $u - u_h$, is the total error

$$u_c - u_h \tag{4.85}$$

the distance of the numerical solution of the engineering model from the physical exact model.

This error determines also the error in the functionals

$$J(u_c) - J(u_h)\,. \tag{4.86}$$

By employing basically the same technique as before this error can be estimated as follows [4],

$$\begin{aligned} J(u_c) - J(u_h) = &- d(u_h; G_h) \\ &+ \frac{1}{2}\left\{\rho(u_h; G_c - G_h) + \rho^*(u_h, G_h)(u_c - u_h)\right\} \\ &- \frac{1}{2}\left\{d(u_h; e_G) + d'(u_h; e_u, G_h)\right\} + \frac{1}{2}\,\mathcal{R}^{(3)} \end{aligned} \tag{4.87}$$

where $e_u = u_c - u_h$ and $e_G = G_c - G_h$.

To steer an adaptive refinement process it is suggested [4], to replace the exact solutions u_c and G_c by an upgrade of the FE-solution. Upgrade means that if the FE-solution is based on bilinear elements then u_h and G_h are interpolated by bi-quadratic functions, $u_h^{(2)}$ and $G_h^{(2)}$, that is

$$G_c - G_h \approx G_h^{(2)} - G_h \qquad u_c - u_h \approx u_h^{(2)} - u_h\,. \tag{4.88}$$

Because of the Galerkin orthogonality it is

$$\rho(u_h; G_h^{(2)} - G_h) = \rho(u_h; G_h^{(2)}) \tag{4.89}$$

$$\rho^*(u_h, G_h; u_h^{(2)} - u_h) = \rho^*(u_h, G_h; u_h^{(2)}) . \tag{4.90}$$

Neglecting higher-order terms gives

$$J(u_c) - J(u_h) \approx \eta_h + \eta_m \tag{4.91}$$

where

$$\eta_h = \frac{1}{2} \{\rho(u_h; G_h^{(2)}) + \rho^*(u_h, G_h; u_h^{(2)})\} \tag{4.92}$$

$$\eta_m = -d(u_h; G_h) . \tag{4.93}$$

are estimates for the two types of errors.

4.10 Pollution

Pollution is the effect that the accuracy of the FE-solution on a certain patch is adversely affected by disturbances whose cause lies *outside* the region itself.

Such an effect can be observed for example in Fig. 4.11 where the influence function for the shear force N_{yx} of the plate

$$N_{yx} = \int_0^l \sigma_{yx} \, dx \tag{4.94}$$

is approximated with finite elements. The sliding movement, $u_x = 1$, which is the influence function, can only be poorly resolved on the mesh and so the poor resolution in cross-section $A - A$ leads to a large drift at the top of the shear wall where the value of the FE-Green's function is 2.3 while the exact value is 1.0. This is pollution: a singularity at a lower level, in cross-section $A - A$, makes its presence felt on the top level.

The upper part of the FE-plate slides to the right but additionally it performs a slight rigid body rotation and so if we would make the plate higher and higher we could produce *arbitrarily large errors* in the displacement field of the Green's function!

There are three types of pollution [3]:

1. *Pollution due to local unsmoothness of the right-hand side p. This type of pollution is essentially limited in the neighborhood of the points of unsmoothness of p.*
2. *Pollution due to local non-uniformity of the mesh.*

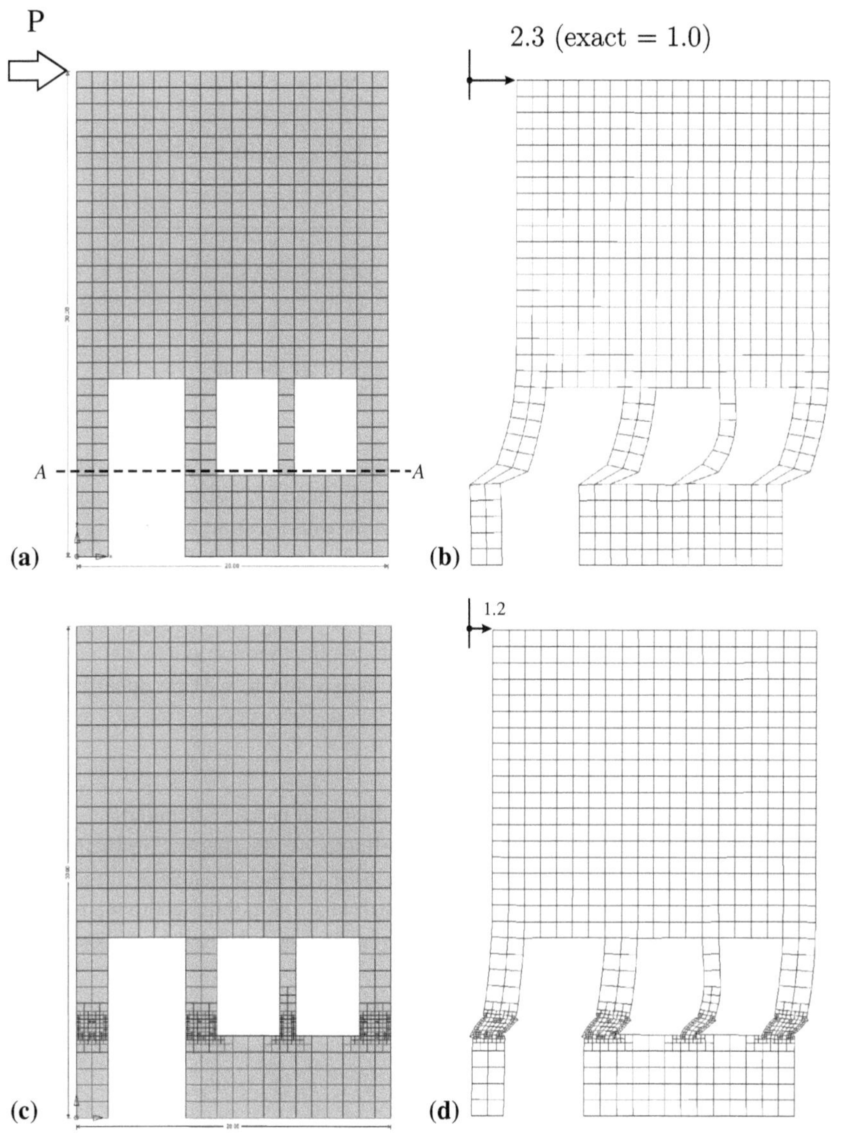

Fig. 4.11 Drift of the influence function for the shear force N_{yx} in cross-section $A - A$, **a** shear wall, **b** FE-solution G_h, **c** refined model, **d** improved FE-solution G_h

3. *Pollution due to unsmoothness in the Green's function which is related also to the unsmoothness of the coefficients in the differential equation.*

Pollution produced by discontinuous loads is negligible in practice. An engineer is well trained to anticipate possible side effects of such loads p.

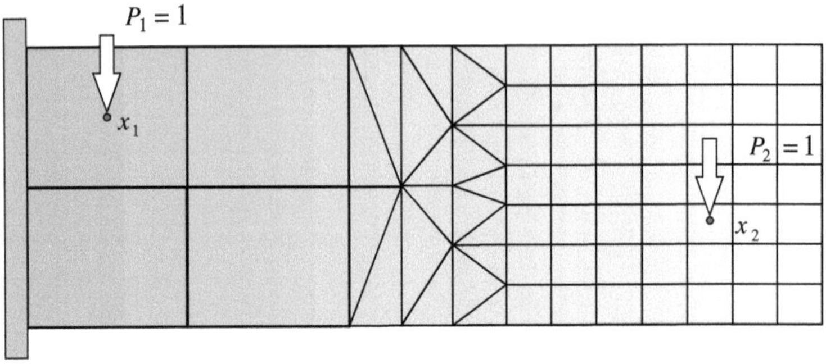

Fig. 4.12 Uneven mesh but the reciprocal FE-displacements are the same

The second effect also is more or less predictable: a very uneven mesh—large elements followed by sequence of small elements—leads to pollution. Though there is a curious side effect built into Betti's theorem: duality straightens things out, is "egalitarian", the richer do not get richer. A finely detailed mesh neighboring to a very coarse mesh cannot profit from its richness because the messages it sends out into the adjacent coarse mesh cannot exceed in accuracy the messages the coarse mesh sends into the fine mesh, see Fig. 4.12. Maxwell's theorem states that two point loads, one on the fine mesh, and one on the coarse mesh, will produce the same effects at the opposite points

$$\delta_{12} = \delta_{21} \tag{4.95}$$

and so not much is gained if only one part of the mesh is finely detailed.

But what about cooperation in adaptive refinement? In adaptive refinement we normally refine the mesh only in the loaded zone (in the absence of singularities on the boundary) and we do *not* refine the mesh near the points where we evaluate the solution. This is only done in goal-oriented refinement. Why then does standard adaptive refinement succeed?

The reason becomes apparent when we look at the error equations

$$\begin{aligned} u(\boldsymbol{x}) - u_h(\boldsymbol{x}) &= \int_\Omega G(\boldsymbol{y}, \boldsymbol{x})(p(\boldsymbol{y}) - p_h(\boldsymbol{y}))\, d\Omega_{\boldsymbol{y}} \\ &= \int_\Omega (G(\boldsymbol{y}, \boldsymbol{x}) - G_h(\boldsymbol{y}, \boldsymbol{x}))\, p(\boldsymbol{y})\, d\Omega_{\boldsymbol{y}}\,. \end{aligned} \tag{4.96}$$

The mesh refinement in the loaded zone makes that the error $p - p_h$ gets smaller and according to the first equation this error gets processed by the *exact* kernel which is accurate at the observation point irrespective of the size of the mesh. The second equation explains why this effect is independent of the non-refinement at the evaluation point $\boldsymbol{x}$, because if the solution u lies in V_h—or very nearly so if $p - p_h$

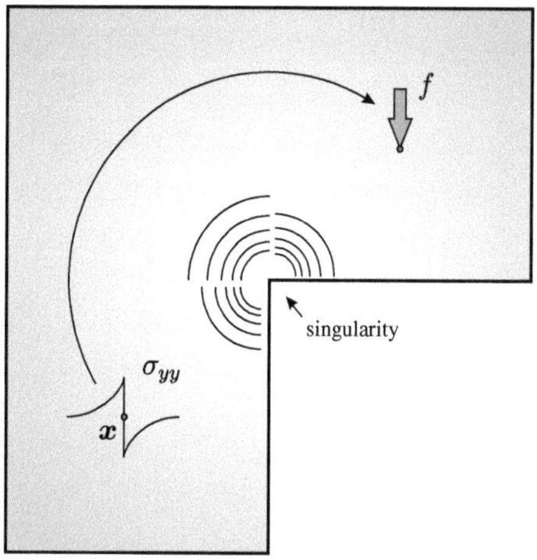

Fig. 4.13 The singularity at the corner point has a negative effect on the information that flows from the source point $\boldsymbol{x}$ to the point load f

is small[1]—then the error in the Green's function for $u(\boldsymbol{x})$ is orthogonal to p and the coarse grain nature of the Green's function for $u(\boldsymbol{x})$ at the evaluation point cannot spoil the picture; it may be only a poor approximation, but its error gets unnoticed.

When we would test Maxwell's theorem with two point loads at $\boldsymbol{x}_1$ and $\boldsymbol{x}_2$ respectively then in goal-oriented refinement the error estimates—if the Green's function would have finite energy—would be symmetric

displacement at $\boldsymbol{x}_1$ caused by point load at $\boldsymbol{x}_2$

$$|\Delta_{12} - \Delta_{12}^h| \leq \underbrace{\|G[\boldsymbol{x}_1] - G_h[\boldsymbol{x}_1]\|_E}_{error\, in\, GF} \underbrace{\|G[\boldsymbol{x}_2] - G_h[\boldsymbol{x}_2]\|_E}_{error\, in\, sol.} \tag{4.97}$$

and

displacement at $\boldsymbol{x}_2$ caused by point load at $\boldsymbol{x}_1$

$$|\Delta_{21} - \Delta_{21}^h| \leq \underbrace{\|G[\boldsymbol{x}_2] - G_h[\boldsymbol{x}_2]\|_E}_{error\, in\, GF} \underbrace{\|G[\boldsymbol{x}_1] - G_h[\boldsymbol{x}_1]\|_E}_{error\, in\, sol.} \tag{4.98}$$

where the bracket is to indicate that this is the Green's function for the point $\boldsymbol{x}_1$ or $\boldsymbol{x}_2$ respectively, $G[\boldsymbol{x}_1] = G(\boldsymbol{y}, \boldsymbol{x}_1)$. In calculating the norm the derivatives are taken with respect to $\boldsymbol{y}$ and integration is also done in this coordinate.

To summarize:

The principal cause for pollution though is the singular behavior of the Green's functions in the neighborhood of corner points. The error in the element next to the singularity spreads

[1] We assume the problem to be well behaved in the sense that $\|u - u_h\| \leq \|p - p_h\|$.

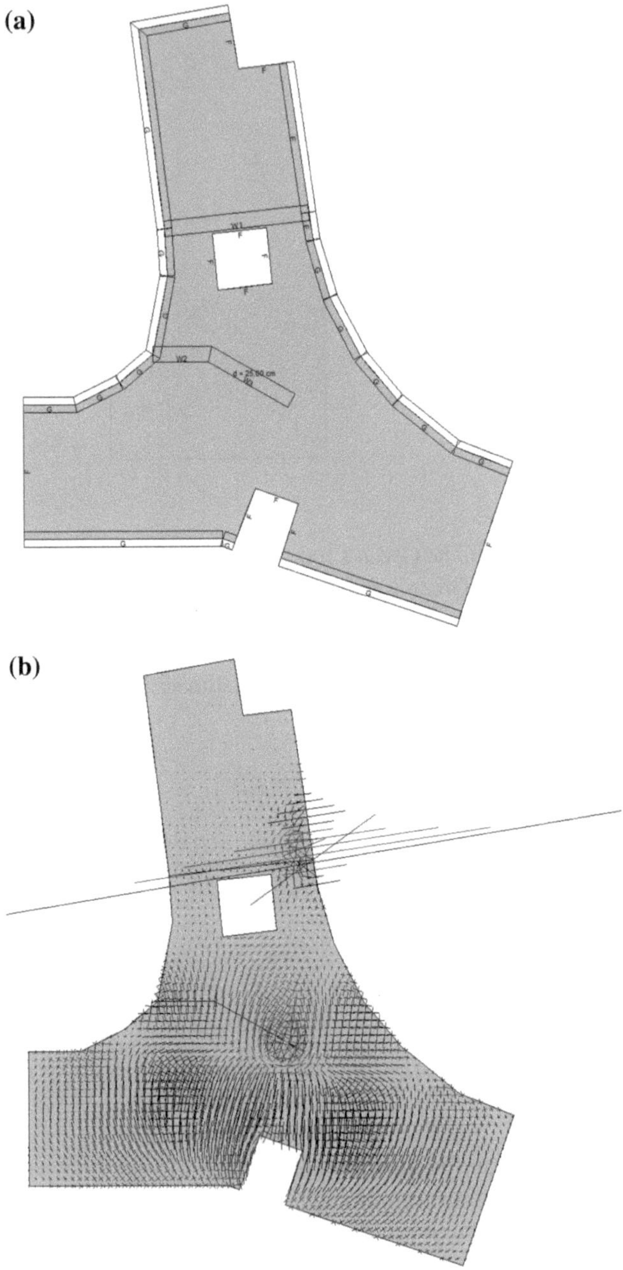

Fig. 4.14 A slightly misplaced node produces a singularity in the principal moments of the slab, **a** slab, **b** principal moments

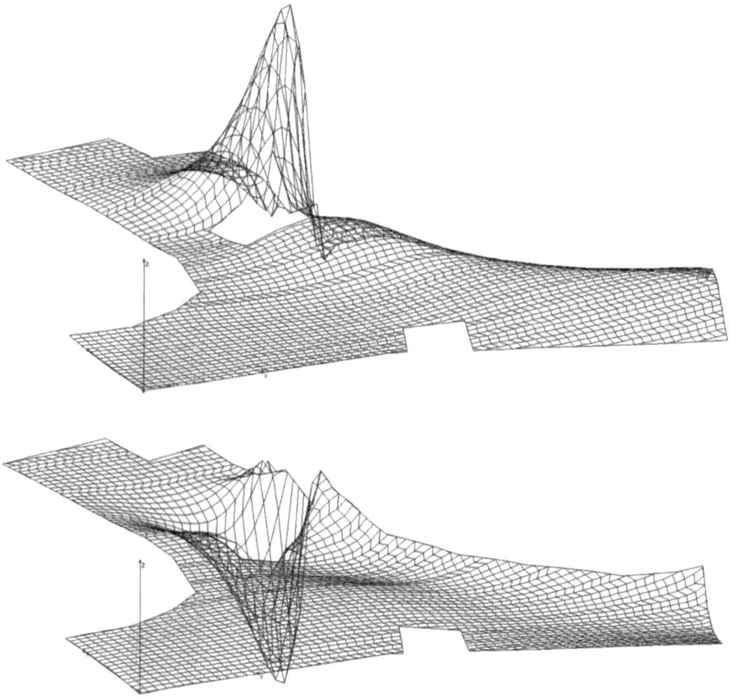

Fig. 4.15 The influence functions for the edge moment m_{nn} at two nearly identical points close to the singularity are widely different

> over the entire mesh and can be the dominant component of the error in the majority of the elements [3].

Note that the authors, *Babuska* and *Strouboulis*, attribute the error to the Green's function, not the solution because it is the Green's functions which produces the output. If the Green's functions are poorly resolved in the neighborhood of singular points then they spoil the picture. The drift of the Green's function manifests itself as the error of the FE-solution.

Imagine a point load f acts at a node and the stress σ_{yy} it produces at a certain point $\boldsymbol{x}$, see Fig. 4.13, is to be calculated. To do this the FE-program applies a vertical unit dislocation at the source point $\boldsymbol{x}$ and it measures by how much the point load is displaced by this dislocation. If the FE-program could determine this value u exactly it would output the exact value for σ_{yy} but the drift of the Green's function produces an error

$$\sigma_{yy}(\boldsymbol{x}) = f \cdot u \qquad \text{exact} \qquad (4.99)$$

$$\sigma^h_{yy}(\boldsymbol{x}) = f \cdot (u + \text{drift}) \qquad \text{FEM} \qquad (4.100)$$

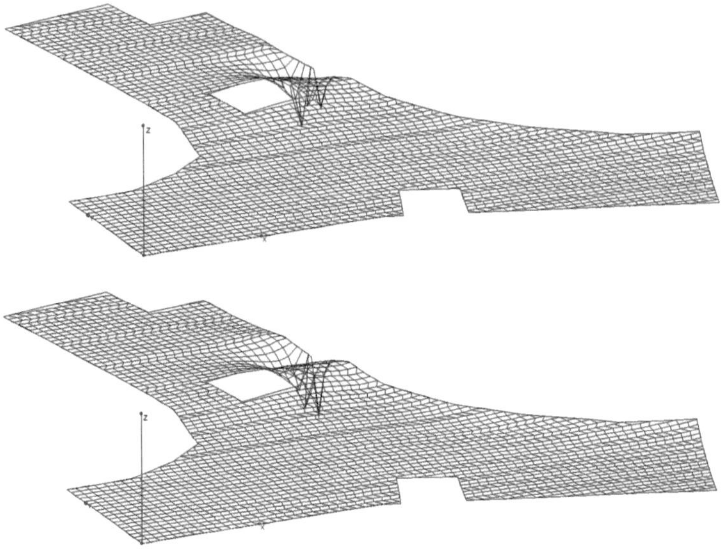

Fig. 4.16 The same two influence functions after the singularity has been removed by placing the node at the correct position

Recall that a Greens function can be split into a fundamental solution $g(\mathbf{y}, \mathbf{x})$ and a regular part $u_R(\mathbf{y}, \mathbf{x})$

$$u(\mathbf{x}) = \int_\Omega G(\mathbf{y}, \mathbf{x})\, p(\mathbf{y})\, d\Omega_y = \int_\Omega (g(\mathbf{y}, \mathbf{x}) + u_R(\mathbf{y}, \mathbf{x}))\, p(\mathbf{y})\, d\Omega_y \qquad (4.101)$$

and so it is the task of the regular part to provide the fit between the fundamental solution and the problem domain. This is why also Green's function have to cope with the singularities on the boundary, see Sect. 3.16.

Pollution may be controlled by employing a mesh which is sufficiently refined in the appropriate places for example in the neighborhood of the singular points for the Green's function. Pollution is negligible if the following three conditions apply *simultaneously*

- the solution u is smooth
- the mesh is uniform
- there are no singularities on the boundary

Pollution is silent. The problem with pollution is that it often goes unnoticed because the drift of the solution and the Green's functions produced by singularities is frequently smooth and so the engineer who looks for suspicious oscillations in the solution has no idea that the output on the screen is biased, is nearly uniformly shifted in one direction.

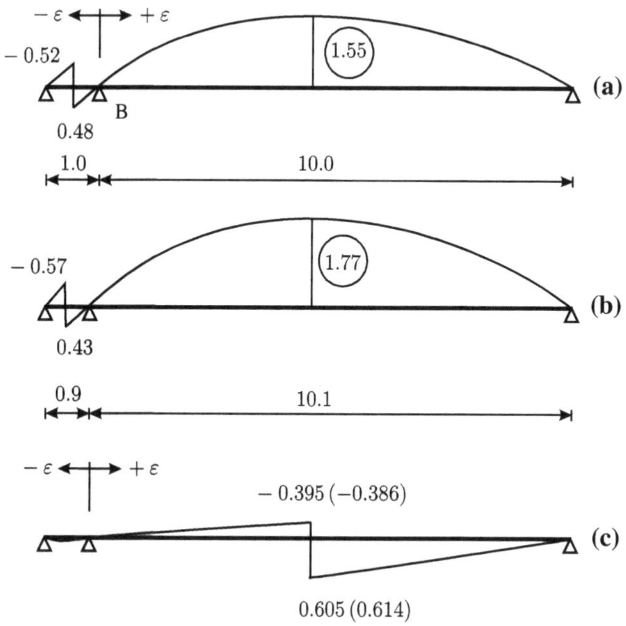

Fig. 4.17 Sensitivity of the Green's function for a shear force. A slight change, ± 0.1 ($= \varepsilon$), in the position of the intermediate support B has a relatively large effect on the influence function, 1.77 versus 1.55 at the center, while the same maneuver hardly has any effect if the source point is the center of the second span

4.11 Gauss Points and Green's Functions

So singularities cause pollution, produce a drift in the solution, see Figs. 4.14, 4.15 and 4.16. But then one may rightly ask why are the stresses or shear forces at the Gauss points (relatively) accurate? How does a program manage to produce accurate Green's functions for the stresses at the Gauss points although the singularities of these Green's functions are *much* higher than the singularities on the boundary. They are for example of the order r^{-2} and r^{-3} for the bending moments (second order) and for the shear force (third order) respectively in a Kirchhoff plate (biharmonic equation)?

To understand this phenomenon imagine that a point load is applied at the center of a plate (shear wall). The associated displacement field has a singularity of the order $O(\ln r)$ and the stresses behave as $1/r$. To an engineer it is evident that given a certain distance from the source point it is no longer possible to distinguish between a true point load and the effects produced by an equivalent (tightly packed) surface load. That is given a certain distance from the source point the Green's function is "back on track", has shed off its singular nature and behaves as any other regular solution.

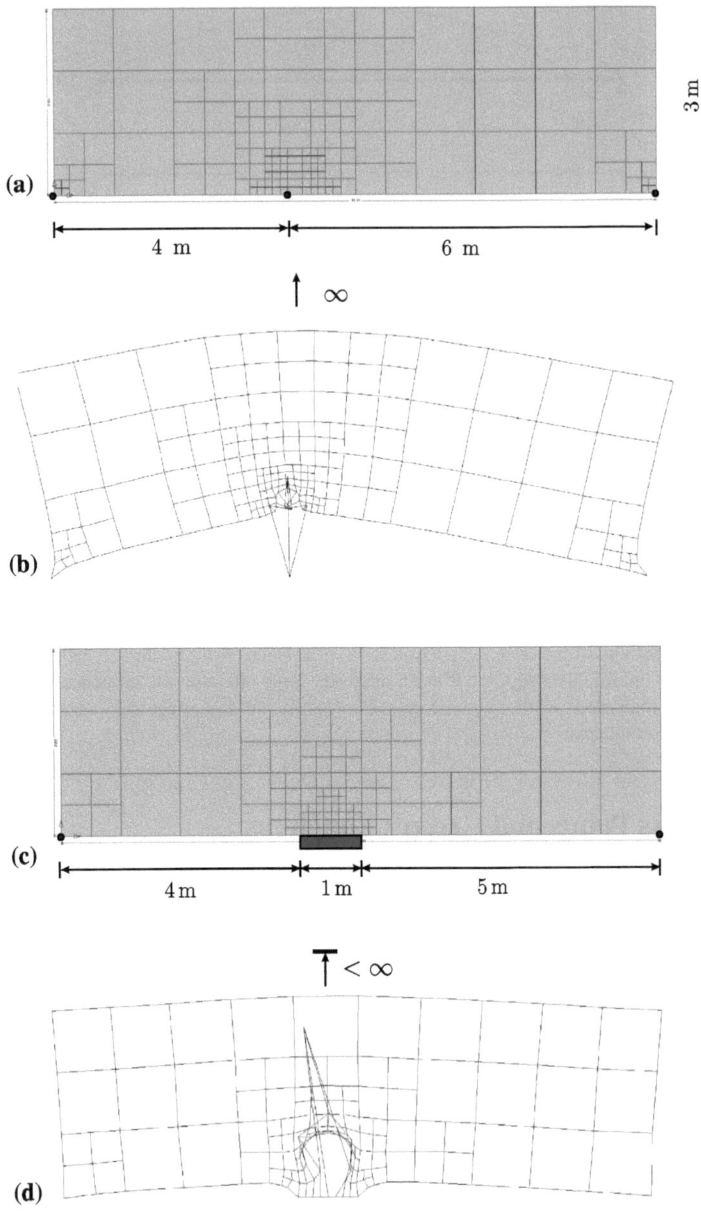

Fig. 4.18 Plate **a** on point supports **b** a dislocation 0.1 m above the point support ultimately produces a nearly infinite displacement while **c** a line support can better cope with such a dislocation, **d** the displacement field remains bounded

What is necessary for this to happen though is that the Green's function has enough "room" to perform the transition. At Gauss points far away from the boundary this is obviously the case. The effects produced by the two opposite peaks, $\pm\infty$, of a dislocation tend to annihilate in the distance but if the dislocation becomes a one-sided swing as it happens when the source point moves to the boundary or encounters an intermediate fixed point as in Fig. 2.19 then the balance gets disturbed: the effects at distant points no longer cancel so fast because the opposite half swing is missing.

Eventually singularities from conflicting boundary conditions add to the singularity in the Green's function. Then the "self-healing forces", the intelligence built into the solutions of elliptic partial differential equations no longer can cope with the situation—in particular if the kinematics is hampered by a low degree approximation of the displacement field—and so the disturbances generated on the boundary spread over the whole domain.

The 1-D problem of a two span beam, see Fig. 4.17, tries to exemplify the situation. The short first span represents the negative effect a singularity on the boundary has on the solution. The influence function for the shear force $V(x)$ at the center of the short span is very sensitive with regard to the position of the intermediate support B; slight variations $x_B \pm \varepsilon$ in the coordinate of the support lead to relatively large changes in the influence function while the influence function for the same quantity in the second span is hardly affected by such alterations.

The evaluation of the solution near the boundary per se poses no problem. The problem is the evaluation of the solution near singular points as point supports or corner points. Singular stresses means that dislocations (= influence functions for stresses) close to the singular point result in infinite displacements, see Fig. 4.18b, while the same dislocations produce only (relatively) mild effects, see 4.18d, if the boundary is fixed, if the point support is replaced by a line support—even if it is only a small stretch.

References

1. Bangerth W, Rannacher R (2003) Adaptive finite element methods for differential equations. Birkhäuser Verlag, Basel
2. Materna D (2009) Structural and sensitivity analysis for the primal and dual problems in the physical and material spaces, Ph.D. thesis, University of Dortmund
3. Babuska I, Strouboulis T (2001) The finite element method and its reliability. Oxford University Press, Oxford
4. Braack M, Ern A (2003) A posteriori control of modeling errors and discretization errors. Multiscale Model Simul 1:221–238

Chapter 5
Modeling Error

In this chapter we are concerned with the question how the influence of modeling errors on the output values can be assessed quantitatively. How do displacements or stresses change if the coefficients of the differential equation change? In a wider perspective this concerns the question in which way a solution depends on the coefficients of the differential equation or how precise a model must be to give reasonable answers.

Any change in the underlying equation or change of an FE-model translates into a change $\boldsymbol{K} \rightarrow \boldsymbol{K} + \boldsymbol{\Delta K}$ of the stiffness matrix and its inverse. The Woodbury-Morrison-formula allows to compute the inverse of the modified matrix by doing a correction to the original inverse. We will apply instead a "direct" method to calculate the solution vector $\boldsymbol{u}_c$ of the modified system and we will tie this to the *force method* of structural analysis which provides a natural and intuitive way to handle such problems. Basically in this approach it is not the stiffness matrix $\boldsymbol{K}$ which gets modified but the right-hand side, $\boldsymbol{f} \rightarrow \boldsymbol{f} + \boldsymbol{f}^{+}$.

While the correction $\boldsymbol{\Delta K}$ to the stiffness matrix can be quite small its effects on the influence functions are global—each entry of the inverse of the stiffness matrix $\boldsymbol{K}^{-1}$ changes—and so to trace the changes in the output due to such modifications a complete reanalysis of the influence functions is necessary, that is integration must be done again from 0 to l, if $\Omega = (0, l)$ is an interval or else over the full domain Ω in higher dimensions.

But there exists an alternative formulation where integration needs only to be done over the defective element to predict which effects a modification in the element stiffness has on all other points in the domain. This integral, the d-form $d(G, u_c)$ is the strain energy product between the Green's function and the modified solution u_c. It is a weak influence function. Various techniques are discussed how best to calculate this d-form in the context of the FE-method. By solving small auxiliary problems the effects caused by changes in an element stiffness can be calculated exactly.

In sensitivity analysis the effects of such changes $\boldsymbol{K} \rightarrow \boldsymbol{K} + \boldsymbol{\Delta K}$ are studied systematically. One question in particular concerns the effects which changes in model parameters have on the displacements and stresses, etc., at a given point. The

F. Hartmann, *Green's Functions and Finite Elements*,
DOI: 10.1007/978-3-642-29523-2_5,

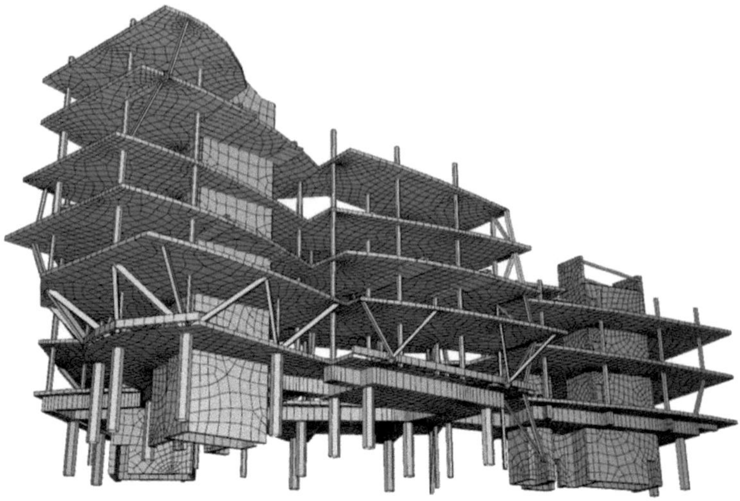

Fig. 5.1 What a stiffness matrix looks like if you ask an engineer [1]

other question is which effects a partial loss of stiffness in one particular element has on all other points and nodes and the relevant functionals at these points. These questions too can be studied by an analysis of the d-form and the contributions to the d-form coming from the Green's functions and the original solution. Based on such estimates adaptive refinement can now be steered by the discretization error and the modeling error simultaneously.

5.1 Linear Algebra

In FE-analysis even the most complex structure, see Fig. 5.1, is reduced at the end to a (possibly quite large) stiffness matrix $\boldsymbol{K}$ which depends on a multitude of model parameters and so even the smallest change to the coefficients of the differential equation of a single structural element requires an update, $\boldsymbol{K} \to \boldsymbol{K} + \boldsymbol{\Delta K}$, of the stiffness matrix. But while the update may regard only one or two entries k_{ij} such an update affects the inverse of the stiffness matrix as a whole. *All* coefficients of the inverse change if one coefficient k_{ij} changes.

The vector of nodal displacement of the modified system $\boldsymbol{u}_c$ is the solution of the system

$$(\boldsymbol{K} + \boldsymbol{\Delta K})\, \boldsymbol{u}_c = \boldsymbol{f} \tag{5.1}$$

while the vector $\boldsymbol{f}$ on the right-hand side is the same vector as in the original system

$$\boldsymbol{K}\, \boldsymbol{u} = \boldsymbol{f}\,. \tag{5.2}$$

This implies

$$\boldsymbol{u}_c = (\boldsymbol{K} + \Delta \boldsymbol{K})^{-1} \boldsymbol{K} \boldsymbol{u} \tag{5.3}$$

and

$$\boldsymbol{K}\,(\boldsymbol{u}_c - \boldsymbol{u}) = -\Delta \boldsymbol{K} \boldsymbol{u}_c \tag{5.4}$$

or

$$\boldsymbol{u}_c - \boldsymbol{u} = -\boldsymbol{K}^{-1} \Delta \boldsymbol{K} \boldsymbol{u}_c \tag{5.5}$$

and

$$(\boldsymbol{I} + \boldsymbol{K}^{-1} \boldsymbol{\Delta}\, \boldsymbol{K})\, \boldsymbol{u}_c = \boldsymbol{u} \tag{5.6}$$

as well.

The last equation means that a system which evolves from $\boldsymbol{K}$ to $\boldsymbol{K} + \boldsymbol{\Delta K}$ is not free in its transition from $\boldsymbol{u} \rightarrow \boldsymbol{u}_c$ but rather that the new state $\boldsymbol{u}_c$ must be compatible with the previous state $\boldsymbol{u}$, must be a solution of (5.6). *A system must know its own history*!

So theoretically one could split the stiffness matrix into a series of sub matrices

$$\boldsymbol{K}_1 = \boldsymbol{\Delta}\, \boldsymbol{K}_1 \tag{5.7}$$
$$\boldsymbol{K}_2 = \boldsymbol{\Delta}\, \boldsymbol{K}_1 + \boldsymbol{\Delta}\, \boldsymbol{K}_2 \tag{5.8}$$
$$\ldots = \ldots \tag{5.9}$$
$$\boldsymbol{K}_n = \boldsymbol{\Delta}\, \boldsymbol{K}_1 + \boldsymbol{\Delta}\, \boldsymbol{K}_2 + \cdots \boldsymbol{\Delta}\, \boldsymbol{K}_n = \boldsymbol{K} \tag{5.10}$$

and watch how the final state $\boldsymbol{u}_n$ evolves out of the previous states

$$(\boldsymbol{I} + \boldsymbol{K}_i^{-1} \boldsymbol{\Delta}\, \boldsymbol{K}_{i+1})\, \boldsymbol{u}_{i+1} = \boldsymbol{u}_i\,. \tag{5.11}$$

To asses which consequences the step $\boldsymbol{K} \rightarrow \boldsymbol{K} + \boldsymbol{\Delta K}$ has for a particular functional $J(u)$ we start with the basic formula

$$J(u_h) = \boldsymbol{j}^T\, \boldsymbol{u} = \boldsymbol{g}^T\, \boldsymbol{K}\, \boldsymbol{u} = \boldsymbol{g}^T\, \boldsymbol{f} \tag{5.12}$$

where $\boldsymbol{u}$ is the nodal vector of the FE-solution u_h, $\boldsymbol{g}$ is the nodal vector of the Green's function of the functional J and $\boldsymbol{j} = \boldsymbol{K}\, \boldsymbol{g}$ is the vector of nodal forces of the Green's function.

It follows that (5.4) must only be multiplied from the left with the nodal vector $\boldsymbol{g}$ to register by how much the functional changes when a matrix $\boldsymbol{\Delta K}$ is added to $\boldsymbol{K}$

$$J(u_c^h) - J(u_h) = \boldsymbol{g}^T\, \boldsymbol{K}\, (\boldsymbol{u}_c - \boldsymbol{u}) = -\boldsymbol{g}^T\, \Delta \boldsymbol{K}\, \boldsymbol{u}_c. \tag{5.13}$$

Basically what happens is that the extra stiffness $\boldsymbol{\Delta K}$ leads to additional nodal forces $\boldsymbol{\Delta K}\, \boldsymbol{u}_c$ which shift the value of the functional from $J(u_h)$ to $J(u_c^h)$.

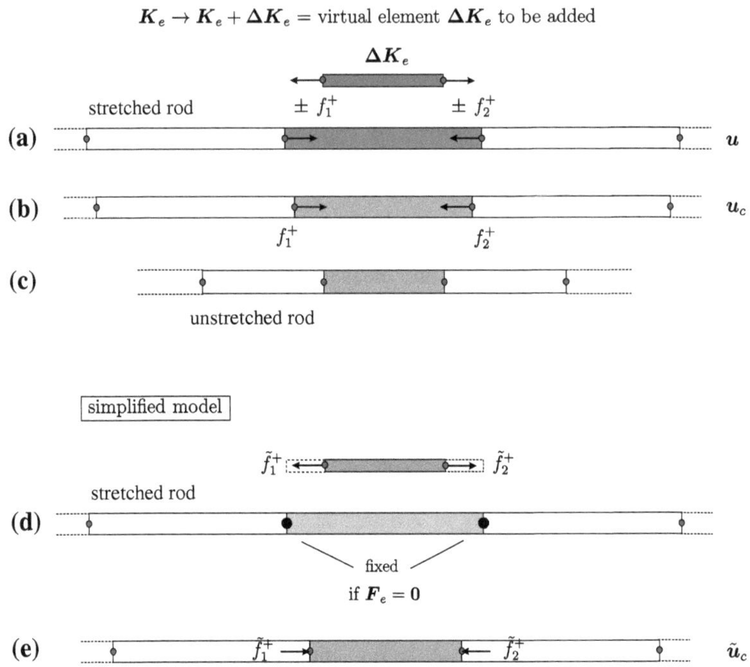

Fig. 5.2 Rod stretched by forces f_i (not shown) and an increase in the stiffness of an element. **a** Two pairs of forces $\pm f_i^+$ provide the fit between the added unstretched virtual element and the deformed rod; **b** the new shape $\boldsymbol{u}_c$ is the response to the forces $f_i + f_i^+$; **d** if $\boldsymbol{F}_e = \boldsymbol{0}$ then the $\tilde{f}_i^+$ must go it alone and stretch the element over a longer distance; **e** consequently the effect is overestimated the rod gets compressed too much (when the $f_i + \tilde{f}_i^+$ are applied the two fixed nodes are again released)

and the change in the value of a functional can be calculated with the formula

$$J(\boldsymbol{e}) = J(\boldsymbol{u}_c) - J(\boldsymbol{u}) = -\boldsymbol{g}^T \, \Delta \boldsymbol{K} \, \boldsymbol{u}_c. \tag{5.21}$$

The vectors $\boldsymbol{g}$ and $\boldsymbol{g}_c$ are the nodal vectors of the original and modified Green's functions respectively—modified because with $\boldsymbol{K} \to \boldsymbol{K} + \boldsymbol{\Delta K}$ also the Green's functions change.

The aim is it to determine $J(\boldsymbol{e})$ *without* calculating the full vector $\boldsymbol{u}_c$ explicitly. In this regard we have two options:

- Substitute for $\boldsymbol{u}_c$ the vector $\boldsymbol{u}$.
- Calculate only those parts of $\boldsymbol{u}_c$ which belong to the nodes of $\boldsymbol{\Delta K}$.

Substituting $\boldsymbol{u}$ for $\boldsymbol{u}_c$ amounts, as we will see, to setting the flexibility matrix $\boldsymbol{F}_e$ to zero, that is the forces $\tilde{\boldsymbol{f}}^+$ are too large, they come from (5.18) and the vector $\tilde{\boldsymbol{u}}$ in

$$J(\tilde{\boldsymbol{u}}_c) = \boldsymbol{g}^T \, \tilde{\boldsymbol{u}}_c \tag{5.22}$$

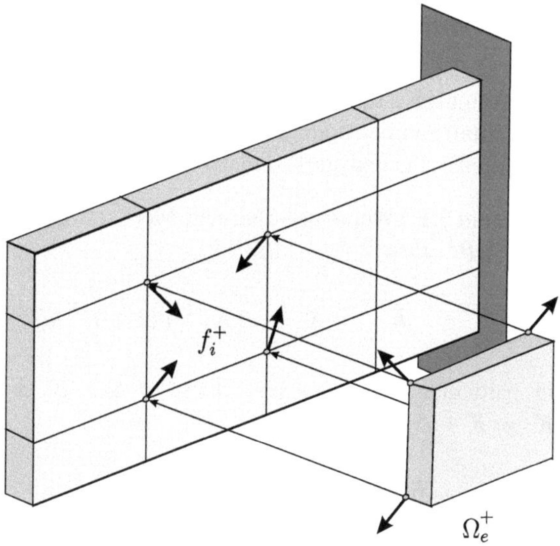

Fig. 5.3 A change $\boldsymbol{K} + \boldsymbol{\Delta K}$ means that an element Ω_e^+ with the stiffness $\boldsymbol{\Delta K}$ is attached to the structure

is the solution of the system $\boldsymbol{K}\,\tilde{\boldsymbol{u}}_c = \boldsymbol{f} + \tilde{\boldsymbol{f}}^+$ and what the formula

$$J(\tilde{\boldsymbol{e}}) = \boldsymbol{g}^T\,\boldsymbol{\Delta K}\,\tilde{\boldsymbol{u}}_c \tag{5.23}$$

measures is the difference

$$J(\tilde{\boldsymbol{e}}) = J(\tilde{\boldsymbol{u}}_c) - J(\boldsymbol{u}) \tag{5.24}$$

between the approximate $\tilde{\boldsymbol{u}}_c$ and the original $\boldsymbol{u}$.

Another option is to calculate by a local analysis the missing parts of $\boldsymbol{u}_c$. Missing means that we only need those parts of $\boldsymbol{u}_c$ which are in contact with the element whose stiffness changes. We will make some suggestions how this could be done.

Of course the central question is: when does it make sense to proceed in this fashion? Certainly not if the modifications are for good because then a reformulation of the stiffness matrix of the structure would be much faster and less cumbersome than this technique.

This technique is valuable in the sense of an inspection method when we want to judge how changes in the stiffness of certain members influence the overall response of a system.

But before discussing all this in more details we will first study which techniques we have to extrapolate the new state $\boldsymbol{u}_c$ from the previous state $\boldsymbol{u}$.

5.3 Woodbury-Sherman-Morrison Formula

This formula is a technique by which the inverse of a slightly modified matrix (rank-k correction) can be computed by doing a corresponding correction to the inverse of the original, the undisturbed matrix.

Theorem 5.1 (Woodbury-Sherman-Morrison formula) *Let* $\boldsymbol{K}_c = \boldsymbol{K} + \Delta \boldsymbol{K} = \boldsymbol{K} + \boldsymbol{A}\,\boldsymbol{B}^T$, *then*

$$\boldsymbol{K}_c^{-1} = \boldsymbol{K}^{-1} - \left[\boldsymbol{K}^{-1}\boldsymbol{A}\,(\boldsymbol{I} + \boldsymbol{B}^T \boldsymbol{K}^{-1} \boldsymbol{A})^{-1}\, \boldsymbol{B}^T\, \boldsymbol{K}^{-1}\right]. \tag{5.25}$$

In particular if the matrices $\boldsymbol{A} \equiv \boldsymbol{a}$ and $\boldsymbol{B} \equiv \boldsymbol{b}$ are column vectors then $\boldsymbol{K}_c = \boldsymbol{K} + \Delta \boldsymbol{K} = \boldsymbol{K} + \boldsymbol{a}\,\boldsymbol{b}^T$,

$$\boldsymbol{K}_c^{-1} = \boldsymbol{K}^{-1} - \gamma^{-1}\, \boldsymbol{K}^{-1}\, \boldsymbol{a}\, \boldsymbol{b}^T\, \boldsymbol{K}^{-1} \qquad \gamma = 1 + \boldsymbol{a}^T\, \boldsymbol{K}^{-1}\, \boldsymbol{b}. \tag{5.26}$$

5.3.1 One Entry on the Diagonal Changes, $k_{ii} + \Delta k$

For a start we first assume that the correction is done only on the main diagonal in row i, to the coefficient $k_{ii} \to k_{ii} + \Delta k$ is added a term Δk. In this case the matrix $\Delta \boldsymbol{K}$ is, up to the entry Δk in row i and column i, the null-matrix, and so

$$\boldsymbol{a} = \sqrt{\Delta k}\, \boldsymbol{e}_i = \boldsymbol{b} \qquad \Delta \boldsymbol{K} = \boldsymbol{a}\, \boldsymbol{b}^T \qquad \gamma = 1 + \Delta k\, k_{ii}^{(-1)}. \tag{5.27}$$

Hence we obtain for $\boldsymbol{u}_c = \boldsymbol{K}_c^{-1} \boldsymbol{f}$ the result

$$\begin{aligned} \boldsymbol{u}_c &= \boldsymbol{K}^{-1} \boldsymbol{f} - \gamma^{-1}\, \boldsymbol{K}^{-1}\, \boldsymbol{a}\, \boldsymbol{b}^T\, \boldsymbol{K}^{-1} \boldsymbol{f} \\ &= \boldsymbol{u} - \gamma^{-1} \Delta k\, \boldsymbol{K}^{-1}\, \boldsymbol{e}_i\, \boldsymbol{e}_i^T\, \boldsymbol{u} = \boldsymbol{u} - \gamma^{-1}\, \Delta k\, u_i\, \boldsymbol{K}^{-1}\, \boldsymbol{e}_i \\ &= \boldsymbol{u} - \underbrace{\gamma^{-1}\, u_i\, \Delta k}_{force}\, \boldsymbol{c}_i, \qquad \text{(no sum over } i) \end{aligned} \tag{5.28}$$

where u_i is the ith component of $\boldsymbol{u}$ and $\boldsymbol{c}_i$ is column i of the inverse $\boldsymbol{K}^{-1}$. Column $\boldsymbol{c}_i$ is the (discrete) Green's function for u_i. Because of Betti this vector lists the influence a unit force at node x_i has on the displacements u_j at the other nodes. So $u_i\, \Delta k\, \gamma^{-1}$ must be a force. It acts at node x_i and it effects additional displacements at the other nodes x_j in accordance with the entry c_{ji} in row j of the vector $\boldsymbol{c}_i$.

5.3.2 The Inverse of the Updated Stiffness Matrix K_c

Equation (5.28) can also be applied to calculate the columns $\boldsymbol{c}_j^c$ of the inverse $\boldsymbol{K}_c^{-1}$. Column $\boldsymbol{c}_j$ of $\boldsymbol{K}^{-1}$ is the solution of the system $\boldsymbol{K}\,\boldsymbol{c}_j = \boldsymbol{e}_j$, and so with

$$\boldsymbol{K}^{-1}\,\boldsymbol{e}_i\,\boldsymbol{e}_i^T\,\boldsymbol{c}_j = \boldsymbol{K}^{-1}\,\boldsymbol{e}_i\,k_{ij}^{(-1)} = \boldsymbol{c}_i\,k_{ij}^{(-1)} \tag{5.29}$$

it follows that the new columns

$$\boldsymbol{c}_j^c = \boldsymbol{c}_j - \gamma^{-1}\,k_{ij}^{(-1)}\,\Delta k\,\boldsymbol{c}_i, \qquad \text{(no sum over } i\text{)} \tag{5.30}$$

are the previous columns plus a certain multiple of column $\boldsymbol{c}_i$. All the columns get shifted in the *same* direction though to a varying degree because the shift depends on the entries $\kappa_{ij}^{(-1)}$ in row i of $\boldsymbol{K}^{-1}$.

Equation (5.30) implies that

$$\boldsymbol{K}_c^{-1} = \boldsymbol{K}^{-1} - \gamma^{-1}\,\Delta k\,\boldsymbol{c}_i \otimes \boldsymbol{c}_i \qquad \text{(no sum over } i\text{).} \tag{5.31}$$

Example 5.1 To better understand the logic behind the Woodbury-Sherman-Morrison formula we try it on a small problem. The taut rope in Fig. 5.4 first hangs free and later node #2 is hooked up to a spring which has the stiffness

$$\Delta k = \frac{0.1}{5} = 0.02 \qquad [\text{kN/m}]. \tag{5.32}$$

This effectively means that a term $\Delta k = 0.1/5$ is added to the entry k_{22} of the stiffness matrix, similar to Fig. 5.5. So if the rope previously had the shape $\boldsymbol{u}$ then with the additional support of the spring it will become the shape

$$\boldsymbol{u}_c = \boldsymbol{u} - \underbrace{0.01786 \cdot u_2}_{force} \cdot \begin{bmatrix} 3.00 \\ 6.00 \\ 4.00 \\ 2.00 \end{bmatrix} \tag{5.33}$$

where the column vector is the column $\boldsymbol{c}_2$ of the inverse stiffness matrix $\boldsymbol{K}^{-1}$—it is the shape of the rope when a unit force $f_2 = 1$ is applied at the node #2—and the term

$$0.01786 = \gamma^{-1}\,\Delta k = \frac{1}{1 + \Delta k\,k_{22}^{(-1)}} = \frac{1}{1 + 0.02 \cdot 6.00} \cdot 0.02 \tag{5.34}$$

is—as will be explained in the next paragraph—the coupling force X_1 between the rope and the spring if there would be a gap of one unit, $u_2 = 1$, between the end of

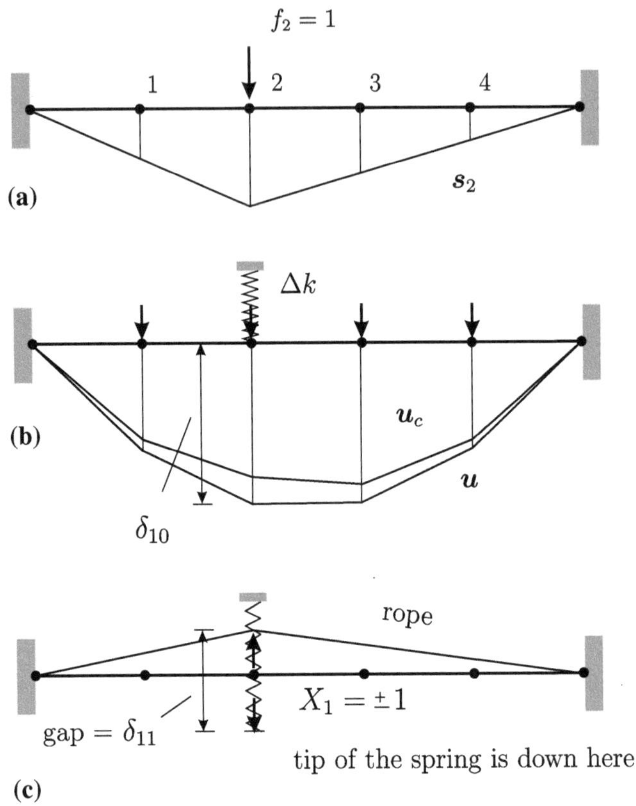

Fig. 5.4 A spring is attached to the taut rope. **a** Green's function for node #2; **b** the deflection before ($\boldsymbol{u}$) and after ($\boldsymbol{u}_c$) is attached; **c** application of the force method

$$\boldsymbol{K} = \frac{1}{5}\begin{bmatrix} 2 & -1 & 0 & 0 \\ -1 & 2 & -1 & 0 \\ 0 & -1 & 2 & -1 \\ 0 & 0 & -1 & 2 \end{bmatrix} \quad \Rightarrow \quad \boldsymbol{K}^{-1} = \begin{bmatrix} 4.00 & 3.00 & 2.00 & 1.00 \\ 3.00 & 6.00 & 4.00 & 2.00 \\ 2.00 & 4.00 & 6.00 & 3.00 \\ 1.00 & 2.00 & 3.00 & 4.00 \end{bmatrix}$$

$$\boldsymbol{K} = \frac{1}{5}\begin{bmatrix} 2 & -1 & 0 & 0 \\ -1 & \textcircled{2.1} & -1 & 0 \\ 0 & -1 & 2 & -1 \\ 0 & 0 & -1 & 2 \end{bmatrix} \quad \Rightarrow \quad \boldsymbol{K}^{-1} = \begin{bmatrix} 3.83 & 2.68 & 1.78 & 0.89 \\ 2.68 & 5.36 & 3.57 & 1.79 \\ 1.78 & 3.57 & 5.71 & 2.86 \\ 0.89 & 1.79 & 2.86 & 3.93 \end{bmatrix}$$

$$\boldsymbol{u} = \boldsymbol{K}^{-1}\boldsymbol{f}$$

Fig. 5.5 Change one coefficient in the stiffness matrix and the whole inverse changes

the spring and the rope. And so if the gap is of size u_2 a force

$$X_1\, u_2 = 0.01786\, u_2 \tag{5.35}$$

is necessary to attach the spring to the rope, to close the gap, and so $X_1\, u_2$ is the part of the load that is carried by the spring.

5.4 Direct Formulations

The Woodbury-Sherman-Morrison formula is not very transparent. It works like magic. You can apply it, and it is guaranteed to work, but you do not see what is going on.

Also, the primary aim of our analysis is not to find the inverse of the modified matrix $\boldsymbol{K} + \boldsymbol{\Delta K}$ but to find the solution $\boldsymbol{u}_c$ of the modified system $(\boldsymbol{K} + \boldsymbol{\Delta K})\, \boldsymbol{u}_c = \boldsymbol{f}$ when the solution $\boldsymbol{u}$ of the original system $\boldsymbol{K}\, \boldsymbol{u} = \boldsymbol{f}$ is given. This leads, see Sect. 5.1, to the system

$$(\boldsymbol{I} + \boldsymbol{K}^{-1}\, \Delta \boldsymbol{K})\, \boldsymbol{u}_c = \boldsymbol{u}. \tag{5.36}$$

In the next sections we will show that the force method is the proper framework for analyzing this problem.

5.5 Force Method

The idea of a force X_1 which we introduced for illustrative purposes to explain the result of the Woodbury-Sherman-Morrison formula is the approach of the *force method*. This method is applied by structural engineers to solve *statically indeterminate* structures when the equilibrium conditions alone do not suffice to determine the support reactions and the internal actions.

Engineers solve such problems by reducing the structure to a statically determinate *cut-back structure* and adding to this system a set of so-called *redundant* forces X_i

$$\text{statically indeterminate structure} = \text{cut-back structure} + X_1, X_2, \ldots, X_n$$

which essentially means that they strip the structure of all "unnecessary" constraints (one step more and the structure would collapse) and they then reattach the removed parts via coupling forces $\pm X_i$.

To this procedure corresponds the transition from $\boldsymbol{K}$ to $\boldsymbol{K} + \boldsymbol{\Delta K}$. The original matrix $\boldsymbol{K}$ plays the role of the stiffness matrix of the cut-back structure and by adding additional constraints to the cut-back structure—imagine these as additional "springs"—the stiffness matrix $\boldsymbol{K}$ becomes the matrix $\boldsymbol{K} + \boldsymbol{\Delta K}$.

Let us see how a structural engineer solves the previous problem with the force method: first he applies the load to the unrestrained, freely hanging rope and he calculates the displacement $\delta_{10} := u_2$ at the node x_2 caused by the load. Next he applies two opposite forces $X_1 = \pm 1$, one at the rope and one at the spring (which at this stage of the analysis are not connected) and he measures the relative displacement δ_{11} between the two points caused by the pair $X_1 = \pm 1$

$$\delta_{11} = \underbrace{k_{22}^{(-1)}}_{rope} + \underbrace{\frac{1}{\Delta k}}_{spring} = \frac{1 + \Delta k\, k_{22}^{(-1)}}{\Delta k} \tag{5.37}$$

and so for the gap δ_{10} (produced by the load) to close the unknown force X_1 must satisfy the equation

$$X_1\, \delta_{11} + \delta_{10} = 0 \tag{5.38}$$

or

$$X_1 = -\frac{\delta_{10}}{\delta_{11}} = -\frac{u_2\, \Delta k}{1 + \Delta k\, k_{22}^{(-1)}} = -\gamma^{-1}\, u_2\, \Delta k. \tag{5.39}$$

This additional force X_1 applied at node #2 will result in an additional displacement of the rope so that the updated displacement vector of the rope is the vector

$$\boldsymbol{u}_c = \boldsymbol{u} + X_1\, \boldsymbol{c}_2 = \boldsymbol{u} - \gamma^{-1}\, u_2\, \Delta k\, \boldsymbol{c}_2 \tag{5.40}$$

where $\boldsymbol{c}_2$ is the vector of nodal displacements when a unit force $f_2 = 1$ is acting at node #2. This equation is the same expression as in the *Woodbury-Sherman-Morrison* formula (5.28).

The important point here is not how X_1 is calculated but the important point is that the analysis of the statically indeterminate structure can be done by applying an additional force X_1 to the cut-back structure and this means:

The analysis of the statically indeterminate system can be done solely with the Green's functions of the cut-back structure.

This is the key point! Let the functional $J(w) = u(x)$ the deflection at the mid-point of the rope then the final value u is the deflection u_0 of the cut-back structure plus the deflection u_1 produced by the force X_1

$$u(x) = u_0(x) + u_1(x) = \int_0^l G(y, x)\, p(y)\, dy + G(l, x)\, X_1 \tag{5.41}$$

where $G(y, x)$ is the influence function for $u(x)$ of the cut-back system.

In this chapter the original system, stiffness matrix $\boldsymbol{K}$, plays the role of the cut-back structure and the modified system, the cracked system $\boldsymbol{K} + \boldsymbol{\Delta K}$, plays the role of the statically indeterminate structure. The X_i are the forces f_i^+ which are applied additionally to the *original* system (matrix $\boldsymbol{K}$) to produce the deformation of the cracked system

$$\boldsymbol{u}_c = \boldsymbol{K}^{-1}(\boldsymbol{f} + \boldsymbol{f}^+). \tag{5.42}$$

This is the beautiful (albeit hidden) idea of the force method. *The force method does not change the stiffness matrix, it changes the right-hand side.*

5.5.1 Notation

To begin with let us first assume that the stiffness of an element Ω_e changes and that this can be described by adding a matrix $\boldsymbol{\Delta K}_e$ to the element matrix

$$\boldsymbol{K}_e \rightarrow \boldsymbol{K}_e + \boldsymbol{\Delta K}_e. \tag{5.43}$$

Let $\boldsymbol{F} = [c_{ij}]$ denote the flexibility matrix of the system, that is the inverse of the global stiffness matrix $\boldsymbol{K}$. Parallel to $\boldsymbol{\Delta K}_e$ we introduce the matrix $\boldsymbol{F}_e$ which contains only those entries c_{ij} of $\boldsymbol{F}$ which belong to the nodes of the element Ω_e and we introduce the matrix

$$\boldsymbol{F}_{e+} := \boldsymbol{F}_e + \boldsymbol{\Delta K}_e^{-1} \tag{5.44}$$

which is the sum of $\boldsymbol{F}_e$ plus the inverse of the matrix $\Delta \boldsymbol{K}_e$ (we assume for a moment that it exists).

Some of the formulations that follow are done on the element level and some are done on the global level. We willingly do not distinguish this in the notation because it would unnecessarily have complicated the notation.

So either $\boldsymbol{\Delta K}_e, \boldsymbol{F}_e, \boldsymbol{F}_{e+}$ have the same size as the element matrix $\boldsymbol{K}_e$ or they have the same size as the global stiffness matrix $\boldsymbol{K}$. In the latter case it is understood that this is done by filling these matrices up with zeros.

Eventually there will also be formulations where the entries k_{ij} which are subject to change do not belong to a contiguous block of an element matrix. Then the notations above are to be understood correspondingly, but essentially they retain their meaning. The important point is to realize that $\boldsymbol{F}_{e+}$ is a "local" flexibility matrix.

We denote the elements of the inverse $\boldsymbol{F}_{e+}^{-1}$ by the small Greek letter κ

$$\boldsymbol{F}_{e+}^{-1} = [\kappa_{ij}] \tag{5.45}$$

because the inverse of a flexibility matrix is a stiffness matrix.

5.5.2 Changes on the Diagonal

Let us first assume that only one entry on the diagonal changes, $k_{ii} \to k_{ii} + \Delta k$, then $\boldsymbol{\Delta K}$ has only one entry, Δk, and so it follows

$$\boldsymbol{I} + \boldsymbol{K}^{-1}\,\boldsymbol{\Delta K} = \boldsymbol{I} + \Delta k\,\boldsymbol{C}_i \tag{5.46}$$

where the matrix $\boldsymbol{C}_i = [\boldsymbol{0}, \boldsymbol{0}, \ldots, \boldsymbol{c}_i, \boldsymbol{0}, \boldsymbol{0}]$ contains only one non-zero column, the column $\boldsymbol{c}_i$ of the inverse $\boldsymbol{K}^{-1}$. The inverse of this particular matrix is

$$(\boldsymbol{I} + \boldsymbol{K}^{-1}\,\boldsymbol{\Delta K})^{-1} = (\boldsymbol{I} + \Delta k\,\boldsymbol{C}_i)^{-1} = \boldsymbol{I} - \frac{\Delta k}{1 + c_{ii}\,\Delta k}\,\boldsymbol{C}_i \tag{5.47}$$

where c_{ii} is the component in row i of column $\boldsymbol{c}_i$ and so the displacement vector of the modified system is

$$\boldsymbol{u}_c = \boldsymbol{u} - \frac{\Delta k}{1 + c_{ii}\,\Delta k}\,\boldsymbol{c}_i \tag{5.48}$$

which is the same result as in (5.28).

5.5.2.1 Two Entries on the Diagonal Change

The case that two diagonal elements are changing

$$k_{ii} \to k_{ii} + \Delta k_{ii} \qquad k_{jj} \to k_{jj} + \Delta k_{jj} \tag{5.49}$$

that is

$$\boldsymbol{I} + \boldsymbol{K}^{-1}\,\boldsymbol{\Delta K} = \boldsymbol{I} + \Delta k_{ii}\,\boldsymbol{C}_i + \Delta k_{jj}\,\boldsymbol{C}_j, \tag{5.50}$$

can be handled as well with the force method, see Fig. 5.6.

Now there are two unknowns, X_i and X_j, and the system for these two unknowns is

$$\begin{bmatrix} c_{ii} + \Delta k_{ii}^{-1} & c_{ij} \\ c_{ji} & c_{jj} + \Delta k_{jj}^{-1} \end{bmatrix} \begin{bmatrix} X_i \\ X_j \end{bmatrix} + \begin{bmatrix} u_i \\ u_j \end{bmatrix} = \begin{bmatrix} 0 \\ 0 \end{bmatrix} \tag{5.51}$$

or

$$\boldsymbol{F}_{e+}\,\boldsymbol{x} + \boldsymbol{u} = \boldsymbol{0} \quad \leftarrow \quad \text{gaps must close.} \tag{5.52}$$

The gaps u_i and u_j between the old structure (sans springs) and the springs must close when the ends of the springs are attached to the structure. The matrix $\boldsymbol{F}_{e+}$ times the vector $\boldsymbol{x} = \{X_i, X_j\}^T$ are the relative displacements between the ends of the springs and the structure produced by the two coupling forces X_i and X_j and these forces must be so tuned that the two gaps close simultaneously that is

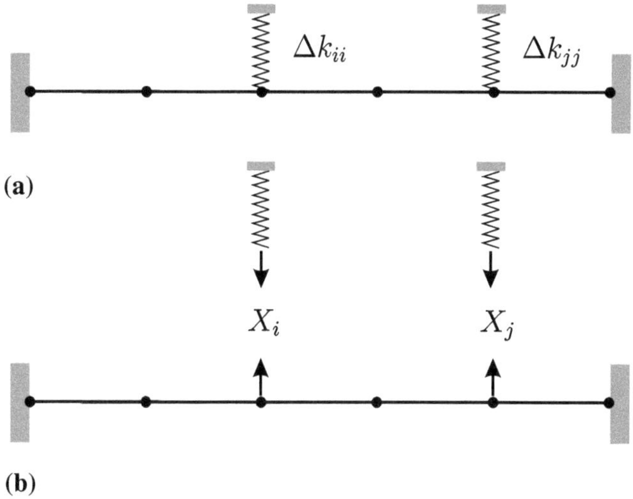

Fig. 5.6 Two springs are attached to the taut rope. **a** Modified system; **b** analysis of the problem with the force method

$$\boldsymbol{x} = \begin{bmatrix} X_i \\ X_j \end{bmatrix} = -\boldsymbol{F}_{e+}^{-1}\,\boldsymbol{u}. \tag{5.53}$$

The two forces X_i and X_j which pull on the structure result in additional displacements, $\boldsymbol{c}_i\, X_i$ and $\boldsymbol{c}_j\, X_j$, ($\boldsymbol{c}_i$ is the displacement of the structure if a unit force $f_i = 1$ acts in the direction of u_i) and so the displacement vector of the modified structure becomes

$$\boldsymbol{u}_c = \boldsymbol{u} + \boldsymbol{c}_i\, X_i + \boldsymbol{c}_j\, X_j. \tag{5.54}$$

This result is identical with

$$\boldsymbol{u}_c = \boldsymbol{u} + \boldsymbol{K}^{-1}\boldsymbol{x} = \boldsymbol{K}^{-1}\boldsymbol{f} + \boldsymbol{K}^{-1}\boldsymbol{x} = \boldsymbol{K}^{-1}(\boldsymbol{f} + \boldsymbol{f}^{+}) \tag{5.55}$$

where we have introduced the vector $\boldsymbol{f}^{+} := \boldsymbol{x}$ to denote the coupling forces.

Remark 5.2 The fact that in the system (5.51) appears the reciprocal Δk_{ii}^{-1} of the change Δk_{ii} is not without risk because for values of Δk_{ii} close to zero the reciprocal is nearly infinite and only infinitely small forces $X_i \sim 0$ are allowed to come in contact with such infinite flexibilities.

5.5.3 *The Inverse*

Because of

$$\boldsymbol{f}^{+} = -\boldsymbol{F}_{e+}^{-1}\,\boldsymbol{u}, \tag{5.56}$$

see (5.53), the last equation can be written as

$$\boldsymbol{u}_c = \boldsymbol{u} - \boldsymbol{K}^{-1} \boldsymbol{F}_{e+}^{-1} \boldsymbol{u} = (\boldsymbol{I} - \boldsymbol{V})\, \boldsymbol{u} \tag{5.57}$$

and so the inverse of the matrix in (5.36) is

$$(\boldsymbol{I} - \boldsymbol{K}^{-1}\, \boldsymbol{\Delta K})^{-1} = \boldsymbol{I} - \boldsymbol{V}. \tag{5.58}$$

This result immediately implies

$$\boldsymbol{u}_c = (\boldsymbol{I} - \boldsymbol{V})\, \boldsymbol{u} = (\boldsymbol{I} - \boldsymbol{V})\, \boldsymbol{K}^{-1}\, \boldsymbol{f} \tag{5.59}$$

and so

$$\boldsymbol{K}_c^{-1} = (\boldsymbol{I} - \boldsymbol{V})\, \boldsymbol{K}^{-1} \tag{5.60}$$

must be the inverse of the updated stiffness matrix $\boldsymbol{K}_c$.

The nearly vacant matrix $\boldsymbol{V}$ contains only two non-zero columns

$$\boldsymbol{V} = \{\boldsymbol{0}, \ldots, \boldsymbol{0}, \boldsymbol{v}_i, \boldsymbol{0}, \ldots, \boldsymbol{0}, \boldsymbol{v}_j, \boldsymbol{0}, \ldots \boldsymbol{0}\} = \boldsymbol{K}^{-1} \boldsymbol{F}_{e+}^{-1} \tag{5.61}$$

which are

$$\boldsymbol{v}_i = \kappa_{ii}\, \boldsymbol{c}_i + \kappa_{ji}\, \boldsymbol{c}_j \qquad \boldsymbol{v}_j = \kappa_{ij}\, \boldsymbol{c}_i + \kappa_{jj}\, \boldsymbol{c}_j \tag{5.62}$$

where the coefficients $\kappa_{ij} = f_{ij}^{(-1)}$ are the entries in the inverse of the 2×2 flexibility matrix $\boldsymbol{F}_{e+}$ in (5.51).

5.6 Example

We apply these formulas to the frame in Fig. 5.7 to see which steps are required if the 4×4 stiffness matrix $\boldsymbol{K}_e$ (the bending part) of a frame element gets an update, $\boldsymbol{K}_e \to \boldsymbol{K}_e + \boldsymbol{\Delta K}_e$. For simplicity it is assumed that the relevant nodal degrees of freedom are numbered 1, 2, 3, 4, see Fig. 5.7.

1. First $\boldsymbol{K}\, \boldsymbol{u} = \boldsymbol{f}$ is solved for the original displacement vector $\boldsymbol{u}$.
2. Then the 4×4 flexibility matrix is constructed

$$\boldsymbol{F}_{e+} = \boldsymbol{F}_e + \Delta \boldsymbol{K}_e^{-1} = \underbrace{\begin{bmatrix} c_{11} & c_{12} & c_{13} & c_{14} \\ c_{21} & c_{12} & c_{13} & c_{14} \\ c_{31} & c_{12} & c_{13} & c_{14} \\ c_{41} & c_{12} & c_{13} & c_{14} \end{bmatrix}}_{\text{entries in } \boldsymbol{K}^{-1}} + \boldsymbol{\Delta K}_e^{-1}. \tag{5.63}$$

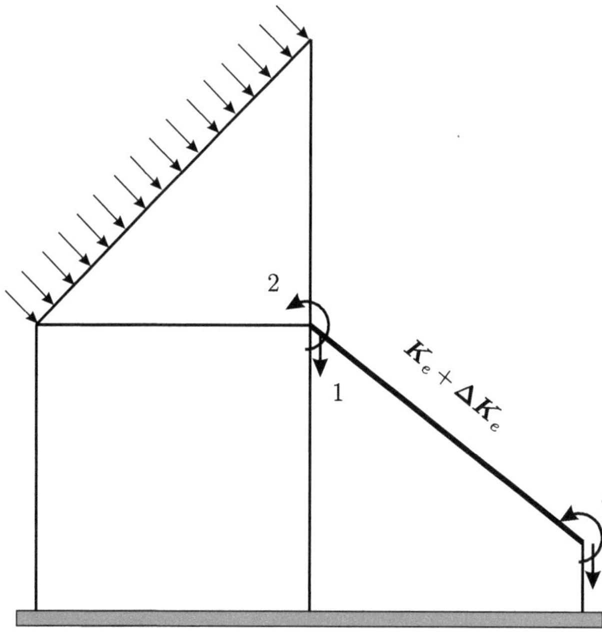

Fig. 5.7 Modification in the stiffness matrix of a single beam element

To calculate the entries c_{ij} in the first column a unit force $f_1 = 1$ will be applied to the structure and the displacements at the stations $u_i, i = 1, 2, 3, 4$ will be measured; these values form the column #1. And this is repeated with $f_i, i = 2, 3, 4$ to find all four columns. To the matrix $\boldsymbol{F}_e$ is added the matrix $\boldsymbol{\Delta K}_e^{-1}$, the inverse of $\boldsymbol{\Delta K}_e$. (We presume for a moment that $\boldsymbol{\Delta K}_e^{-1}$ exists).

3. The inverse $\boldsymbol{F}_{e+}^{-1} = [\kappa_{ij}]$ of the (4×4) matrix $\boldsymbol{F}_{e+}$ is calculated and then the four forces

$$f_i^+ := -\sum_{j=1}^{4} \kappa_{ij} u_j \qquad i = 1, 2, 3, 4 \tag{5.64}$$

all other $f_i^+, i > 4$ are zero.

4. Finally the system

$$\boldsymbol{K} \boldsymbol{u}_c = \boldsymbol{f} + \boldsymbol{f}^+ \tag{5.65}$$

is solved for the new displacement vector $\boldsymbol{u}_c$.

5.6.1 What It Means

When the (unknown) displacement vector $\boldsymbol{u}_c$ of the modified system

$$\boldsymbol{K}_c \, \boldsymbol{u}_c = \boldsymbol{f} \tag{5.66}$$

is substituted into the original system

$$\boldsymbol{K} \, \boldsymbol{u}_c = \boldsymbol{f} + \boldsymbol{f}^+ \tag{5.67}$$

then out pops the vector $\boldsymbol{f}^+ = \{f_1^+, f_2^+, f_3^+, f_4^+, 0, 0, \ldots, 0\}^T$, see (5.64) and signals that $\boldsymbol{u}_c$ is the response of the original system to "updated" equivalent nodal forces $\boldsymbol{f} + \boldsymbol{f}^+$. In mechanical term are the forces f_i^+ the coupling forces between the original structure and the added element Ω_e^+ with its added stiffness $\boldsymbol{\Delta K}_e$. The modifications $\boldsymbol{K}_e \to \boldsymbol{K}_e + \boldsymbol{\Delta K}_e$ in an element Ω_e can be thought of—so we would explain it to a layman—as attaching an additional element Ω_e^+ to the structure, see Fig. 5.3. This element is in size and shape identical with Ω_e and it is connected to Ω_e at the nodes where the two elements interchange the forces $\pm f_i^+$. And the reaction of the original structure $\boldsymbol{K}$ to these nodal forces $\boldsymbol{f}^+$ plus the original forces $\boldsymbol{f}$ is the displacement vector $\boldsymbol{u}_c$.

Because of $\boldsymbol{K} \, \boldsymbol{u} = \boldsymbol{f}$ and $\boldsymbol{e} = \boldsymbol{u}_c - \boldsymbol{u}$ is (5.67) equivalent to

$$\boldsymbol{K} \, \boldsymbol{e} = \boldsymbol{f}^+ \tag{5.68}$$

which corresponds to the variational problem for the displacement increment $e = u_c - u$ due to a change in the coefficients of the differential equation

$$a(e, \varphi_i) = -d(u_c, \varphi_i) \qquad \forall \, \varphi_i \tag{5.69}$$

with the switch

$$-\, d(u_c, \varphi_i) = (f_i^+, \varphi_i) \tag{5.70}$$

from virtual interior energy to virtual exterior work, see Sect. 5.8.1. The vector $\boldsymbol{f}^+$ is then called the "pseudo-load vector" corresponding to the shift $\boldsymbol{K} \to \boldsymbol{K} + \boldsymbol{\Delta K}$.

Remark 5.3 The added stiffness $\boldsymbol{\Delta K}_e$ can also be negative. Mathematically there is no difference between adding or removing a stiffness.

Example 5.2 Imagine the uppermost (bilinear) element of the plate in Fig. 5.8a near the support cracks, $\boldsymbol{K}_e \to \boldsymbol{K}_e + \boldsymbol{\Delta K}_e$. The displacement vector of the cracked model is the solution of the system

$$\boldsymbol{K} \, \boldsymbol{u}_c = \boldsymbol{f} + \boldsymbol{f}^+ \tag{5.71}$$

where the non-zero components of the vector $\boldsymbol{f}^+$ are the equivalent nodal forces f_i^+, see Fig. 5.8b, which act at the nodes of the cracked element.

The forces f_i^+ are calculated as follows:

1. $\boldsymbol{K} \, \boldsymbol{u} = \boldsymbol{f}$ is solved for the displacement vector $\boldsymbol{u}$.
2. The 4×4 matrix

$$\boldsymbol{F}_{e+} = \boldsymbol{F}_e + \boldsymbol{\Delta K}_e^{-1} \tag{5.72}$$

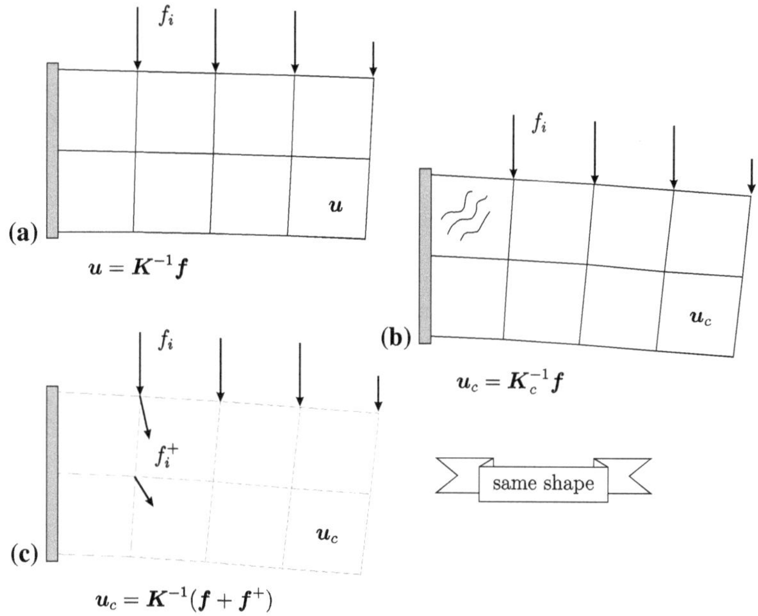

Fig. 5.8 The uncracked model assumes the shape $\boldsymbol{u}_c$ of the cracked model if additional nodal forces f_i^+ are applied

is formed, (the entries of $\boldsymbol{F}_e$ are calculated as in (5.63)), it is inverted, $\boldsymbol{F}_{e+}^{-1} = [\kappa_{ij}]$ and the f_i^+ are calculated

$$f_i^+ = -\sum_{j=1}^{4} \kappa_{ij} u_j \tag{5.73}$$

3. Finally

$$\boldsymbol{K}\,\boldsymbol{u}_c = \boldsymbol{f} + \boldsymbol{f}^+ \tag{5.74}$$

is solved for $\boldsymbol{u}_c$.

5.6.2 The Inverse of $\boldsymbol{\Delta K}_e$

Most often the matrix $\boldsymbol{\Delta K}_e = -\alpha\,\boldsymbol{K}_e$ is a multiple $(0 < |\alpha| \leq 1)$ of the singular element matrix $\boldsymbol{K}_e$, and so it is not invertible.

5.6.2.1 Adding Springs

One possible remedy would be to attach small springs to the ends of the element which would eliminate the rigid body motions and render the stiffness matrix of the element, say, a bar element

$$\boldsymbol{K}_e^+ = \frac{EA}{l_e}\begin{bmatrix} 1+\varepsilon & -1 \\ -1 & 1+\varepsilon \end{bmatrix} \qquad 0 < \varepsilon \tag{5.75}$$

invertible. Of course this is a delicate maneuver and it gives only approximate results but if the focus is more on the sensitivity of a system than on sheer numbers, one may be justified in proceeding this way. It may be good enough to find the directions into which a system evolves.

5.6.2.2 Two Step Approach

An accurate procedure is the following where the calculation is split into two steps.

1. First springs $\varepsilon > 0$ are added to the 4×4 element matrix $\boldsymbol{K}_e \to \boldsymbol{K}_e^+$

$$k_{11}^+ = k_{11} + \varepsilon \qquad k_{33}^+ = k_{33} + \varepsilon, \tag{5.76}$$

all other values remain the same, $k_{ij}^+ = k_{ij}$, to make $\boldsymbol{\Delta K}_e$ invertible

$$\boldsymbol{\Delta K}_e^+ = -\alpha\, \boldsymbol{K}_e^+. \tag{5.77}$$

Next $\boldsymbol{\Delta K}_e^+$ is inverted and the 4×4 flexibility matrix $\boldsymbol{F}^{(1)}$ is calculated

$$\boldsymbol{F}_{e+}^{(1)} := \boldsymbol{F}_e + (\boldsymbol{\Delta K}_e^+)^{-1} \tag{5.78}$$

and its inverse

$$(\boldsymbol{F}_{e+}^{(1)})^{-1} = [\kappa_{ij}^{(1)}]. \tag{5.79}$$

(The matrix $\boldsymbol{F}_e$ is, because of our numbering scheme, the upper 4×4 block in the inverse $\boldsymbol{K}^{-1}$).
This will provide the four vectors

$$\boldsymbol{v}_i^{(1)} = -\sum_{j=1}^{4} \kappa_{ij}^{(1)}\, \boldsymbol{c}_j \qquad i = 1, 2, 3, 4 \tag{5.80}$$

which allow to calculate the intermediate solution

$$\boldsymbol{u}^{(1)} = \boldsymbol{u} - \sum_{i=1}^{4} u_i \, \boldsymbol{v}_i^{(1)} = \boldsymbol{u} - \sum_{i,j=1}^{4} \kappa_{ij}^{(1)} \, u_i \, \boldsymbol{c}_j \tag{5.81}$$

and the first four columns, $i = 1, 2, 3, 4,$

$$\boldsymbol{c}_i^{(1)} = \boldsymbol{c}_i - \sum_{j=1}^{4} c_{ij} \, \boldsymbol{v}_j^{(1)} = \boldsymbol{c}_i - \sum_{j,k=1}^{4} c_{ik} \, \kappa_{kj}^{(1)} \, \boldsymbol{c}_j \tag{5.82}$$

of the inverse $(\boldsymbol{K}_c^{(1)})^{-1}$ of the intermediate stiffness matrix

$$\boldsymbol{K}_c^{(1)} = \boldsymbol{K} + \boldsymbol{\Delta K}_e^{+}. \tag{5.83}$$

These columns will be needed in the following step. (The entries c_{ij} are the elements of the inverse $\boldsymbol{K}^{-1}$).

2. In the next step the spring elements at the end points are removed, that is $\boldsymbol{K}_c^{(1)}$ is modified by adding a correction matrix $\boldsymbol{\Delta K}^{(2)}$

$$\boldsymbol{K}_c = \boldsymbol{K} + \boldsymbol{\Delta K}_e^{+} + \boldsymbol{\Delta K}^{(2)} = \boldsymbol{K}_c^{(1)} + \boldsymbol{\Delta K}^{(2)} \tag{5.84}$$

with only two non-zero elements on the diagonal,

$$\Delta k_{33}^{(2)} = \Delta k_{11}^{(2)} = -\varepsilon \tag{5.85}$$

and so the flexibility matrix

$$\boldsymbol{F}_{e+}^{(2)} := \boldsymbol{F}_e^{(1)} + (\boldsymbol{\Delta K}^{(2)})^{-1} \tag{5.86}$$

($\boldsymbol{F}_0^{(1)}$ is the upper 4×4 block in the inverse of $\boldsymbol{K}_c^{(1)}$) is invertible

$$(\boldsymbol{F}_{e+}^{(2)})^{-1} = [\kappa_{ij}^{(2)}] \tag{5.87}$$

and the four vectors

$$\boldsymbol{v}_i^{(2)} = \sum_{j=1}^{4} \kappa_{ij}^{(2)} \, \boldsymbol{c}_j^{(1)} \qquad i = 1, 2, 3, 4 \tag{5.88}$$

can be calculated to produce the final solution

$$\boldsymbol{u}_c = \boldsymbol{u}^{(1)} - \sum_{i=1}^{4} u_i^{(1)} \, \boldsymbol{v}_i^{(2)} = \boldsymbol{u}^{(1)} - \sum_{i,j=1}^{4} \kappa_{ij}^{(2)} \, u_i^{(1)} \, \boldsymbol{c}_j^{(1)}. \tag{5.89}$$

Whether this procedure is advantageous depends on the circumstances. In any case it is a local procedure. Regardless of how large $\boldsymbol{K}$ is, we only have to invert twice a 4×4 matrix and calculate the 2×4 columns $\boldsymbol{c}_i$ and $\boldsymbol{c}_i^{(1)}$, which are the first four columns in the inverses of the matrices $\boldsymbol{K}^{-1}$ and $\boldsymbol{K}_c^{(1)}$ (because of our numbering scheme).

The first set $\boldsymbol{c}_i, i = 1, 2, 3, 4$ could be calculated by applying unit forces $f_i = 1$ at the pertinent nodes. The response of the structure to these four forces are the columns $\boldsymbol{c}_i$. The second set can be retrieved from the first set via the formula

$$\boldsymbol{c}_i^{(1)} = \boldsymbol{c}_i - \sum_{j=1}^{4} c_{ij}\, \boldsymbol{v}_j^{(1)} = \boldsymbol{c}_i - \sum_{j=1}^{4} c_{ij} \sum_{k=1}^{4} \kappa_{jk}^{(1)}\, \boldsymbol{c}_k \qquad i = 1, \ldots, 4. \tag{5.90}$$

5.6.2.3 Notation

The symbol

$$\boldsymbol{E}_\kappa = \boldsymbol{0}_{(n\times n)} + \left[\kappa_{ij}\right] \tag{5.91}$$

denotes a matrix which contains the matrix κ_{ij} as a single block (or eventually sprinkled over $\boldsymbol{E}_\kappa$ if the indices are not contiguous) but otherwise is empty. In the following we will use two matrices

$$\boldsymbol{E}_{\kappa^{(1)}} = \boldsymbol{0}_{(n\times n)} + \left[\kappa_{ij}^{(1)}\right] \qquad \boldsymbol{E}_{\kappa^{(2)}} = \boldsymbol{0}_{(n\times n)} + \left[\kappa_{ij}^{(2)}\right] \tag{5.92}$$

of this kind. The formula (5.81) for $\boldsymbol{u}^{(1)}$ becomes in this notation

$$\boldsymbol{u}^{(1)} = \boldsymbol{u} - \boldsymbol{E}_{\kappa^{(1)}}\, \boldsymbol{C}\, \boldsymbol{u} \qquad \boldsymbol{C} = \boldsymbol{K}^{-1} \tag{5.93}$$

and the formula (5.82) for the intermediate inverse

$$\boldsymbol{C}^{(1)} = \boldsymbol{C} - \boldsymbol{C}\, \boldsymbol{E}_{\kappa^{(1)}}\, \boldsymbol{C} \tag{5.94}$$

and the formula (5.89) for the final product

$$\boldsymbol{u}_c = \boldsymbol{u}^{(1)} - \boldsymbol{E}_{\kappa^{(2)}}\, \boldsymbol{C}^{(1)} \boldsymbol{u}^{(1)}. \tag{5.95}$$

5.6.3 Example

The rod in Fig. 5.9 may serve as a test example. The stiffness matrix of the rod with a uniform cross section is

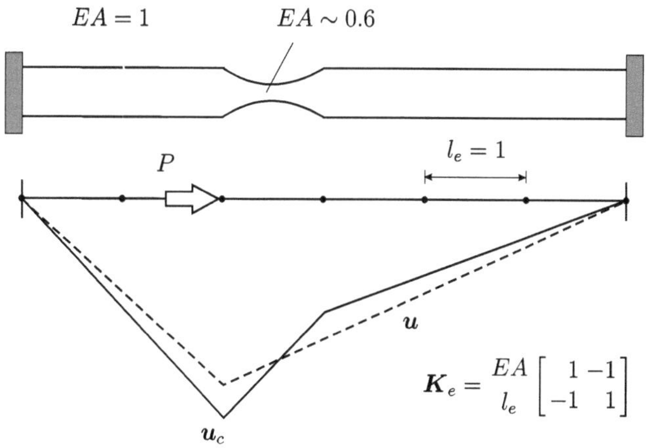

Fig. 5.9 Horizontal displacement of the rod; approximate analysis with linear elements; *dashed line* is result for uniform rod

$$
\boldsymbol{K} = \begin{bmatrix} 2 & -1 & 0 & 0 & 0 \\ -1 & 2 & -1 & 0 & 0 \\ 0 & -1 & 2 & -1 & 0 \\ 0 & 0 & -1 & 2 & -1 \\ 0 & 0 & 0 & -1 & 2 \end{bmatrix} \qquad \boldsymbol{K}^{-1} = \frac{1}{6}\begin{bmatrix} 5 & 4 & 3 & 2 & 1 \\ 4 & 8 & 6 & 4 & 2 \\ 3 & 6 & 9 & 6 & 3 \\ 2 & 4 & 6 & 8 & 4 \\ 1 & 2 & 3 & 4 & 5 \end{bmatrix} \tag{5.96}
$$

and the displacement vector is

$$
\boldsymbol{u} = \boldsymbol{K}^{-1}\begin{bmatrix} 0 \\ 1 \\ 0 \\ 0 \\ 0 \end{bmatrix} = \begin{bmatrix} 0.6667 \\ 1.3333 \\ 1.0000 \\ 0.6667 \\ 0.3333 \end{bmatrix}. \tag{5.97}
$$

The single element with its concave contraction is modeled with a single element which has only 40 % of the longitudinal stiffness of a standard element, so that $\boldsymbol{\Delta K}_e = -0.6\,\boldsymbol{K}_e$.

5.6.3.1 Springs

To make the matrix $\boldsymbol{\Delta K}_e$ invertible an extra term $\varepsilon = 0.01$, which corresponds to 1 % of the stiffness of K_e, is added on the diagonal

$$
\boldsymbol{\Delta K}_e^{+} = -0.6\begin{bmatrix} 1 & -1 \\ -1 & 1 \end{bmatrix} + \varepsilon\,\boldsymbol{I}_{(2\times 2)} \qquad (\boldsymbol{\Delta K}_e^{+})^{-1} = -\begin{bmatrix} 49.58 & 50.42 \\ 50.42 & 49.58 \end{bmatrix}.
$$

The flexibility matrix and its inverse are

$$\boldsymbol{F}_{e+} = \frac{1}{6}\begin{bmatrix} 8 & 6 \\ 6 & 9 \end{bmatrix} - \begin{bmatrix} 49.58 & 50.42 \\ 50.42 & 49.58 \end{bmatrix} \quad \boldsymbol{F}_{e+}^{-1} = \begin{bmatrix} -1.1770 & 1.1849 \\ 1.1849 & -1.1732 \end{bmatrix} \tag{5.98}$$

and $-\boldsymbol{F}_{e+}^{-1}$ is multiplied with the nodal displacements u_2 and u_3 to retrieve the additional forces f_i^+

$$\boldsymbol{f}^+ = -\boldsymbol{F}_{e+}^{-1}\begin{bmatrix} u_2 \\ u_3 \end{bmatrix} = \begin{bmatrix} 0.3845 \\ -0.4066 \end{bmatrix} \tag{5.99}$$

which are added to the nodal forces $\boldsymbol{f}$ to produce the (approximate) new vector $\boldsymbol{u}_c^{\sim}$

$$\boldsymbol{u}_c^{\sim} = \boldsymbol{K}^{-1}(\boldsymbol{f} + \boldsymbol{f}^+) = \begin{bmatrix} 0.7197 \\ 1.4394 \\ 0.7745 \\ 0.5164 \\ 0.2582 \end{bmatrix} \qquad \boldsymbol{u}_c = \begin{bmatrix} 0.7333 \\ 1.4667 \\ 0.8000 \\ 0.5333 \\ 0.2667 \end{bmatrix} \tag{5.100}$$

which is quite close to the exact solution $\boldsymbol{u}_c$.

5.6.3.2 Two Step Approach

In the two-step approach we would now calculate the columns $\boldsymbol{c}_2^{(1)}$ and $\boldsymbol{c}_3^{(1)}$ of the inverse $\boldsymbol{C}^{(1)} = (\boldsymbol{K}_c^{(1)})^{-1}$ and form the second flexibility matrix

$$\begin{aligned} \boldsymbol{F}_{e+}^{(2)} &= \boldsymbol{F}_{e+}^{(1)} + (-\varepsilon\, \boldsymbol{I})^{-1} = \boldsymbol{F}_{e+}^{(1)} - \frac{1}{\varepsilon}\boldsymbol{I} \\ &= \begin{bmatrix} 1.4394 & 0.7745 \\ 0.7745 & 1.7621 \end{bmatrix} - \frac{1}{0.01}\begin{bmatrix} 1 & 0 \\ 0 & 1 \end{bmatrix} = \begin{bmatrix} -98.5606 & 0.7745 \\ 0.7745 & -98.2379 \end{bmatrix} \end{aligned} \tag{5.101}$$

calculate its inverse

$$(\boldsymbol{F}_{e+}^{(2)})^{-1} = [\kappa_{ij}^{(2)}] = \begin{bmatrix} -0.0101 & -0.0001 \\ -0.0001 & -0.01002 \end{bmatrix} \tag{5.102}$$

and calculate the final solution

$$\boldsymbol{u}_c = \boldsymbol{u}_c^{\sim} - \boldsymbol{E}_\kappa^{(2)}\boldsymbol{C}^{(1)}\boldsymbol{u}_c^{\sim} = \begin{bmatrix} 0.7197 \\ 1.4394 \\ 0.7745 \\ 0.5164 \\ 0.2582 \end{bmatrix} - \begin{bmatrix} -0.0137 \\ -0.0273 \\ -0.0255 \\ -0.0170 \\ -0.0085 \end{bmatrix} = \begin{bmatrix} 0.7333 \\ 1.4667 \\ 0.8000 \\ 0.5333 \\ 0.2667 \end{bmatrix}$$

which is exact.

5.6.3.3 Accuracy

With a spring constant of $\varepsilon = 0.01$ the largest relative error in the spring model is about 3 %, with $\varepsilon = 0.001$ it is 0.3 % and with $\varepsilon = 0.0001$ it becomes 0.03 %.

For practical purposes the second step therefore may not be needed if ε is chosen to be sufficiently small. Of course there is always the danger that ε is too small and that the system blows up, becomes singular, but we would leave the choice of an appropriate ε to the engineer. Often an engineer in studying the response of a structure is not interested in pure numbers but more in the sensitivity of a structure, see Fig. 5.10, in answers to questions such as "what happens if?".

5.6.3.4 Technical Remark

The technique is also applicable to non-symmetric matrices $\boldsymbol{K}$ and $\boldsymbol{\Delta K}$ and also to single changes in off-diagonal terms, say $k_{23} \rightarrow k_{23} + \Delta k_{23}$, though in the latter case there is the handicap that the matrix

$$\boldsymbol{\Delta K} = \begin{bmatrix} 0 & \Delta k_{23} \\ 0 & 0 \end{bmatrix} \tag{5.103}$$

is no longer invertible. In this situation the starting point would be a slightly shifted matrix

$$\boldsymbol{\Delta K}^{+} = \begin{bmatrix} 0.01 & \Delta k_{23} \\ 0 & 0.01 \end{bmatrix} = \boldsymbol{\Delta K} + 0.01\,\boldsymbol{I}_{(2\times 2)} \tag{5.104}$$

and in the second step the extra terms would be removed again. This procedure provides the exact result, though it was often found that for engineering purposes the second step was not necessary.

5.6.4 Collapse

If the flexibility matrix

$$\boldsymbol{F}_{e+} = \boldsymbol{F}_e + \boldsymbol{\Delta K}_e^{-1} \tag{5.105}$$

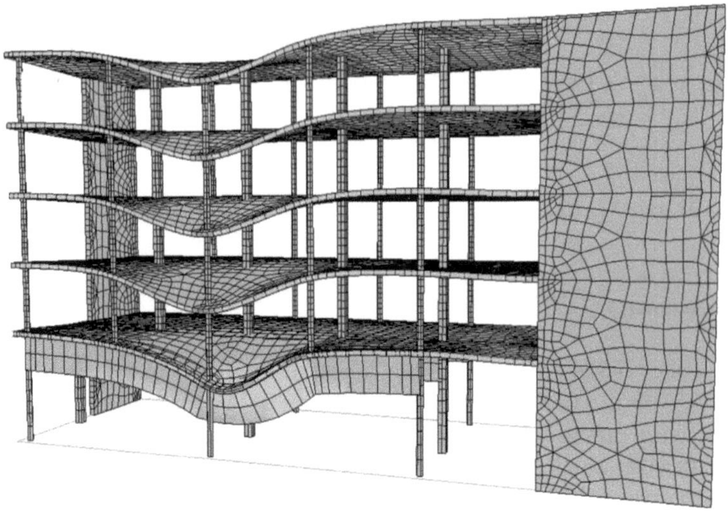

Fig. 5.10 Influence function for the bending moment in a girder

is not invertible then the modifications in the element matrix, $\boldsymbol{K}_e \rightarrow \boldsymbol{K}_e + \Delta \boldsymbol{K}_e$, are too excessive, the changes $\boldsymbol{\Delta K}_e^{-1}$ to $\boldsymbol{F}_e$ render the system—or at least a part of it—unstable.

Note that the removal of an element, $\boldsymbol{\Delta K}_e = -\boldsymbol{K}_e$, does not render $\boldsymbol{F}_e$ singular because $\boldsymbol{F}_e$ is "more" than the inverse of $\boldsymbol{K}_e^{-1}$ alone (again assumed that $\boldsymbol{K}_e^{-1}$ exists).

5.7 Functionals

What is the effect of local changes in an element stiffness on displacements or stresses? Recall that there are two ways to evaluate a functional

$$J(\boldsymbol{u}) = \begin{cases} \boldsymbol{j}^T \boldsymbol{u} \\ \boldsymbol{g}^T \boldsymbol{f} \end{cases} . \tag{5.106}$$

So one can either set up and solve the modified system

$$(\boldsymbol{K} + \boldsymbol{\Delta K})\, \boldsymbol{u}_c = \boldsymbol{f} \tag{5.107}$$

for the new vector $\boldsymbol{u}_c$—how this can be done with as little effort as possible have we discussed above—and simply compare the results

$$J(\boldsymbol{e}) = J(\boldsymbol{u}_c) - J(\boldsymbol{u}) = (\boldsymbol{u}_c - \boldsymbol{u})^T \boldsymbol{j} \tag{5.108}$$

or solve the modified system

$$(\boldsymbol{K} + \boldsymbol{\Delta K})\, \boldsymbol{g}_c = \boldsymbol{j} \tag{5.109}$$

for the new Green's function vector $\boldsymbol{g}_c$ and compare the results

$$J(\boldsymbol{e}) = J(\boldsymbol{u}_c) - J(\boldsymbol{u}) = (\boldsymbol{g}_c - \boldsymbol{g})^T\, \boldsymbol{f}. \tag{5.110}$$

But the force method offers an alternative formulation to the last formula. Instead of modifying $\boldsymbol{g}$ we modify the right-hand side $\boldsymbol{f}$

$$J(\boldsymbol{e}) = J(\boldsymbol{u}_c) - J(\boldsymbol{u}) = \boldsymbol{g}^T(\boldsymbol{f} + \boldsymbol{f}^+) - \boldsymbol{g}^T\, \boldsymbol{f} = \boldsymbol{g}^T\, \boldsymbol{f}^+ \tag{5.111}$$

The elements of the force vector

$$\boldsymbol{f}^+ = -\boldsymbol{F}_{e+}^{-1}\, \boldsymbol{u} \tag{5.112}$$

are the unknowns $X_i = f_i^+$ in the force method which provide the fit between the nodes of the structure and the nodes of the attached element Ω_e^+; they close, so to speak, the gap between Ω_e^+ and the structure and so they come in pairs, see Fig. 5.3.

According to (5.13) the value $J(\boldsymbol{e})$ can also be expressed as

$$J(\boldsymbol{e}) = -\boldsymbol{g}^T\, \boldsymbol{\Delta K}_e\, \boldsymbol{u}_c \tag{5.113}$$

where $\boldsymbol{\Delta K}_e$ is the modification to the element matrix. This implies that

$$\boldsymbol{F}_{e+}^{-1}\, \boldsymbol{u} = \boldsymbol{\Delta K}_e\, \boldsymbol{u}_c \tag{5.114}$$

5.7.1 The Gradient of a Functional

To predict the new value $J(\boldsymbol{u}_c)$ of a functional one needs to know either the vector $\boldsymbol{u}_c$ or the vector $\boldsymbol{g}_c$ or the vector $\boldsymbol{f}^+$

$$J(\boldsymbol{u}_c) = \boldsymbol{u}_c^T\, \boldsymbol{j} = \boldsymbol{g}_c^T\, \boldsymbol{f} = J(\boldsymbol{u}) + \boldsymbol{g}^T\, \boldsymbol{f}^+. \tag{5.115}$$

But the gradient $\nabla J(\boldsymbol{u})$ of a functional, the direction into which a functional will evolve, instead can be calculated exactly at the current position, that is with the vectors $\boldsymbol{u}$ and $\boldsymbol{g}$. In sensitivity analysis this is known as the *adjoint method of sensitivity analysis*. We will later discuss this in more details, see Sect. 5.14. Here we only want to give the main result because with the gradient one can do a first order Taylor analysis of $J(\boldsymbol{u})$

$$J(\boldsymbol{u}_c) = J(\boldsymbol{u}) + \nabla J(\boldsymbol{u})\, \boldsymbol{d}\, \boldsymbol{p} + \cdots \tag{5.116}$$

which corresponds to a linearization of the functional.

Let $\boldsymbol{p}$ a vector of m design variables. In frame analysis this would be the list of the bending stiffness EI_i of the single beam elements

$$\boldsymbol{p} = \{EI_1, EI_2, \ldots, EI_m\}^T . \tag{5.117}$$

The functionals then depend on $\boldsymbol{u}$ and $\boldsymbol{p}$ and evaluating a functional can be seen as evaluating $J(\boldsymbol{u}, \boldsymbol{p})$ under a certain side condition

$$functional \quad J(\boldsymbol{u}, \boldsymbol{p}) \quad side\ condition \quad \boldsymbol{K}(\boldsymbol{p})\,\boldsymbol{u} = \boldsymbol{f} . \tag{5.118}$$

The vector $\boldsymbol{f}$ usually does not depend on $\boldsymbol{p}$ and so the equation $\boldsymbol{K}(\boldsymbol{p})\,\boldsymbol{u} = \boldsymbol{f}$ implies that

$$\boldsymbol{K},_{p_i}\,\boldsymbol{u} + \boldsymbol{K}\,\boldsymbol{u},_{p_i} = \boldsymbol{0} \qquad i = 1, 2, \ldots, m \tag{5.119}$$

and in the same sense $J(\boldsymbol{u}, \boldsymbol{p}) = \boldsymbol{u}^T\,\boldsymbol{j}$ implies that

$$J,_{p_i} = \boldsymbol{u},_{p_i}^T\,\boldsymbol{j} = \boldsymbol{u},_{p_i}^T\,\boldsymbol{K}\,\boldsymbol{g} = \boldsymbol{g}^T\,\boldsymbol{K}\,\boldsymbol{u},_{p_i} = -\boldsymbol{g}^T\,\boldsymbol{K},_{p_i}\,\boldsymbol{u}. \tag{5.120}$$

So that the expression

$$J(\boldsymbol{e}) = J(\boldsymbol{u}_c) - J(\boldsymbol{u}) \simeq \sum_i J,_{p_i}\,dp_i = -\sum_i \boldsymbol{g}^T\,\boldsymbol{K},_{p_i}\,dp_i\,\boldsymbol{u} \tag{5.121}$$

is a first-order approximation to $J(\boldsymbol{e})$. It should not be too difficult to see that this is an approximation of the exact equation (5.113) when we allow the change in only one p_i and read $\boldsymbol{\Delta K}_e = \boldsymbol{K},_{p_i}\,dp_i$.

5.8 Weak Formulations and the d-Term

We started this chapter with linear algebra, that is we started directly with the finite element formulation and we studied how modifications of a stiffness matrix lead to changes in the solution vector and how functionals are affected by these modifications. In the following we want to do a step back and study these modifications and the consequences they entail by looking at the original weak formulation of the boundary value problem. How does a weak formulation react to changes in the coefficients of a differential equation?

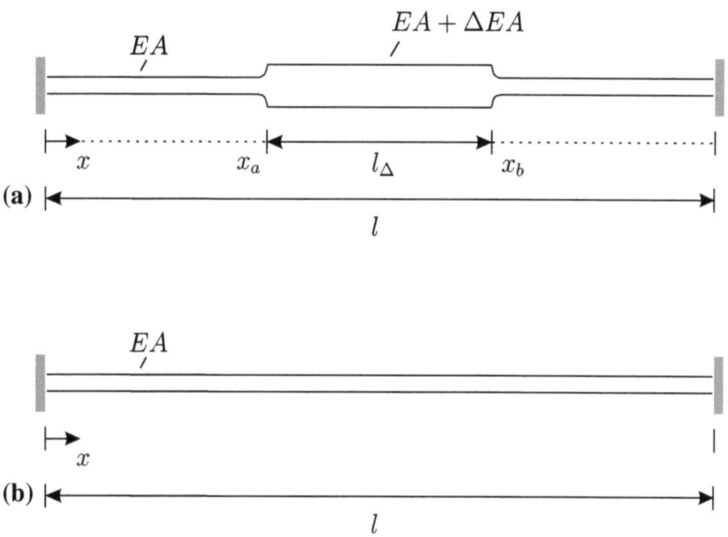

Fig. 5.11 Stepped bar. **a** Exact model; **b** simplified model

5.8.1 Linear Problems

In a stepped bar as in Fig. 5.11 the coefficient EA in the differential equation for the longitudinal displacement of the bar

$$-\,EA\,u'' = p \tag{5.122}$$

is only piecewise constant. We want to study which effect such a local modification of the coefficient has on the solution of the differential equation.

To this end we consider two equations: the base model with the uniform stiffness EA

$$-\,EA\,u'' = p \tag{5.123}$$

and the stepped bar

$$-\,EA_c\,u_c'' = p \tag{5.124}$$

with the piecewise constant coefficient

$$EA_c = \begin{cases} EA & x \notin [x_a, x_b] \\ EA + \Delta EA & x \in [x_a, x_b] \end{cases} . \tag{5.125}$$

The bilinear form of the original model is

$$a(u, v) = \int_0^l EA\, u'\, v'\, dx \tag{5.126}$$

while the bilinear form of the modified problem

$$\begin{aligned} a_c(u, v) &= \int_0^l EA_c\, u'\, v'\, dx = \int_0^l (EA + \Delta EA)\, u'\, v'\, dx \\ &= \int_0^l EA\, u'\, v'\, dx + \int_{x_a}^{x_b} \Delta EA\, u'\, v'\, dx \\ &= a(u, v) + d(u, v), \end{aligned} \tag{5.127}$$

contains an additional symmetric and bilinear term

$$d(u, v) = \int_{x_a}^{x_b} \Delta EA\, u'\, v'\, dx. \tag{5.128}$$

Consequently the difference between the original weak problem

$$a(u, v) = (p, v) \qquad \forall\, v \in \mathcal{V} \tag{5.129}$$

and the weak form of the modified problem

$$a(u_c, v) + d(u_c, v) = (p, v) \quad \forall\, v \in \mathcal{V} \tag{5.130}$$

is

$$a(u_c - u, v) + d(u_c, v) = 0 \quad \forall\, v \in \mathcal{V} \tag{5.131}$$

or if $e := u_c - u$ is introduced as the modeling error

$$a(e, v) = -d(u_c, v) \quad \forall\, v \in \mathcal{V}. \tag{5.132}$$

This is the weak formulation for the modeling error e. The right-hand side is an expression of internal energy which is of course equivalent to an expression of external virtual energy. To find this expression we note that to the differential operator $-\Delta EA\, u''$ belongs on the interval $[x_a, x_b]$, where ΔEA lives, the identity

$$\mathcal{G}\,(u_c, v) = \underbrace{\int_{x_a}^{x_b} -\Delta E A\, u_c''\, v\, dx + [\Delta E A_c\, u_c'\, v]_{x_a}^{x_b}}_{\text{ext. energy}} - d(u_c, v) = 0 \tag{5.133}$$

or

$$\mathcal{G}\,(u_c, v) = (-f^+, v) - d(u_c, v) = 0 \tag{5.134}$$

for short where[1]

$$(-f^+, v) := \int_{x_a}^{x_b} -\Delta E A\, u_c''\, v\, dx + [\Delta E A_c\, u_c'\, v]_{x_a}^{x_b} \tag{5.135}$$

so that (5.132) is the same as

$$a(e, v) = (f^+, v) \quad \forall v \in \mathcal{V} \tag{5.136}$$

or in the FE-context

$$\boldsymbol{K e} = \boldsymbol{f}^+ \tag{5.137}$$

where $f_i^+ = (f^+, \varphi_i)$. The equivalent nodal forces f_i^+ are exactly the forces which we encountered in the force method, see (5.68). Here the vector $\boldsymbol{f}^+$ is called the "pseudo-load vector".

5.8.2 The Error in Functionals

Next let us study how we can assess the changes in typical functionals such as

$$J(u) = u(x) \qquad J(u) = \sigma(x) \tag{5.138}$$

due to the change in the coefficients of a differential equation. While functionals for displacement terms do not change with a change in the coefficients, functionals for force terms, as $J(u) = \sigma(x)$ do change

$$J(u) = \sigma(x) = E\, u'(x) \qquad \rightarrow \qquad J_c(u) = \sigma_c(x) = E_c\, u'(x). \tag{5.139}$$

We therefore first concentrate on functionals of the type $J(u) = u(x)$.

The important point to note is that—regardless of what happens to the coefficients—the loading p does not change. This implies that a shake test with a virtual displacement v—before and after any modification of the coefficients—

[1] The negative sign in $(-f^+, v)$ allows to write $-d(u_c, v) = (f^+, v)$.

results in the *same* virtual exterior work (p, v) and so also the virtual interior work must be the same as before

$$\delta W_i(u_c, v) = a(u_c, v) + d(u_c, v) = (p, v) = \delta W_e(p, v) \tag{5.140}$$
$$\delta W_i(u, v) = a(u, v) = (p, v) = \delta W_e(p, v), \tag{5.141}$$

which means that

$$\delta W_i(u_c, v) - \delta W_i(u, v) = a(u_c, v) + d(u_c, v) - a(u, v) = 0 \tag{5.142}$$

or

$$a(e, v) = -d(u_c, v) \qquad e = u_c - u. \tag{5.143}$$

Now we need only to substitute for v the Green's function G

$$J(e) = a(e, G) = -d(u_c, G) \tag{5.144}$$

and we have found the central equation for quantifying the modeling error in a functional

$$J(e) = -d(u_c, G)\,. \tag{5.145}$$

Two points are noteworthy: (i) this formula is a weak influence function (a *divergence* form of an influence function) and (ii) the strain energy product, the integral, only extends over the patch Ω_p of the domain, the part where $EA \to EA + \Delta EA$.

We add a third remark: the subscript c can be swapped without changing the result

$$J(e) = a(e, G_c) + d(e, G_c) = -d(u_c, G_c) + d(e, G_c) = -d(u, G_c), \tag{5.146}$$

that is both formulations, (5.145) and (5.146), are equivalent. Here we have used (5.143) with $v = G_c$.

More formulations are possible, we only mention that the two equations

$$a(u_c, G_c) + d(u_c, G_c) = J(u_c) \tag{5.147}$$
$$a(u, G) = J(u), \tag{5.148}$$

where $J(u)$ is assumed to be a displacement functional, and

$$a(u_c, G_c) = a(e, G_c) + a(u, G_c) = -d(u_c, G_c) + a(u, G_c) \tag{5.149}$$

imply that

$$a(u, G_c) = J(u_c) \tag{5.150}$$

which is equivalent to

$$\boldsymbol{g}_c^T \boldsymbol{K} \boldsymbol{u} = \boldsymbol{g}_c^T \boldsymbol{f} = J(u_c) \tag{5.151}$$

in which form the result looks more familiar.

5.8.2.1 Weak Influence Functions

With regard to functionals for force terms such as $J(u) = \sigma(x)$ the basic situation is that only strong formulations (Betti)

$$J_c(u_c) - J(u) = \sigma_c(x) - \sigma(x) = \int_0^l (G_c(y,x) - G(y,x))\, p(y)\, dy \qquad (5.152)$$

allow to quantify effects while weak formulations return zero

$$a(u_c, G_c) + d(u_c, G_c) = 0 \neq \sigma_c(x) \qquad (5.153)$$

$$a(u, G) = 0 \neq \sigma(x) \qquad (5.154)$$

and not $\sigma_c(x)$ or $\sigma(x)$; here the two Green's function G_c and G represent the necessary dislocations.

But we have the interesting result that we can apply (5.145) also to force terms. Before we give a proof we give an example.

5.8.2.2 Example

The non-uniform continuous beam in Fig. 5.12 which is subjected to a constant lateral load p is chosen as model problem. The bending stiffness in the first span is $EI_c = 90{,}626\,\text{kNm}^2$ and in the second span it is half that value, $EI = 45{,}313\,\text{kNm}^2$.

In an engineering handbook the bending moment at the center of the first span is listed as $M(x) = -7.81\,\text{kNm}$ if the beam has a uniform stiffness $EI = 45{,}313\,\text{kNm}^2$. What is the real value of $M(x)$ given that the bending stiffness is not uniform?

The strain energies of the uniform and the non-uniform beam differ in the first span by the term

$$d(u, v) = \Delta EI \int_0^{l_1} u''\, v''\, dx \qquad \Delta EI = 45{,}313. \qquad (5.155)$$

Therefore the error in the bending moment is

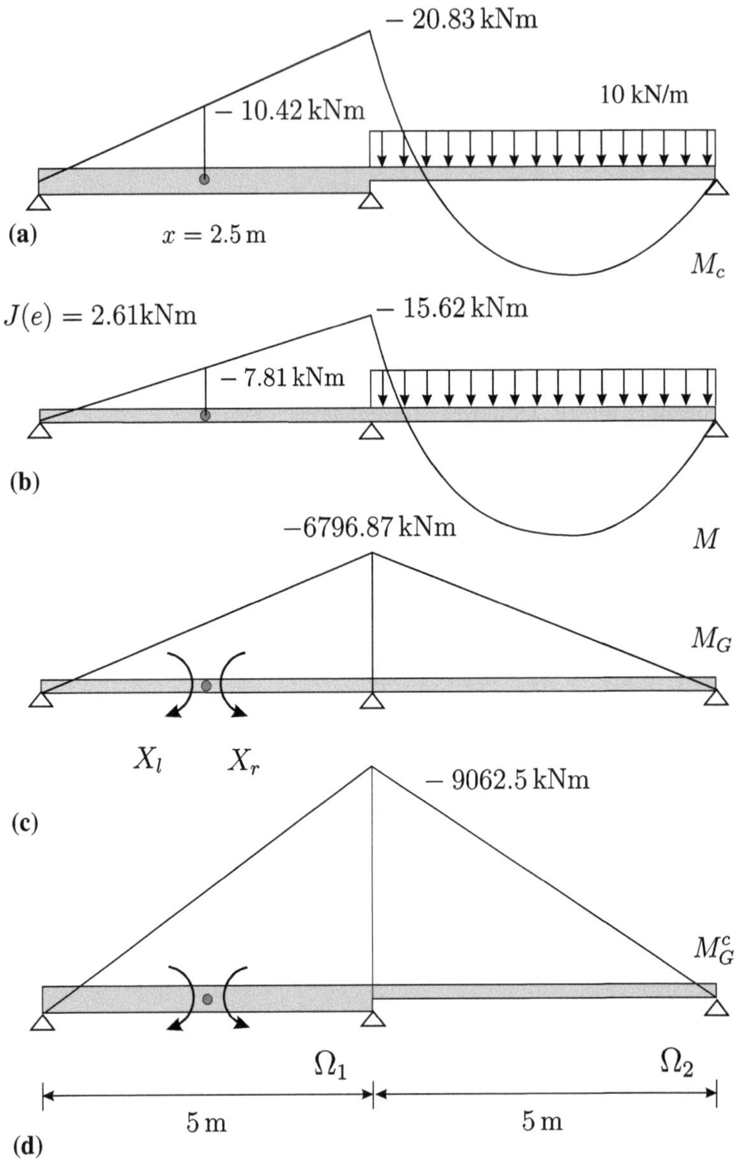

Fig. 5.12 Continuous beam. **a** Bending moments of the non-uniform beam and **b** the uniform beam; **c** bending moment of the Green's function G (uniform beam) and **d** non-uniform beam

$$J(e) = M_c(x) - M(x) = -d(u, G_c) = -\Delta EI \int_0^{l_1} u''\, G_c''\, dx$$

$$= -\Delta EI \int_0^5 \frac{M}{EI}\,\frac{M_G^c}{EI_c} dx = -2.61\,\mathrm{kNm} \qquad (5.156)$$

so that the true value $M_c(x)$ of the bending moment is

$$M_c(x) = M(x) + J(e) = -7.81 - 2.61 = -10.42\,\text{kNm}. \tag{5.157}$$

which is the correct value. The function u in (5.156) is the deflection of the uniform beam and $G_c(y, x)$ is the Green's function for the bending moment $M_c(x)$ in the non-uniform beam.

Remark 5.4 For completeness we mention, see Fig. 5.12, that

$$a(u, G) = EI \int_0^{10} u''\, G''\, dy = \int_0^{10} \frac{M\, M_G}{EI}\, dy = 0 \tag{5.158}$$

$$a(u_c, G) = EI \int_0^{10} u_c''\, G''\, dy = \int_0^{5} \frac{M_c\, M_G}{EI_c}\, dy + \int_5^{10} \frac{M_c\, M_G}{EI}\, dy = 0 \tag{5.159}$$

with $M_c = -EI_c\, u_c''$ and $M_c = -EI\, u_c''$ in the first and the second span respectively while $M_G = -EI\, G''$ along the whole beam. If these two equations were not zero then the first derivatives, $u'(x)$ and $u_c'(x)$ respectively, were discontinuous at x. The two couples, $X_l = X_r = X$, on both sides of x, see Fig. 5.12c, would perform exterior virtual work δW_e

$$\delta W_e = X_r \cdot u'(x_+) - X_l \cdot u'(x_-) = X \cdot (u'(x_+) - u'(x_-)) \neq 0 \tag{5.160}$$

to which would correspond a non-vanishing virtual interior work $\delta W_e = \delta W_i = a(u, G)$ or $a(u_c, G)$ respectively.

The engineer determines the Green's function G for $M(x)$ as follows: he inserts a hinge at the point x, applies two moments $X_l = X_r = X$, first of size $X = 1$, finds the solution G and then he scales the X in such a way that the slope of the Green's function has the required discontinuity, $G_-' - G_+' = 1$, at x.

5.8.2.3 Is the Principle of Virtual Forces Applicable to Forces?

But why can *changes* in force terms such as $M(x)$ be calculated with weak influence functions when weak influence functions for force terms per se give no answer, return zero?

The difference is obviously that in (5.156) the integral only extends over a *part* of the structure.

Let us recap the formulation. The derivation of the equation

$$a(e, v) = -d(u_c, v) \tag{5.161}$$

rests on purely formal arguments (algebra). For all functions e, u_c and v which are sufficiently regular, which lie in the trial space $\mathcal{V}$, this identity is established.

Let G the Green's function for $J(u)(x) = M(x)$. The limit of the sequence

$$\lim_{\varepsilon \to 0} a(G, u)_{\Omega_\varepsilon} = 0 \tag{5.162}$$

where Ω_ε is the punctured domain—the hole gets smaller and smaller—is zero but if a sequence of smooth functions G_ε converges to G the limit of the integrals is

$$\lim_{\varepsilon \to 0} a(G_\varepsilon, u) = J(u)(x) \tag{5.163}$$

so that with $e = u_c - u$ instead of u also

$$\lim_{\varepsilon \to 0} a(G_\varepsilon, e) = J(e)(x). \tag{5.164}$$

Equation (5.161) holds true for any of the functions G_ε as well

$$a(e, G_\varepsilon) = -d(e, G_\varepsilon) \tag{5.165}$$

and therefore also the limits of the two sides must be the same

$$J(e)(x) = \lim_{\varepsilon \to 0} a(e, G_\varepsilon) = -\lim_{\varepsilon \to 0} d(u_c, G_\varepsilon) \tag{5.166}$$

and we may assume that

$$\lim_{\varepsilon \to 0} d(u_c, G_\varepsilon) = d(u_c, G) \tag{5.167}$$

so that the equation

$$J(e) = \lim_{\varepsilon \to 0} a(G_\varepsilon, e) = -\lim_{\varepsilon \to 0} d(u_c, G_\varepsilon) = -d(u_c, G) \tag{5.168}$$

is established.

What could spoil the assumption (5.167) is that $d(u_c, G)$ could be zero, in analogy to

$$\lim_{\varepsilon \to 0} a(u, G_\varepsilon) \neq a(u, G) = 0 \tag{5.169}$$

but $d(u_c, G)$ cannot be zero. We know that the integral over the full length of the two-span beam in Fig. 5.12 is zero, see (5.159),

$$a(u_c, G) = EI \int_0^l u_c''(y)\, G''(y, x)\, dx = 0 \tag{5.170}$$

or

$$
\begin{aligned}
0 = a(u_c, G)_{\Omega_1} + a(u_c, G)_{\Omega_2} &= \frac{\Delta EI}{EI}\, a(u_c, G)_{\Omega_1} + \frac{\Delta EI}{EI}\, a(u_c, G)_{\Omega_2} \\
&= d(u_c, G) + \frac{\Delta EI}{EI}\, a(u_c, G)_{\Omega_2} \qquad (5.171)
\end{aligned}
$$

and this implies

$$
d(u_c, G) = -\frac{\Delta EI}{EI}\, a(u_c, G)_{\Omega_2} \qquad (5.172)
$$

which is only zero if Ω_2 is zero, in which case the change $EI \rightarrow EI + \Delta EI$ affects all parts of the beam, $l = \Omega_1$, no part is spared, $\Omega_2 = 0$, and in this case the bending moment distribution does not change at all, $M_c(x) = M(x)$, and so $J(e) = M_c(x) - M(x) = -d(u_c, G) = 0$.

Remark 5.5 For any pair of coefficients $\{EI, EI_c\}$ the two solutions u and u_c of

$$
EIu^{IV} = p \qquad EI_c u_c^{IV} = p \qquad (5.173)
$$

have the same bending moments $M = -EI\, u'' = -EI_c\, u_c'' = M_c$.

By the same line of reasoning it can be shown that for functionals of force terms holds

$$
a(u, G_c) = J(e). \qquad (5.174)
$$

Namely we have

$$
a(u, G_c) + d(u, G_c) = J_c(u) = 0 \qquad (5.175)
$$

and it is

$$
a(u, G_c) = -d(u, G_c) = -d(u_c, G) = J(e). \qquad (5.176)
$$

Applied to the beam where the left side of this equation is

$$
\begin{aligned}
a(u, G_c) &= EI \int_0^l u''\, G_c''\, dy = \int_0^l \frac{M\, M_G^c}{EI_c} = \int_0^5 \frac{M\, M_G^c}{EI_c}\, dy + \int_5^{10} \frac{M\, M_G^c}{EI}\, dy \\
&= \frac{1}{EI_c}\frac{1}{3} \cdot 9062.5 \cdot 15.625 \cdot 5 + \frac{1}{EI}\frac{1}{3} \cdot 9062.5 \cdot 15.625 \cdot 5 \\
&\quad + \frac{1}{EI}\frac{1}{3} \cdot (-9062.5) \cdot 31.25 \cdot 5 = -2.61 \qquad (5.177)
\end{aligned}
$$

this means that

$$
J(e) = M_c - M = -2.61. \qquad (5.178)
$$

Note that in the FE-context the analog of (5.174) gives

$$\boldsymbol{g}_c^T\, \boldsymbol{K}\, \boldsymbol{u} = \boldsymbol{g}_c^T\, \boldsymbol{f} = J_c(u_h) \tag{5.179}$$

and not $J(e)$.

Remark 5.6 Obviously there is a distinctive difference between displacement terms and force terms, see (5.150) and (5.174),

$$a(u, G_c) = \begin{cases} J(u_c) & \text{displacement terms} \\ J(e) & \text{force terms.} \end{cases} \tag{5.180}$$

5.9 The Basic Idea

Let us pause for a moment to reflect on the contents of the formula $J(e) = -d(G, u_c)$. Imagine J is the functional $J(u) = u(x)$. When we multiply $J(e)$ with the number one, the virtual force, the δ which belongs to the functional,

$$J(e) = 1 \cdot (u_c(x) - u(x)) = -d(G, u_c) = -\boldsymbol{g}^T\, \boldsymbol{\Delta K}_e\, \boldsymbol{u}_c \tag{5.181}$$

then we are reminded that the equation expresses an energy balance: the work done by the virtual point load $P = 1$ on acting through the shift $u_c - u$ in the displacement is equal to the virtual strain energy $-d(G, u_c)$ in the element Ω_e between G and u_c. That is changes in functionals, $J(e)$, at points x which possibly lie far, far away can be predicted by integrating over the defective element Ω_e only! Of course the points communicate with Ω_e, they send signals via the Green's functions to Ω_e. But the information is processed on Ω_e alone. It collects the incoming information, it acts in a sense like a CCD (light sensor) which produces a picture, the change $J(e)$ in the functional.

5.9.1 Continuous and Discrete Case

Here we compare the discrete formulation with the continuous formulation. The model problem is a beam on a spring support, see Fig. 5.13a, and the spring stiffness k undergoes a change, $k \rightarrow k + \Delta k$.

Let the stiffness matrix $\boldsymbol{K}$ be of size $n \times n$ and let the last node x_n coincide with the spring support. In the discrete case the new solution vector is

$$\boldsymbol{u}_c = \boldsymbol{u} - \frac{\Delta k\, u_n}{1 + \Delta k\, k_{nn}^{(-1)}}\, \boldsymbol{c}_n \tag{5.182}$$

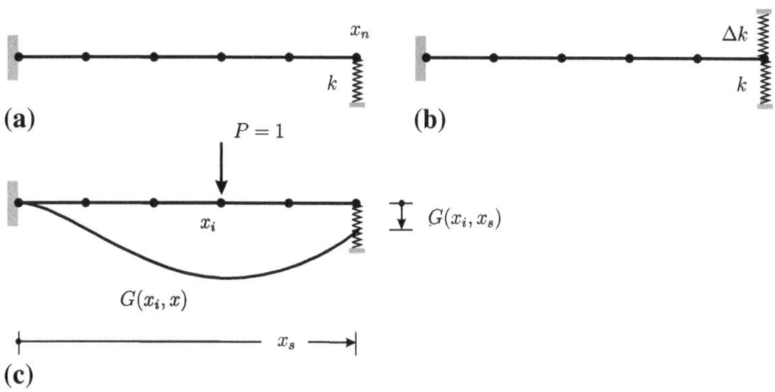

Fig. 5.13 Beam on a spring support. **a** Original system; **b** modified system; **c** Green's function for the deflection $u(x_i)$

where $\boldsymbol{c}_n$ is the last column of the inverse matrix, that is $\boldsymbol{c}_n$ is the discrete Green's function for $u(x_n)$.

It may be guessed that this formula translates in the continuous case to the formula

$$u_c(x) = u(x) - \frac{\Delta k\, u(x_n)}{1 + \Delta k\, G(x_n, x_n)}\, G(x, x_n) \tag{5.183}$$

where

$$G(x_n, x_n) \equiv k_{nn}^{(-1)} \tag{5.184}$$

is the flexibility of the beam at the spring support. To show that this assumption is correct we start with the continuous case where the formula for $u_c(x)$ is

$$u_c(x) = u(x) - d(G[x], u_c). \tag{5.185}$$

The updated strain energy product of the beam + 2 springs is the expression

$$a(u, v) := \int_0^l EI\, u''\, v''\, dx + k\, u(x_n)\, v(x_n) + \underbrace{\Delta k\, u(x_n)\, v(x_n)}_{d(u,v)} \tag{5.186}$$

so that

$$d(G[x], u_c) = \Delta k\, u_c(x_n)\, G(x_n, x). \tag{5.187}$$

In terms of the force method the factor $\Delta k\, u_c(x_n) = X$ is the suspension force in the added spring which contributes the deflection $-X\, G(x_n, x)$ to $u_c(x)$.

The equation for X is

$$X = -\frac{\delta_{10}}{\delta_{11}} = -\frac{u(x_n)}{G(x_n, x_n) + 1/\Delta k} = -\frac{\Delta k\, u(x_n)}{1 + \Delta k\, G(x_n, x_n)} \tag{5.188}$$

so it is found that indeed

$$\begin{aligned} u_c(x) &= u(x) - d(G[x], u_c) = u(x) - X\, G(x_n, x) \\ &= u(x) - \frac{\Delta k\, u(x_n)}{1 + \Delta k\, G(x_n, x_n)}\, G(x_n, x) \end{aligned} \tag{5.189}$$

agrees with (5.183).

5.9.2 Long & Strong and Short & Weak

The formula

$$J(e) = -d(u, G_c) \tag{5.190}$$

is a weak influence function and the integral only extends over the interval $[x_a, x_b]$ where the change $EI \rightarrow EI + \Delta EI$ occurs. Compare this with a formulation based on Betti's theorem

$$J(e) = M_c(x) - M(x) = \sum_{i=1}^{2} \left\{ \int_0^{l_i} (G_c(y, x) - G(y, x))\, p(y)\, dy \right\} \tag{5.191}$$

or in vector notation

$$J(e) = (\boldsymbol{g}_c - \boldsymbol{g})^T \boldsymbol{f} \tag{5.192}$$

where for the same result the integral extends over the *whole structure*, the whole domain, or in this case over *both* spans, $i = 1, 2$. This is what makes strong influence functions "expensive". Even the slightest change in the coefficient of a differential equation will affect the Green's function as a whole, will make the Green's functions change *everywhere*. Modify just one coefficient in the stiffness matrix $\boldsymbol{K}$ and *each* coefficient of the inverse $\boldsymbol{K}^{-1}$ will change, see Fig. 5.5.

Though the distribution of the load, the right hand side, also factors in. Accidentally in this case the extra effort is not larger, because only the second span carries a load, but in the case of a 20 story building under gravity loading it certainly would be helpful if the integral only extends over the element where the coefficient in the differential equation changes.

So the short form, (5.156), which is a weak influence function, must be preferred to (5.191), which is pure Betti, which is a strong influence function.

5.9.3 Estimates

The bilinear form $d(u, G_c)$ is a scalar product and this means that

1. the Cauchy-Schwarz' inequality provides a bound on the energy

$$|d(u, G_c)| \leq d(u, u)\, d(G_c, G_c); \tag{5.193}$$

2. and—as in any scalar product—u and G_c can be orthogonal in which case the product is zero

$$d(u, G_c) = 0 \qquad u \neq 0, \qquad G_c \neq 0 \tag{5.194}$$

   even though neither u nor G_c is zero.

So for modifications in the model parameters to produce significant effects the strains and stresses of the two solutions, u and G_c, must *both* reach elevated levels and be in *sync* or *in parallel* in the region where $d(.,.)$ is defined.

5.10 The Approximation $u_c \approx u$

The problem with the formula

$$J(e) = -d(u_c, G) \tag{5.195}$$

or the equivalent statement

$$J(e) = -d(u, G_c) \tag{5.196}$$

is that either one of the two functions u_c or G_c must be known to predict the change in the functional. But if the modified solution u_c is known then it can be evaluated directly and then there is no need for the formula (5.195).

The natural idea at this point is to substitute for u_c the undisturbed solution u. Adding the identity $d(u, G) - d(u, G) = 0$ to (5.195)

$$J(e) = -d(u, G) + d(u - u_c, G) \tag{5.197}$$

does not change the equation but it allows to conclude that if either (i) the difference $u_c - u \approx 0$ is negligible on Ω_p or (ii) $u - u_c$ and G are orthogonal on Ω_p (in terms of the strain energy product) then the expression

$$J(e) \approx -d(u, G) \tag{5.198}$$

should be a reasonable approximation to $J(e)$.

For a first check of the quality of this predictor we test the formula with the system in Fig. 5.12 and obtain the result

$$J(e) \approx -d(u, G) = \Delta EI \int_0^l \frac{M}{EI} \frac{M_G}{EI_c} dx = 1.94\,\text{kNm} \tag{5.199}$$

instead of the exact value $J(e) = 2.61\,\text{kNm}$. So the approximate value of $M_c(x)$ is

$$M_c(x) \simeq -7.81 - 1.94 = -9.75\,\text{kNm} \tag{5.200}$$

while the true value is $M_c(x) = -10.42\,\text{kNm}$ which is an error of 6 %. This is a small error, but of course there are other examples where the error is much larger. So the question is how the accuracy of the predictor step can be controlled.

In matrix notation the predictor step is the expression

$$J(e) = -\boldsymbol{g}^T \boldsymbol{\Delta K}_e \boldsymbol{u}_c \simeq -\boldsymbol{g}^T \boldsymbol{\Delta K}_e \boldsymbol{u} \tag{5.201}$$

where only those components of $\boldsymbol{g}$ and $\boldsymbol{u}_c$ respectively contribute to the scalar product which are attached to the element $\boldsymbol{K}_e$. Let us call these the "local" components of $\boldsymbol{g}$ and $\boldsymbol{u}$. With this in mind recall, see (5.114), that the force method establishes the following link between the local $\boldsymbol{u}$ and the local $\boldsymbol{u}_c$

$$(\boldsymbol{F}_e + \boldsymbol{\Delta K}_e^{-1})^{-1} \boldsymbol{u} = \boldsymbol{\Delta K}_e \, \boldsymbol{u}_c. \tag{5.202}$$

If the flexibility matrix is set to zero, $\boldsymbol{F}_e = \boldsymbol{0}$, then this results in

$$(\Delta \boldsymbol{K}_e^{-1})^{-1} \boldsymbol{u} = \boldsymbol{\Delta K}_e \, \boldsymbol{u}_c \tag{5.203}$$

or

$$\boldsymbol{\Delta K}_e \, \boldsymbol{u} = \boldsymbol{\Delta K}_e \, \boldsymbol{u}_c \tag{5.204}$$

that is with respect to $\boldsymbol{\Delta K}_e$ the actions of the two vectors are the same. This is the error we commit when we replace $\boldsymbol{u}_c$ by $\boldsymbol{u}$ in (5.201), we neglect the flexibility of the structure, we set $\boldsymbol{F}_e = \boldsymbol{0}$.

The entries in $\boldsymbol{F}_e = [c_{ij}]$ are the flexibilities of the structure at the nodes opposite to the added element Ω_e. Setting these flexibilities to zero effectively means that the coupling forces $\pm f_i^+$ which pull on both sides to close the gap come out too large because the rigid, inflexible nodes of the structure (opposite to Ω_e) do not move, and so the f_i^+ can close the gap only by pulling at the nodes of the element Ω_e.

Let us exemplify this by studying the rope in Fig. 5.4. The strain energy product of the rope with the spring attached is the expression

$$a(u, \hat{u}) + \Delta k \, u(x_i) \, \hat{u}(x_i) = \int_0^l u' \, \hat{u}' dx + \underbrace{\Delta k \, u(x_i) \, \hat{u}(x_i)}_{d(u,\hat{u})} \tag{5.205}$$

and using the approximation $u_c \sim u$ the change in the deflection at a point x is predicted as

$$J(e) = u_c(x) - u(x) \approx -d(G, u) = -\Delta k\, G(x_i, x)\, u(x_i). \tag{5.206}$$

In this formula the term

$$\gamma^{-1} = \frac{1}{1 + \Delta k\, k_{ii}^{(-1)}} = \frac{1}{1 + \Delta k\, G(x_i, x_i)} \tag{5.207}$$

is missing which means that the contribution of the rope, the term $k_{22}^{(-1)}$, has been neglected when δ_{11}, the gap between the rope and the spring produced by the pair of forces $X_1 = \pm 1$, was calculated, s. (5.37).

So erroneously a too large X_1 is applied to close the gap and consequently (5.206) will overshoot, will predict values which are too large.

The following theorem summarizes these observations:

Theorem 5.2 (The error in a first-order Taylor expansion) *Estimating the change in a functional with the formula*

$$J(e) \simeq -\boldsymbol{g}^T\, \boldsymbol{\Delta K}\, \boldsymbol{u} \tag{5.208}$$

as is done in the adjoint method of sensitivity, see Sect. 5.14, and in a first-order Taylor expansion, (5.121), amounts to a disregard for the flexibility of the system opposite to the added element Ω_e^+.

5.11 Linearization

When the load f that pulls on a spring

$$k\, u = f \tag{5.209}$$

doubles in value, the displacement $u = k^{-1}\, f$ will also double in value but if the stiffness drops, $k + \Delta k$ and $\Delta k < 0$, then we cannot later correct this by pumping the lost stiffness back into the system, see Fig. 5.14c. At the end the displacement u of the spring remains larger than before. The response of a spring, as of any linear system, to changes $\pm \Delta k$ in the stiffness is not symmetric.[2]

But when we substitute in the expression

$$J(e) = -d(u, G_c) \tag{5.210}$$

[2] The same holds true when you drive the first half of a distance with 50 mph and the second half with 70 mph. Better to keep a constant speed of 60 mph: you arrive earlier.

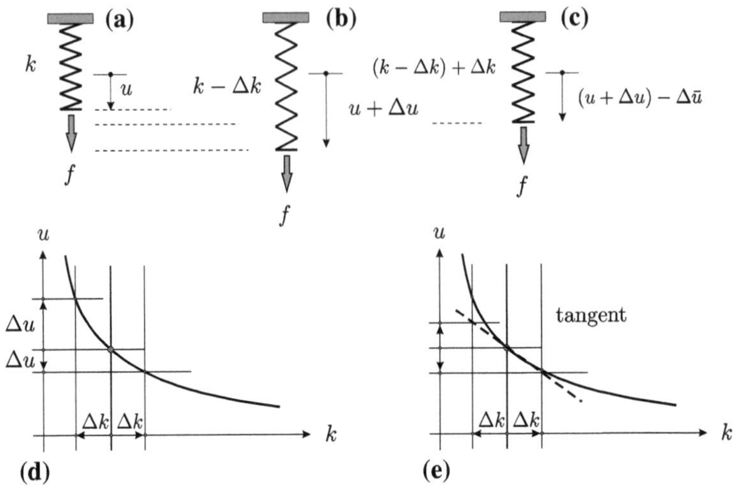

Fig. 5.14 Spring and effects of an increase/decrease in the stiffness k

for G_c the original Green's function then we effectively enforce this behavior—we symmetrize the response of the system to changes $\pm\Delta k$.

To see this let the functional $J(u) = u$ the displacement of the spring. The strain energy product of the spring is $a(u, v) = v\,k\,u$ and so the weak problems of the original (k) and the modified spring ($k + \Delta k$) is

$$a(u, v) = u\,k\,v = f\,v \tag{5.211}$$

$$a(u_c, v) + d(u_c, v) = u_c\,k\,v + u_c\,\Delta k\,v = f\,v\,. \tag{5.212}$$

The "Green's function" is the response of the spring to a unit load, $G = 1/k$,

$$J(u) = a(u, G) = G\,k\,u = \frac{1}{k}\,k\,u = u\,, \tag{5.213}$$

and so by changing the characteristic of a spring, $k \to k + \Delta k$, also the response of the spring changes, $G_c = 1/(k + \Delta k)$, and consequently

$$J(e) = u_c - u = -d(u, G_c) = -G_c\,\Delta k\,u = -\frac{1}{k + \Delta k}\,\Delta k\,u \tag{5.214}$$

which evidently is not symmetric with respect to $\pm\Delta k$ while, when we substitute for G_c the original $G = 1/k$, the response is symmetric

$$J(e) = u_c - u \simeq -d(u, G) = -\frac{1}{k}\,\Delta k\,u. \tag{5.215}$$

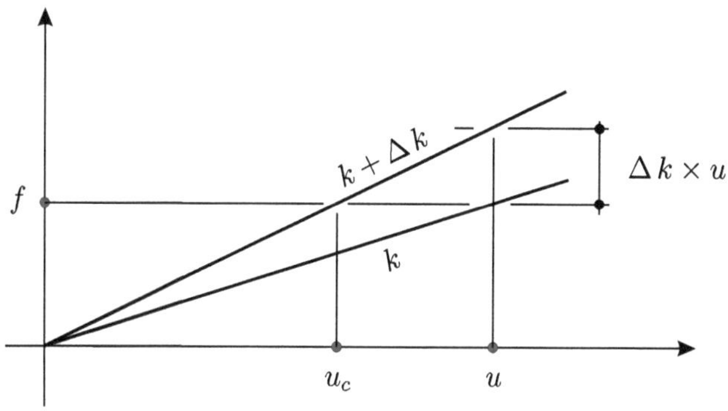

Fig. 5.15 A change in the stiffness k means a change in the slope

So the approximation $G \simeq G_c$ means that the function $1/k$ gets replaced by its tangent

$$u_c - u = u(k + \Delta k) - u(k) = -\frac{1}{k}\,\Delta\,k\,u + \cdots \quad \text{(Taylor series)} \tag{5.216}$$

or

$$u_c \approx u - \frac{1}{k}\,\underbrace{\Delta\,k\,u}_{\text{force}}\,. \tag{5.217}$$

The increase in stiffness $k \to k + \Delta\,k$ makes that the original displacement u produces now the force $(k + \Delta\,k)\,u = f + \Delta\,k\,u$ in the spring and the surplus force must be compensated by an opposite displacement $\Delta\,u_c = \Delta\,k\,u_c/k$ (approx.), see Fig. 5.15.

The Taylor-expansion of a stiffness matrix has the form [2]

$$(\boldsymbol{K} + \boldsymbol{\Delta}\,\boldsymbol{K})^{-1} = \boldsymbol{K}^{-1} - \boldsymbol{K}^{-1}\,\boldsymbol{\Delta K}\,\boldsymbol{K}^{-1} + \cdots\,, \tag{5.218}$$

and the analogy between (5.217) and

$$\boldsymbol{u}_c = (\boldsymbol{K} + \boldsymbol{\Delta K})^{-1}\boldsymbol{f} \approx \boldsymbol{u} - \boldsymbol{K}^{-1}\boldsymbol{\Delta K}\,\boldsymbol{u} \tag{5.219}$$

is evident.

Remark 5.7 Because of the $1/k$-effect the derivative of $u(k)$ is inversely proportional to k^2

$$k\,u = f \quad \Rightarrow \quad u = \frac{1}{k}\,f \quad \Rightarrow \quad u' = -\frac{1}{k^2}\,f, \tag{5.220}$$

and this means that the displacement increments Δu

$$\Delta u = u' \, \Delta k = -\frac{1}{k^2} \, \Delta k \, f \tag{5.221}$$

are the larger the lower the stiffness of the spring.

This also holds true for whole structures. The compensating movements a structure must perform to balance the loss of the stiffness in some members are inversely proportional to the original stiffness in these members. So soft structures are very sensitive to such changes and this implies that the influence functions and coefficients in such structures must be handled with care. A greater safety margin is needed for such structures.

5.12 Engineering Sensitivity Analysis

After these general remarks about functionals and how they are affected by changes in the model parameter let us apply these techniques to two important questions in engineering analysis.

1. Given a point $\boldsymbol{x}$ at which the maximum stresses occur how then would these stresses change if the stiffness in a certain part of the structure changes? Find the parts of the structure which have the most negative influence on the stress state at $\boldsymbol{x}$ and quantify their influence on the point!
2. Given an element Ω_e in which way would cracks in the element influence the results in the structure? How vital is the element for the structure?

5.12.1 *Focus on a Point*

We start with the first problem. The frame in Fig. 5.16 is subjected to wind blowing from the left which generates the bending moments plotted in Fig. 5.16. By how much will the bending moment $M(x)$ at the foot of the vertical frame element change if the bending stiffness EI in one of the frame elements l_i drops by 30 %.

Given the change $EI \to EI + \Delta EI$ in one of the elements l_i the change in the functional $J(u) = M(x)$ amounts to

$$J(e)(x) = M_c(x) - M(x) = -d(G, u_c) = -\Delta EI \int_0^{l_i} u_c'' \, G'' \, dy$$

$$= -\frac{\Delta EI}{EI} \int_0^{l_i} \frac{M_c \, M_G}{EI_c} \, dy \tag{5.222}$$

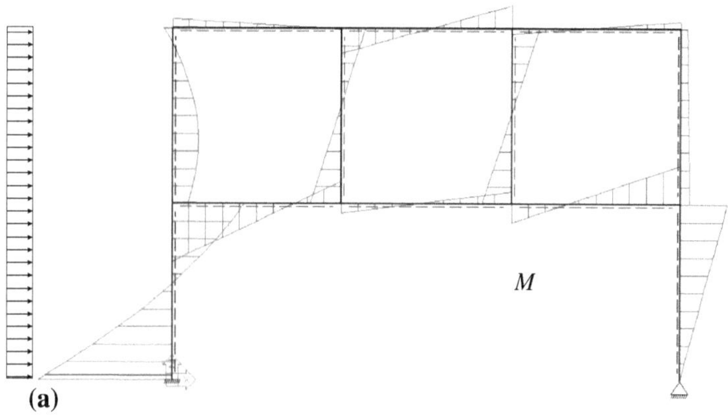

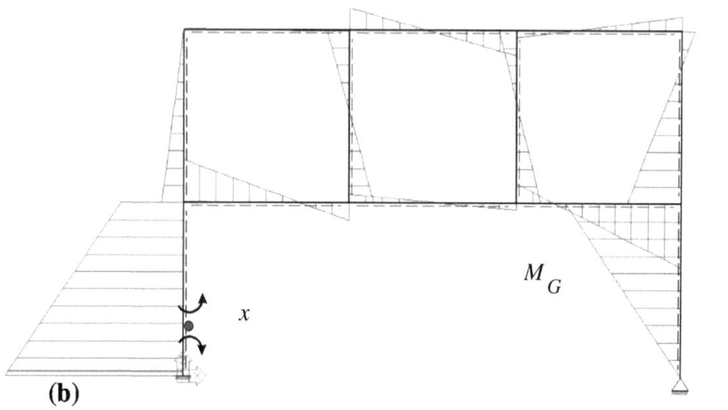

Fig. 5.16 Applied duality. **a** Bending moment M from the wind load; **b** bending moment M_G of the influence function for $M(x)$

where $M_c = -EI_c\, u_c''$ is the bending moment in the afflicted element l_i due to the wind load (after the cracks have developed) and the function $M_G = -EI\, G''$ is the bending moment of the Green's function in this element (before the cracks).

How the moment distribution M_c in the cracked element can be determined will be discussed in Sect. 5.13. Eventually one could approximate M_c with the moment distribution M in the uncracked beam.

The other term in the formula is the bending moment distribution M_G of the Green's function, see Fig. 5.17, which according to Schwarz' inequality

$$|J(e)(x)| \leq \frac{\Delta EI}{EI} \sqrt{\int_0^{l_e} \left(\frac{M_c}{EI_c}\right)^2 dy} \sqrt{\int_0^{l_e} (M_G)^2\, dy} \qquad (5.223)$$

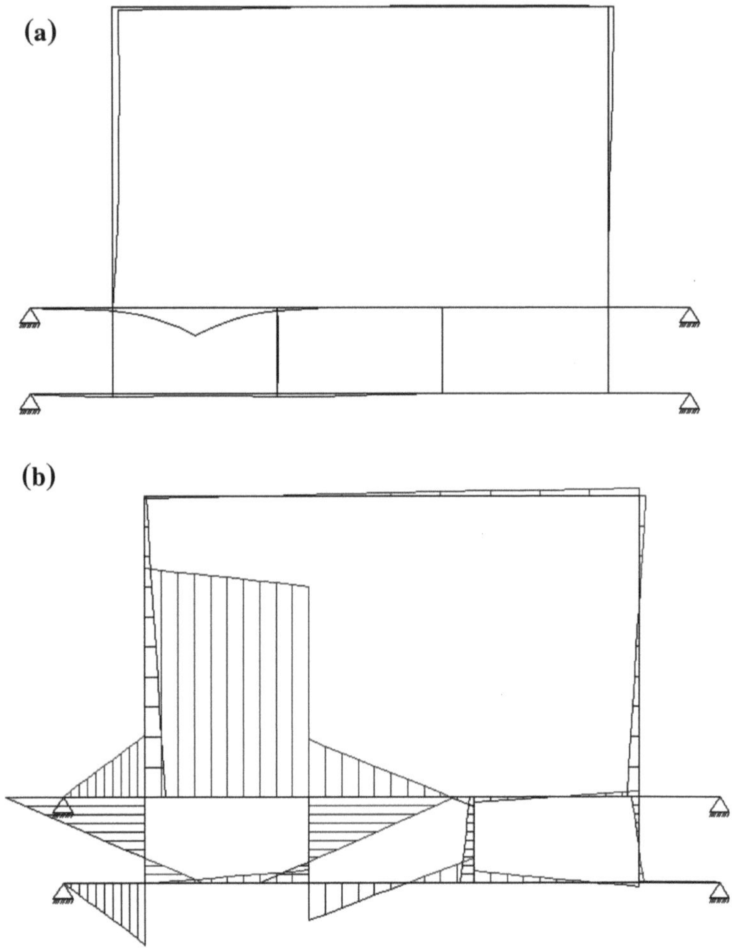

Fig. 5.17 Influence function for a bending moment. **a** Deformation of the system; **b** bending moment M_G of the Green's function

signals how sensitive $J(e)$ is with regard to M_c.

A simple but illustrative example for the role which M_G plays provides the cantilever beam in Fig. 5.18. To the point load at the end of the beam, which generates the influence function for the end deflection, belongs a linear bending moment distribution M_G which has its maximum value at the clamped edge and this means the closer the defective element, $EI \rightarrow EI + \Delta EI$, lies to the edge the larger its influence on the end deflection will be (in reality the distribution of the bending moment M from the load also has some influence on the results—the reasoning would be correct if M were constant).

With regard to the end rotation of the beam the situation is different. The bending moment M_G of the influence function is the same at all points and so we conclude

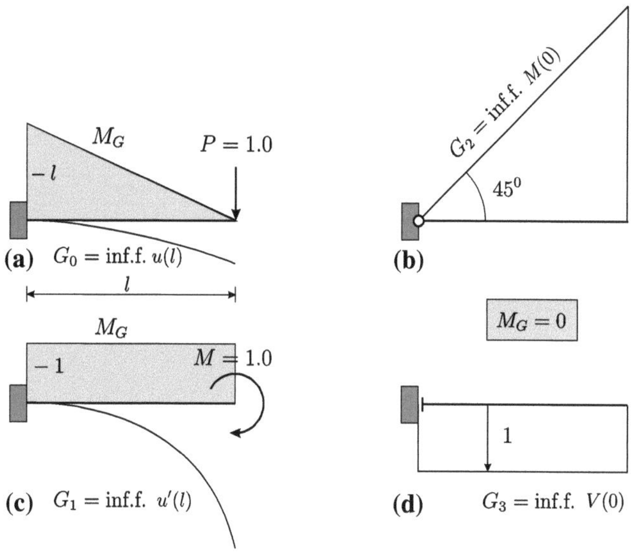

Fig. 5.18 Influence functions and their bending moments M_G. **a** For the end deflection $u(l)$; **b** the bending moment at the clamped edge; **c** for the end rotation $u'(l)$; **d** the shear force $V(0)$

that wherever the cracks will appear in the beam the effect on $u'(l)$ is the same. And the bending moments M_G of the two influence functions for $M(0)$ and $V(0)$ are both zero and this means that cracks will not change these values at all, which is no surprise given that the cantilever is statically determinate.

5.12.1.1 2-D Problems

In 2-D problems the situation essentially is the same. In the case of the Laplacian the d-form is

$$d(G, u_c) = k_\Delta \int_{\Omega_e} (G_{,1}\, u_{c,1} + G_{,2}\, u_{c,2})\, d\Omega_y \qquad (5.224)$$

where k_Δ represents the change, $k \to k + k_\Delta$, in the coefficient of the differential equation $-k\,\Delta u$ in the element Ω_e and so the gradient of the Green's function, see Fig. 5.19, could be used as an indicator when we search for those elements Ω_e or regions where a change $k \to k + k_\Delta$ in the coefficient is the most critical.

In linear elasticity the strain energy is the integral[3]

[3] $\boldsymbol{S} \cdot \boldsymbol{E} = \sigma_{11}\,\varepsilon_{11} + \sigma_{12}\,\varepsilon_{12} + \sigma_{21}\,\varepsilon_{21} + \sigma_{22}\,\varepsilon_{22}$ (scalar product).

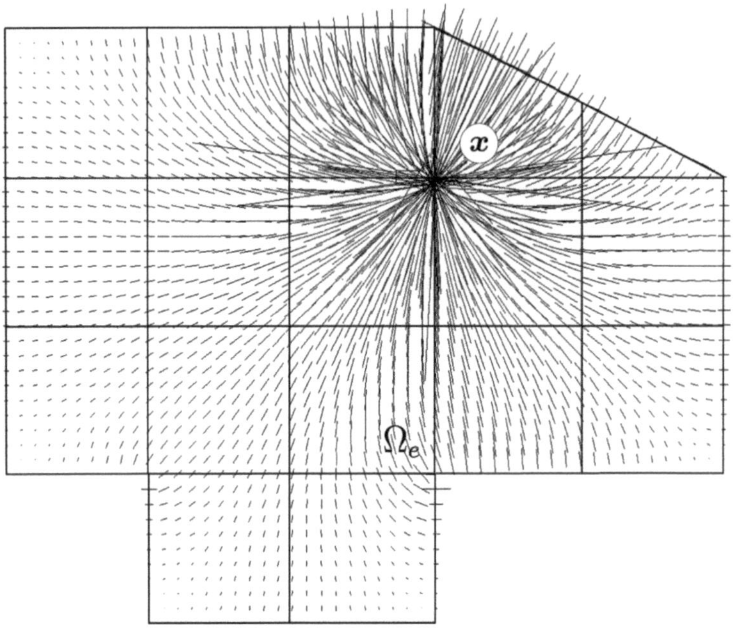

Fig. 5.19 The gradient of the Green's function of the functional $J(u) = u(\boldsymbol{x})$ (Poisson equation) and the subdivision of Ω into different control regions Ω_e. The strong singularity of the gradient at the source point $\boldsymbol{x}$ overshadows all other regions

$$a(\boldsymbol{u}, \boldsymbol{v}) = \int_\Omega \boldsymbol{C}[\boldsymbol{E}(\boldsymbol{u})] \cdot \boldsymbol{E}(\boldsymbol{v})\, d\Omega = \int_\Omega \boldsymbol{S} \cdot \boldsymbol{E}\, d\Omega \tag{5.225}$$

where the dominant term in the elasticity tensor

$$\boldsymbol{C}[\boldsymbol{E}] = E\left(\boldsymbol{E} + \frac{\nu}{1-2\nu}(\mathrm{tr}\boldsymbol{E})\,\boldsymbol{I}\right) = \boldsymbol{S} \tag{5.226}$$

is Young's modulus E (a number) so that defects in an element could be modeled by adding a corrective term to Young's modulus $E \to E + E_\Delta$

$$E_\Delta\, d(\boldsymbol{G}, \boldsymbol{u}_c) = \frac{E_\Delta}{E} \int_{\Omega_e} \boldsymbol{C}[\boldsymbol{E}(\boldsymbol{G})] \cdot \boldsymbol{E}(\boldsymbol{u}_c)\, d\Omega . \tag{5.227}$$

This formula suggests that the stresses associated with the Green's function of the functional $J(u)$ are an indicator of where and when modifications in element stiffness will have an impact on a functional. Of course it could always happen that the stress states of $\boldsymbol{G}$ and $\boldsymbol{u}_c$ are almost orthogonal, $d(\boldsymbol{G}, \boldsymbol{u}_c) \simeq 0$, but by and large the regions where the stresses reach a high level will be the regions to check for possible defects in the material.

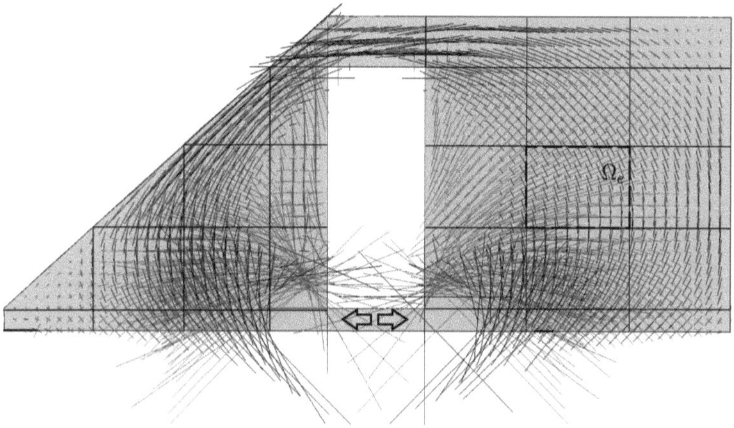

Fig. 5.20 Influence function for σ_{xx} and subdivision of Ω into control regions Ω_e

In Fig. 5.20 is displayed the typical stress distribution of a Green's function (for σ_{xx}). Because the singularity at the source point tends to overshadow stresses at points farther off, see also Fig. 5.19, the neighborhood of the source point must be masked out when the stress distribution resulting from the Green's function is displayed.

For an analysis one would partition the domain Ω into smaller subregions Ω_e, see Fig. 5.20, and one would weight the principal stresses in such a subregion Ω_e with the stresses in the same region resulting from the load.

In plate-bending problems (Kirchhoff plates) the formalism basically is the same. The strain energy product of two functions u and v is the L_2-scalar product of the moment tensor $\boldsymbol{M}$ and the curvature tensor $\boldsymbol{K}$ of the two functions

$$a(u, v) = \int_\Omega \boldsymbol{C}[\boldsymbol{K}(u)] \cdot \boldsymbol{K}(v) \, d\Omega = \int_\Omega \boldsymbol{M} \cdot \boldsymbol{K} \, d\Omega \tag{5.228}$$

where

$$\boldsymbol{C}[\boldsymbol{K}] = K\,((1-\nu)\boldsymbol{K} + \nu\,(\mathrm{tr}\boldsymbol{K})\,\boldsymbol{I}) = \boldsymbol{M}. \tag{5.229}$$

The scalar $K = E\,d^3/12(1-\nu^2)$ is the bending stiffness of a plate with thickness d and modulus of elasticity E. The entries in the curvature tensor $\boldsymbol{K} = \boldsymbol{K}(u)$ are the second derivatives, $K_{ij} = u_{,ij}$.

Cracks $K \to K + K_\Delta$ in an element Ω_e would lead to

$$d(G, u_c) = \frac{K}{K_\Delta} \int_{\Omega_e} \boldsymbol{C}[\boldsymbol{K}(G)] \cdot \boldsymbol{K}(u_c) \, d\Omega_y \tag{5.230}$$

so that the principal moments of the Green's function play the role of an indicator to detect the regions where cracks will have the most influence on a given functional $J(u)(\boldsymbol{x})$, see Fig. 5.21.

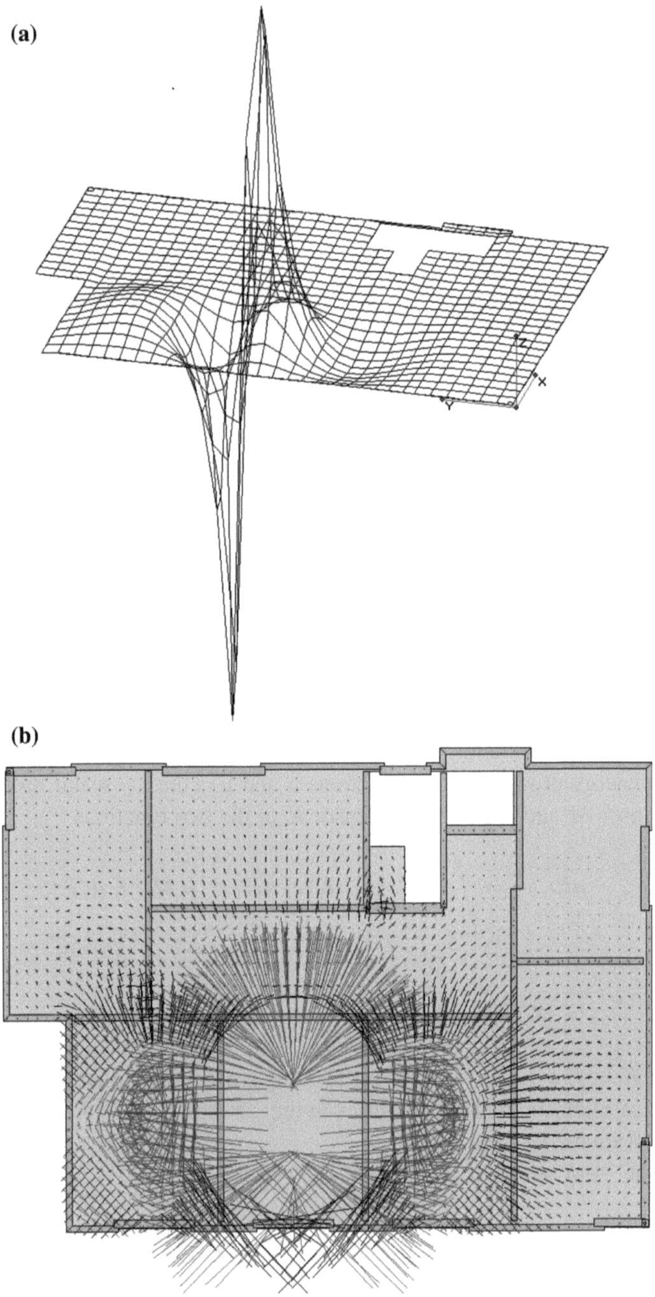

Fig. 5.21 Influence function for the shear force v_x. **a** 3-D plot; **b** principal moments

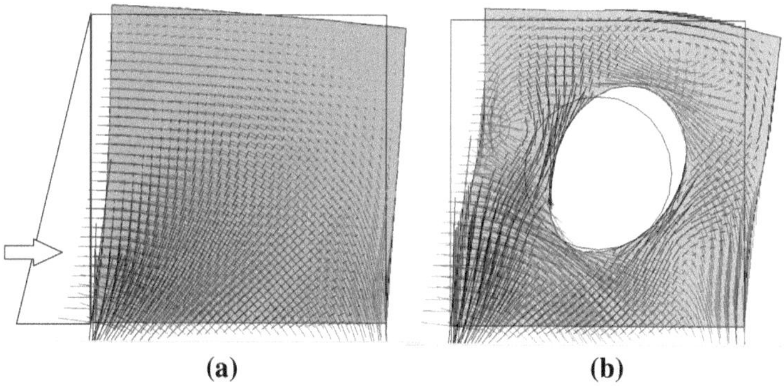

Fig. 5.22 Plate and lateral load. **a** Principal stresses before and **b** after the drilling of a hole

5.12.1.2 Boundary Integrals

In certain situations the d-form is equivalent to a boundary integral.

Let $J(u) = u(\boldsymbol{x})$ and $G(\boldsymbol{y}, \boldsymbol{x})$ the corresponding Green's function and let Ω_e be a patch or a single element of the mesh where the coefficient of the governing equation $-\kappa \Delta u = p$ shifts to a value $\kappa \to \kappa + \kappa_\Delta$. Green's first identity implies that the strain energy product

$$d(u_c, G) = \kappa_\Delta \int_{\Omega_e} \nabla u_c(\boldsymbol{y}) \cdot \nabla G(\boldsymbol{y}, \boldsymbol{x})\, d\Omega_y \tag{5.231}$$

is identical with

$$d(u_c, G) = \kappa_\Delta \int_{\Gamma_e} \frac{\partial G(\boldsymbol{y})}{\partial n} u_c(\boldsymbol{y}, \boldsymbol{x})\, ds_y + c(\boldsymbol{x})\, \kappa_\Delta\, u(\boldsymbol{x}) \tag{5.232}$$

where

$$c(\boldsymbol{x}) = \begin{cases} 1 & \boldsymbol{x} \in \Omega_e \\ \Delta\varphi/2\,\pi & \boldsymbol{x} \in \Gamma_e \\ 0 & \boldsymbol{x} \notin \Omega_e \end{cases} \tag{5.233}$$

is the so called *characteristic function* of Ω_e and $\Delta\varphi$ is the interior angle of Γ_e at the point $\boldsymbol{x}$; at smooth boundary points $\Delta\varphi = \pi$, see Fig. 5.22.

If the source point $\boldsymbol{x}$ lies outside Ω_e then this reduces to

$$d(u_c, G) = \kappa_\Delta \int_{\Gamma_e} \frac{\partial G(\boldsymbol{y})}{\partial n} u_c(\boldsymbol{y}, \boldsymbol{x})\, ds_y. \tag{5.234}$$

Because $\Delta u_c = 0$ on Ω_e[4] and $\boldsymbol{x} \notin \Omega$ only the tractions of u_c and G are pulling on the edge Γ_e and the reciprocal work of these two solutions is the same (Betti)

$$\kappa_\Delta \int_{\Gamma_e} \frac{\partial G(\boldsymbol{y})}{\partial n} u_c(\boldsymbol{y}, \boldsymbol{x})\, ds_{\boldsymbol{y}} = \kappa_\Delta \int_{\Gamma_e} \frac{\partial u_c(\boldsymbol{y})}{\partial n} G(\boldsymbol{y}, \boldsymbol{x})\, ds_{\boldsymbol{y}} \tag{5.235}$$

so that

$$d(u_c, G) = \kappa_\Delta \int_{\Gamma_e} \frac{\partial u_c(\boldsymbol{y})}{\partial n} G(\boldsymbol{y}, \boldsymbol{x})\, ds_{\boldsymbol{y}} \tag{5.236}$$

or what is the same (*the engineer's approach*)

$$d(u, G_c) = \kappa_\Delta \int_{\Gamma_e} \frac{\partial u(\boldsymbol{y})}{\partial n} G_c(\boldsymbol{y}, \boldsymbol{x})\, ds_{\boldsymbol{y}}. \tag{5.237}$$

Let us assume that $\kappa_\Delta = -\kappa$, that is the element or patch Ω_e effectively becomes a void, $\kappa + \kappa_\Delta = \kappa - \kappa = 0$. An engineer would solve this problem by applying the tractions $t = \kappa\, \partial u / \partial n$ of the undisturbed solution u as exterior forces on the edge of the void and predict the effect these forces produce with the modified influence function $G_c(\boldsymbol{y}, \boldsymbol{x})$ of the punched domain $\Omega_c = \Omega - \Omega_e$ and he would add this correction

$$e(\boldsymbol{x}) = -(-\kappa)\, d(u, G_c) = \kappa \int_{\Gamma_e} \frac{\partial u(\boldsymbol{y})}{\partial n} G_c(\boldsymbol{y}, \boldsymbol{x})\, ds_{\boldsymbol{y}} \tag{5.238}$$

to $u(\boldsymbol{x})$ to obtain $u_c(\boldsymbol{x}) = u(\boldsymbol{x}) + e(\boldsymbol{x})$.

Stated differently the change $J(e)$ in any functional can be computed via a boundary integral if the tractions t_c of the modified solution are known on the interface between Ω_e and the rest of the domain Ω.

$$J(e) = \int_{\Gamma_e} t_c(\boldsymbol{y})\, G(\boldsymbol{y}, \boldsymbol{x})\, ds_{\boldsymbol{y}}. \tag{5.239}$$

This is again the idea of the force method. The new solution u_c can be generated on the original physical domain (uniform κ) when to u are added additional effects which come from the boundary layer t_c on the edge of Ω_e.

Remark 5.8 One crucial assumption is of course what happens to the load p that possibly acts on Ω_e when the stiffness $\kappa + \Delta\kappa$ tends to zero. We assume that either $p = 0$ (on Ω_e) or that it tends to zero in sync with $\Delta\kappa \to -\kappa$.

[4] But u_c is not zero on Ω_e.

5.12.2 Focus on an Element

Imagine there is a certain element which looks suspicious and we want to make sure that cracks that eventually develop in this element, $EI \to EI + \Delta EI$, will not lead to a collapse of the structure.

Inquiring for possible negative side effects of these cracks means one must check the normal force, $J = N$, the shear force, $J = V$, the bending moment $J = M$ etc., and each of these functionals at a very large number of points x_i

$$J(e)(x_i) = J(u_c)(x_i) - J(u)(x) = -d(G[x_i], u_c) \qquad i = 1, 2, \ldots \tag{5.240}$$

In this situation it would be much simpler to solve the modified system with the cracked element directly.

But there is an alternative approach which requires less effort. Figuratively speaking we do not have to throw a new stone into the pond at each point x separately to watch the ripples. We can stand still and watch what happens when we drop one stone at the point where we are standing—which is the cracked frame element. There might be situations where this technique has its benefits and we therefore want to detail this method in the following.

Recall that the formula which is used to trace changes in functionals is an expression of internal energy that is it is of the type

$$J(e)(x) = -d(G, u_c) = -\frac{\Delta E}{E} \int_0^{l_e} \sigma(G[x])\, \varepsilon(u_c)\, dy. \tag{5.241}$$

The strains $\varepsilon(u_c)$ of the updated solution u_c are weighted with the stresses produced by $G[x]$, which is the Green's function of the functional J. The point in brackets, $[x]$, is the source point—the point on the left—and the ratio $\Delta E/E$ signals the change in the stiffness; in a beam this would be the term $\Delta EI/EI$.

This expression is, as can be shown, equivalent to

$$J(e)(x) = -d(G, u_c) = -\frac{\Delta E}{E} \int_0^{l_e} J(G_\sigma[y])(x)\, \varepsilon(u_c)\, dy \tag{5.242}$$

where G_σ is the influence function for the stress at the integration point y. G_σ is the movement the (uncracked) structure performs if a unit dislocation is applied at the integration point y. This movement produces an effect J at the point x and because the effect depends on which point y was spread one unit apart $J(G_\sigma[y])(x)$ is a function of the integration point y and so the scalar product with the strain $\varepsilon(u_c)(y)$ provides the value $J(e)(x)$. Note that the functional J appears now on both sides of the equation.

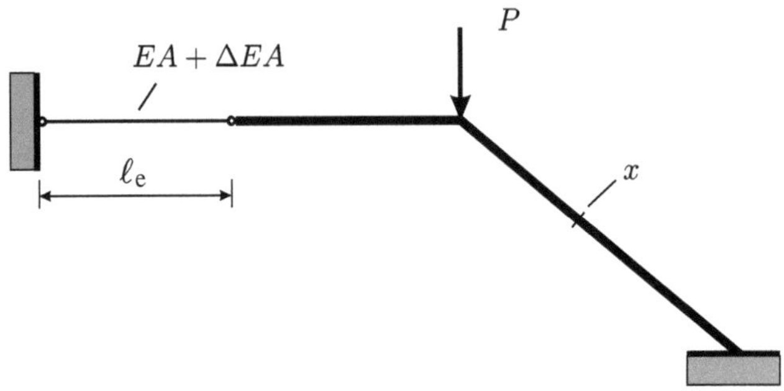

Fig. 5.23 Frame with a truss element

The benefit of this formulation is that only one Green's function is needed. The Green's function under the integral sign, does not depend on the functional. Only the effects studied, the value $J(G_\sigma[y])(x)$, depend on the choice of the functional J and the position x of the observation point.

5.12.2.1 Details

To explicitly detail the single steps of this approach the technique is applied to the frame in Fig. 5.23. The longitudinal stiffness EA of the first frame element drops by some value ΔEA and the task is to predict which effect this change has on the shear force V at the point x, that is on the functional

$$J_V(u)(x) = V(x) = -EI\, u'''(x). \tag{5.243}$$

The analysis proceeds in steps:

(i) The defective frame element is a pin-jointed element in which the strain energy product only depends on the normal force

$$a(u, \hat{u}) = \int_0^{l_e} EA\, u'\, \hat{u}'\, dx = \int_0^{l_e} EA\, \varepsilon\, \hat{\varepsilon}\, dx = \int_0^{l_e} \sigma\, \hat{\varepsilon}\, dx = \int_0^{l_e} \frac{N\, \hat{N}}{EA}\, dx$$

so that

$$J_V(e)(x) = -d(u_c, G) = -\frac{\Delta EA}{EA} \int_0^{l_e} \frac{N_c(y)\, N(G_V[x])(y)}{EA_c}\, dy \tag{5.244}$$

where $N_c(y)$ is the normal force (in the cracked element) due to the applied load p and $N(G_V[x])(y)$ is the normal force at the integration point y produced by the Green's function G_V. The Green's function produces a dislocation of the beam at the source point x in the direction of the shear force V.

(ii) The normal force at the integration point y too is a functional

$$J_N(u)(y) := N(y) = EA\,u'(y) \tag{5.245}$$

so that (5.244) is the same as

$$J_V(e)(x) = -\frac{\Delta EA}{EA}\int_0^{l_e} \frac{N_c(y)\,J_N(G_V[x])(y)}{EA_c}\,dy. \tag{5.246}$$

(iii) The final step: let $G_N(x, y)$ be the Green's function associated with the functional J_N (that is $G_N(x, y)$ is the influence function for the normal force $N(y)$ at the integration point y). The reciprocal values of two functionals with regard to their kernels (= G.F.) are the same

$$J_V(G_N[y])(x) = J_N(G_V[x])(y) \tag{5.247}$$

and this means that (5.246) is equivalent to

$$J_V(e)(x) = -\frac{\Delta EA}{EA}\int_0^{l_e} \frac{N_c(y)\,J_V(G_N[y])(x)}{EA_c}\,dy. \tag{5.248}$$

In this expression the functionals J on both sides are identical.

How is the integral (5.248) calculated? To generate the Green's function $G_N[y]$ the forces

$$j_i(y) = N(\varphi_i^e)(y) = EA\,(\varphi_i^e)'(y) \qquad i = 1, 2 \tag{5.249}$$

have to be applied at the two nodes of the element. Because the shape functions $\varphi_i^e(y)$ are linear the j_i are the same at all source points y. By the same argument the normal force $N_c(y) = N$ due to the load is constant and so with

$$\int_0^{l_e} J_V(G_N[y])(x)\,dy = J_V\left(\int_0^{l_e} G_N[y]\,dy\right)(x) = J_V(G^{Ⓢ})(x) \tag{5.250}$$

(5.248) becomes

$$J_V(e)(x) = -\frac{\Delta EA}{EA}\,\varepsilon_c\,J_V(G^{Ⓢ})(x). \tag{5.251}$$

The function $G^{Ⓢ}$ is the influence function for the *integral value* of N in the element

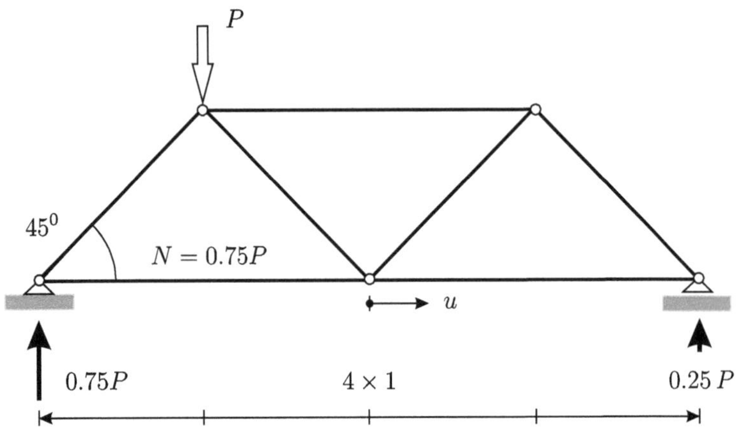

Fig. 5.24 Truss

$$\int_0^{l_e} N(x)\,dx = \int_0^{l} G^{Ⓢ}(x)\,p(x)\,dx. \tag{5.252}$$

It is generated by the two nodal forces, $i = 1, 2,$

$$j_i^{Ⓢ} = \int_0^{l_e} f_i(y)\,dy = \int_0^{l_e} N(\varphi_i^e)(y)\,dy = EA\,\frac{\pm 1}{l_e}\,l_e = \pm EA \tag{5.253}$$

which are independent of l_e. They produce the same strains $\varepsilon = 1$ in short and in long elements.

If the element were free to move then these two end forces would stretch the element to double its length[5] but the nodal forces $j_i^{Ⓢ}$ also must push the neighboring elements aside and so the elongation Δl of the element will be less than l_e. How these forces manage to deform the structure is a telltale sign of how changes $E \to E + \Delta E$ in the element will effect the structure.

Example 5.3 Under the influence of the nodal load P the center node of the truss in Fig. 5.24 moves sideways by

$$u = \frac{N\,l}{EA} = \frac{0.75 \cdot P \cdot 2}{EA} \tag{5.254}$$

units. By how much will this displacement increase if the element which holds the node tight (the left element) cracks, say $\Delta E = -0.5E$? The formula for the change in the functional is

[5] A stress $\sigma = E$ implies because of $\sigma = E\,\varepsilon$ that $\varepsilon = 1$ and so $\Delta l = \varepsilon \cdot l = l$.

$$J(e) = \frac{\Delta EA}{EA}\,\varepsilon_c\, J(G) \tag{5.255}$$

where G is the influence function for the integral value of N in the element—two forces $j_i^{Ⓢ} = \pm EA$ stretch the element (unhindered because the truss is statically determinate) to double its length, $\Delta l = l = 2$—and $J(G) = 2$ is the nodal displacement caused by this stretching. The term

$$\varepsilon_c = \frac{N_c}{E_c\, A} = 0.75\,\frac{P}{E_c\, A} \qquad E_c = E + \Delta E = \frac{1}{2}\, E \tag{5.256}$$

is the strain in the cracked element due to the load (the normal force N in the element does not alter, $N_c = N$) so that the change in the displacement amounts to

$$J(e) = -\frac{-0.5\, EA}{EA}\, 0.75\, \frac{P}{E_c A}\, J(G) = \frac{1}{2}\, u \cdot 2 = u. \tag{5.257}$$

This is the additional displacement, so the total displacement $u_c = u + u$ is twice the previous value.

Certainly one could have guessed the result without applying this elaborate technique but this example is just to demonstrate that the effects observed when the element is stretched to double its length, $\varepsilon = 1$, is an indicator for the change $J(e)$ in the functionals. Of course in statically indeterminate structures and in 2-D and 3-D problems when the elements are tightly packed no element achieves $\varepsilon = 1$ (or $\kappa = 1$ in bending problems). The neighboring elements dampen the movement and so they lessen the influence of cracks. What here appears as a hinderance is actually beneficial because it increases the *redundancy*. The neighboring elements help their brethren out when it cracks under the load.

Example 5.4 A more realistic example is the multi-story frame in Fig. 5.25. A change ΔEA in the stiffness of the center front column amounts to a change

$$J(e) = -\frac{\Delta EA}{EA}\,\varepsilon_c\, J(G^{Ⓢ}) \tag{5.258}$$

in any functional. Here $G^{Ⓢ}$ is the influence function for the integral value of the normal force in the column and ε_c is the strain in the column

$$\varepsilon_c = \frac{\Delta u}{l} \tag{5.259}$$

where Δu is the amount by which the column shortens when EA decreases.

In FE-analysis $G^{Ⓢ}$ would be generated by applying two forces $j_i^{Ⓢ} = \pm EA$ at the end points of the column, see Fig. 5.25 and the plots of $G^{Ⓢ}$ and of the relevant functionals, $J(G^{Ⓢ}) = M(G^{Ⓢ})$ or $J(G^{Ⓢ}) = V(G^{Ⓢ})$, would be used as indicators

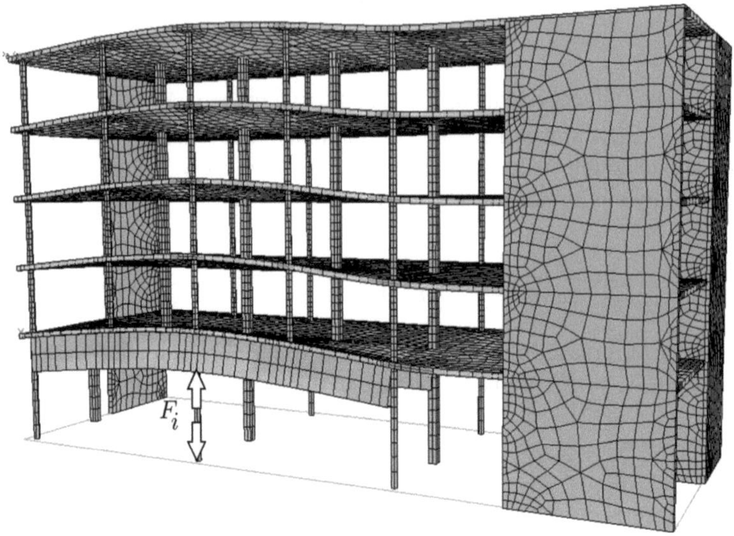

Fig. 5.25 Influence function for the normal force in the column

to assess which functional values—the engineer would say: which internal actions, M, N, V—are the most affected by cracks in the element.

5.12.2.2 Total Loss of Stiffness

If the column breaks down, if it is no longer able to support the structure, then this is equivalent to a total reverse, $\Delta EA = -EA$, in the stiffness and so (5.258) becomes

$$J(e) = \varepsilon_c\, J(G^{Ⓢ})(x) \tag{5.260}$$

where ε_c is the strain in the column if it were drained of all is stiffness that is the Δl in (5.259) is by how much the floor plate sacks when the column topples.

Basically this is the same logic which the engineer applies: To see how much the structure gets stressed without the column the engineer would remove the column and apply the opposite of the normal force N as nodal forces at the end points and add these results to the original load case so that in the end the result is the same

$$J(u_c) = J(u) + J(e). \tag{5.261}$$

Actually the engineer uses the expression (5.260) in the second form

$$J(e) = \varepsilon_c\, J(G^{Ⓢ})(x) = \underbrace{\varepsilon\, J(G_c^{Ⓢ})(x)}_{engineer} \tag{5.262}$$

(recall that $d(u_c, G) = d(u, G_c)$) and he applies integration by parts to the strain energy so that

$$J(e) = \varepsilon\, J(G_c^{Ⓢ})(x) = \sum_i f_i\, G_c^{Ⓢ}(y_i, x) \tag{5.263}$$

becomes an expression of exterior work namely the work done by the nodal forces f_i opposite to the vanished column on acting through the displacements of the same nodes caused by the influence function $G_c^{Ⓢ}$.

So whenever there is a total loss of the stiffness as in the column, $EA \rightarrow EA - EA = 0$, the curve u_c is that curve (or surface in 2-D) which bridges the gap in the structure left open by the now defunct element. The curve u_c can be a very "wild" function—the released nodes can move unhindered—and because the strain energy in the blank fictitious element Ω_e is measured with the full stiffness, $d(u, v) = a(u, v)$, the net effect $J(e)$ can be quite large. It will no longer do to approximate ε_c by ε. The plot of $G^{Ⓢ}$ can only give hints as to which parts of a structure are the most affected by a total loss of the stiffness.

5.12.3 Beams

The strain energy in a beam is

$$a(u, \hat{u}) = \int_0^l EI\, u''\, \hat{u}''\, dx = \int_0^l \frac{M\, \hat{M}}{EI}\, dx \qquad M := -EI\, u'' \tag{5.264}$$

and so a modification $EI \rightarrow EI + \Delta EI = EI_c$ in an element $(0, l_e)$ produces the shift

$$\begin{aligned} J(e) &= -\Delta EI\, d(u_c, G) = -\Delta EI \int_0^{l_e} u_c''\, G''\, dy \\ &= -\frac{\Delta EI}{EI} \int_0^{l} \frac{M_c}{EI_c}\, M_G\, dy \qquad M_G = -EI\, G'' \end{aligned} \tag{5.265}$$

in any linear functional. Applying the previous logic it follows that

$$J(e) = -\frac{\Delta EI}{EI} \int_0^{l_e} \frac{M_c}{EI_c}\, J(G_M[y])(x)\, dy \tag{5.266}$$

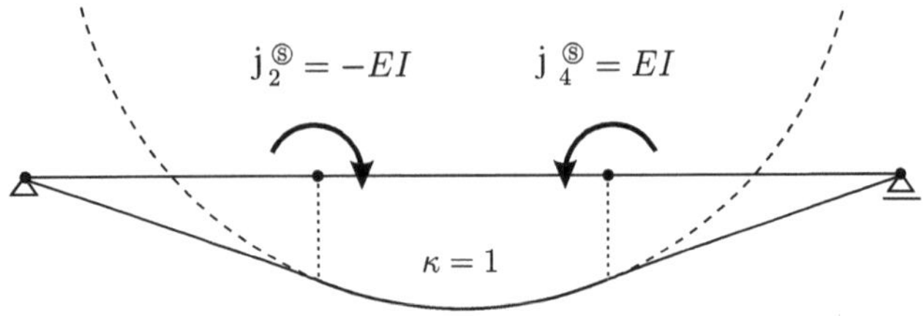

Fig. 5.26 The center element has a constant curvature

where $G_M[y] = G_M(y, x)$ is the influence function for the bending moment $M = -EIu''$ at the integration point y.

While in a bar element the integrands in the strain energy $d(u_c, G)$ were constant they are now linear. This requires some more effort.

To prepare for the following recall that

$$J(u)(x) = \boldsymbol{g}(x)^T \boldsymbol{f} \tag{5.267}$$

where the vector $\boldsymbol{g}$ is the nodal vector of the Green's function G_h associated with the functional $J(u)$ and $\boldsymbol{f}$ are the nodal forces belonging to u.

So if the curvature $\kappa_c = -M_c/EI_c$ in the element is constant then (5.266) is identical with

$$\begin{aligned} J(e) &= \frac{\Delta EI}{EI}\,\kappa_c \int_0^{l_e} J(G_M[y])(x)\,dy = \frac{\Delta EI}{EI}\,\kappa_c \int_0^{l_e} \boldsymbol{g}(x)^T\,\boldsymbol{f}(y)\,dy \\ &= \frac{\Delta EI}{EI}\,\kappa_c\,\boldsymbol{g}(x)^T\,\boldsymbol{f}^{Ⓢ} = \frac{\Delta EI}{EI}\,\kappa_c\,J(G^{Ⓢ}) \end{aligned} \tag{5.268}$$

where $G^{Ⓢ}$ is the influence function for the integral of M in the element.

The influence function $G_M(y, x)$ is generated by the nodal forces

$$j_i = -EI\,\varphi_i''(y) \qquad i = 1, 2, 3, 4 \tag{5.269}$$

where the φ_i are the four shape functions of the beam, see (1.91). So that $G^{Ⓢ}$ is generated by the integrals of these forces

$$j_i^{Ⓢ} = \int_0^{l_e} f_i(y)\,dy = EI\,(0, -1, 0, 1). \tag{5.270}$$

These two couples produce a constant curvature $\kappa = 1$ in the element if the element can deflect unhindered that is if the beam is statically determinate as in Fig. 5.26.

If $\kappa_c = u_c''$ is not constant then it is linear (the shape functions φ_i are cubic)

$$\kappa_c = \kappa_1^c\,\psi_1(x) + \kappa_2^c\,\psi_2(x) = \kappa_1^c\,\frac{l_e - x}{l_e} + \kappa_2^c\,\frac{x}{l_e} \tag{5.271}$$

and it follows

$$J(e) = \frac{\Delta EI}{EI}\int_0^{l_e}(\kappa_1^c\,\psi_1(y) + \kappa_2^c\,\psi_2(y))\,J(G_M[y])(x)\,dy \tag{5.272}$$

so that this expression could be split into two influence functions

$$J(e) = \frac{\Delta EI}{EI}\Big(\kappa_1^c\,J\big(G_{(1)}^{Ⓢ}\big) + \kappa_2^c\,J\big(G_{(2)}^{Ⓢ}\big)\Big) \tag{5.273}$$

where $G_{(1)}^{Ⓢ}$ is generated by the nodal forces

$$j_i^{Ⓢ\,(1)} = \int_0^{l_e} f_i(y)\,\psi_1(y)\,dy = EI\Big(\frac{1}{l_e}, -1, \frac{-1}{l_e}, 0\Big) \tag{5.274}$$

and $G_{(2)}^{Ⓢ}$ by the nodal forces

$$j_i^{Ⓢ\,(2)} = \int_0^{l_e} f_i(y)\,\psi_2(y)\,dy = EI\Big(\frac{-1}{l_e}, 0, \frac{1}{l_e}, 1\Big). \tag{5.275}$$

5.12.4 Poisson Problem

The extension to 2-D problems—in principle—is straightforward. Let u the solution to the problem

$$-\kappa\,\Delta u = p \qquad \text{on } \Omega \qquad u = 0 \qquad \text{on } \Gamma. \tag{5.276}$$

If the coefficient κ changes in an element Ω_e of Ω from a value κ to a value $\kappa + \kappa_\Delta$ then the corresponding change in any linear functional $J(u)$ amounts to

$$J(e)(\boldsymbol{x}) = -d(u_c, G) = -\frac{\kappa_\Delta}{\kappa} \int\limits_{\Omega_e} \nabla u_c \cdot \nabla G \, d\Omega_y$$
$$= -\frac{\kappa_\Delta}{\kappa} \int\limits_{\Omega_e} (u_{c,1}(\boldsymbol{y})\, G_{,1}(\boldsymbol{y}, \boldsymbol{x}) + u_{c,2}(\boldsymbol{y})\, G_{,2}(\boldsymbol{y}, \boldsymbol{x}))\, d\Omega_y \tag{5.277}$$

where G is the Green's function of the functional $J(u)(\boldsymbol{x})$.

When the shape functions are linear this equation can be recast in the form

$$J(e)(\boldsymbol{x}) = -\frac{\kappa_\Delta}{\kappa} \left[J(G_1^{Ⓢ})(\boldsymbol{x})\, u_{,1}(\boldsymbol{y}_c) + J(G_2^{Ⓢ})(\boldsymbol{x})\, u_{,2}(\boldsymbol{y}_c)\right] |\Omega_e| \tag{5.278}$$

where $G_1^{Ⓢ}$ and $G_2^{Ⓢ}$ are the influence functions for the integrals of $u_{,1}$ and $u_{,2}$ respectively over the element Ω_e and the values $u_{c,1}$ and $u_{c,2}$ are the derivatives of the solution u_c at the center $\boldsymbol{y}_c$ of the element.

The function $G_1^{Ⓢ}$ is generated by applying the nodal forces

$$j_i^{Ⓢ} = \int\limits_{\Omega_e} \varphi_{i,x}\, d\Omega \tag{5.279}$$

and for $G_2^{Ⓢ}$ the derivatives are taken with respect to y.

5.12.4.1 Elasticity

In 2-D elasticity the strain energy integral is the integral

$$a(\boldsymbol{u}, \boldsymbol{v}) = \int\limits_{\Omega} (\sigma_{11}\, \varepsilon_{11} + 2\, \sigma_{12}\, \varepsilon_{12} + \sigma_{22}\, \varepsilon_{22})\, d\Omega \tag{5.280}$$

and so, given a change E_Δ in Young's modulus, the corresponding expression for linear elements would be

$$J(\boldsymbol{e})(\boldsymbol{x}) = -\frac{E_\Delta}{E} (J(\boldsymbol{G}_{11}^{Ⓢ})(\boldsymbol{x})\, \varepsilon_{11}(\boldsymbol{y}_c) + 2\, J(\boldsymbol{G}_{12}^{Ⓢ})(\boldsymbol{x})\, \varepsilon_{12}(\boldsymbol{y}_c)$$
$$+ J(\boldsymbol{G}_{22}^{Ⓢ})(\boldsymbol{x})\, \varepsilon_{22}(\boldsymbol{y}_c))\, |\Omega_e| \tag{5.281}$$

where the ε_{ij} are the constant strains and the Green's function $\boldsymbol{G}_{ij}^{Ⓢ}$ (which are displacement fields = vector valued functions) are the influence functions for the integral values of the stresses

$$\int_{\Omega_e} \sigma_{kl}(\boldsymbol{x})\, d\Omega = \int_{\Omega} \boldsymbol{G}_{kl}^{Ⓢ}(\boldsymbol{x}) \cdot \boldsymbol{p}(\boldsymbol{x})\, d\Omega \tag{5.282}$$

each of which would be generated by the application of $2 \cdot 3$ nodal forces

$$j_i^{Ⓢ\, kl} = \int_{\Omega_e} \sigma_{kl}(\boldsymbol{\Phi}_i)\, d\Omega_{\boldsymbol{y}}. \tag{5.283}$$

The $\boldsymbol{\Phi}_i$ are the six displacement fields corresponding to horizontal and vertical unit displacements respectively of the nodes. If we would let Poisson's ratio $\nu = 0$ then the number of nodal forces would reduce to three, but we would still have to solve three additional load cases to determine the three functions $\boldsymbol{G}_{kl}^{Ⓢ}(\boldsymbol{x})$. Whether this is worth the effort or whether it is not just simpler to modify the system directly to watch for the effects of $E \to E + E_{\Delta}$ depends on the circumstances and the aim of the analysis.

This approach has one advantage namely that influence functions can be generated which signal how important an element is for the safety of a structure. Basically the message is that the negative influence which an element that cracks has is proportional to by how much the element is able to spread uniformly, $\varepsilon_{xx} = 1, \varepsilon_{yy} = 1$ and $\varepsilon_{xy} = 1$ in all three "directions". The more the neighboring elements can hinder these movements the less important the element is.

Of course additionally the stress distribution due to the load in the element factors in but in the sense of signals or weights the influence functions carry the message.

5.12.5 Kirchhoff Plates

The strain energy of a Kirchhoff plate is the expression

$$\begin{aligned} a(u, \hat{v}) &= \int_{\Omega} m_{ij}\, \hat{\kappa}_{ij}\, d\Omega \\ &= \int_{\Omega} (m_{11}\, \hat{\kappa}_{11} + 2\, m_{12}\, \hat{\kappa}_{12} + m_{22}\, \hat{\kappa}_{22})\, d\Omega. \end{aligned} \tag{5.284}$$

It has the same structure as the equations before and so

$$\begin{aligned} J(\boldsymbol{e})(\boldsymbol{x}) = &- \frac{K_{\Delta}}{K}\, (J(G_{11}^{Ⓢ})(\boldsymbol{x})\, \kappa_{11}(\boldsymbol{y}_c) + 2\, J(G_{12}^{Ⓢ})(\boldsymbol{x})\, \kappa_{12}(\boldsymbol{y}_c) \\ &+ J(G_{22}^{Ⓢ})(\boldsymbol{x})\, \kappa_{22}(\boldsymbol{y}_c))\, |\Omega_e| \end{aligned} \tag{5.285}$$

would be the analogous expression where the three functions $G_{ij}^{Ⓢ}$ are the Green's function for the integral values of the three moments, m_{11}, m_{12}, m_{22}

$$\int_{\Omega_e} m_{ij}(\boldsymbol{x})\, d\Omega = \int_{\Omega} G_{ij}^{Ⓢ}(\boldsymbol{x})\, p(\boldsymbol{x})\, d\Omega. \tag{5.286}$$

5.12.6 Analysis

Let us shortly recap the logic we applied. The typical influence function for $J(e)$ has the form

$$J(e)(x) = \int_0^{l_e} J(G_\sigma[y])(x)\, \varepsilon_c(y)\, dy. \tag{5.287}$$

First it is assumed that $\varepsilon_c(y) = \varepsilon_c$ is constant

$$J(e)(x) = \varepsilon_c \int_0^{l_e} J(G_\sigma[y])(x)\, dy \tag{5.288}$$

and then the functional is placed outside the integral

$$J(e)(x) = \varepsilon_c\, J\Big(\int_0^{l_e} G_\sigma[y]\, dy\Big)(x) \tag{5.289}$$

which maneuver implicitly introduces the Green's function $G^{Ⓢ}$ for the integral of σ

$$J(u)(x) = \varepsilon_c\, J(G^{Ⓢ})(x). \tag{5.290}$$

This influence function

$$\int_0^{l_e} \sigma(x)\, dx = \int_0^{l} G^{Ⓢ}(x)\, p(x)\, dx \tag{5.291}$$

is generated by applying the integrals

$$j_i^{Ⓢ} = \int_0^{l} \sigma(\varphi_i)(x)\, dx \tag{5.292}$$

of the shape functions as equivalent nodal forces.

If $\varepsilon_c(y)$ is not constant then we have at least an estimate

$$|J(e)(x)| \leq \max_{(0,l_e)} |\varepsilon_c| \, |J(G^{Ⓢ})(x)|. \tag{5.293}$$

In 2-D and 3-D problems σ is a tensor $\boldsymbol{S} = [\sigma_{ij}]$ and so the estimate becomes

$$|J(\boldsymbol{e})(\boldsymbol{x})| \leq \sum_{i,j=1}^{n} \max_{\Omega_e} |\varepsilon_{ij}| \, |J(\boldsymbol{G}_{ij}^{Ⓢ})(\boldsymbol{x})| \qquad n = 2, 3 \tag{5.294}$$

where

$$\boldsymbol{G}_{11}^{Ⓢ} \quad \boldsymbol{G}_{12}^{Ⓢ} \quad \boldsymbol{G}_{22}^{Ⓢ} \qquad (n = 2) \tag{5.295}$$

are the influence functions for the integral values of the stresses

$$\int_{\Omega_e} \sigma_{ij}(\boldsymbol{x}) \, d\Omega = \int_{\Omega} \boldsymbol{G}_{ij}^{Ⓢ}(\boldsymbol{x}) \cdot \boldsymbol{p}(\boldsymbol{x}) \, d\Omega. \tag{5.296}$$

The forces $j_k^{Ⓢ}$ which generate these influence functions would stretch—if unimpeded by the neighboring elements—the element Ω_e uniformly in horizontal ($\varepsilon_{11} = 1$) and vertical direction ($\varepsilon_{22} = 1$) respectively while the third set would produce a shear deformation ($\varepsilon_{12} = 1$).

The more these movements can spread unhindered over a domain Ω the larger the effect a change in the element stiffness will have on the results.

5.13 Equations for the Unknown Stresses on Ω_e

The formula

$$J(e) = -d(u_c, G) \tag{5.297}$$

allows to calculate the change in any internal action so why not apply this formula to calculate the unknown strains or stresses of the modified solution u_c within the defective element Ω_e itself?

To be specific let us choose the frame in Fig. 5.27 as a test problem. The parameters of the beam are

$$EI = 3{,}320 \, \text{kNm}^2 \qquad EA = 996{,}000 \, \text{kN}. \tag{5.298}$$

When the longitudinal stiffness of the vertical pier drops by 50 %

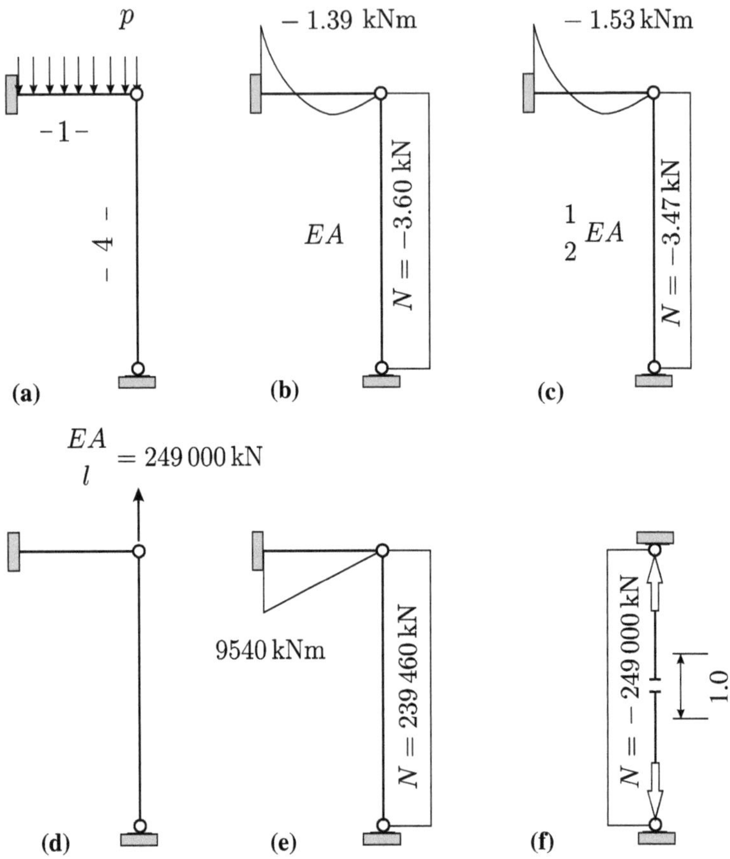

Fig. 5.27 Frame. **a** System and load; **b** internal actions M and N; **c** the same quantities when EA drops to half its values; **d** nodal force for FE-influence function G for N; **e** internal actions of G; **f** fixed end forces resulting from a unit dislocation of an internal point

$$EA_c = EA + \Delta EA = 996{,}000 - 0.5 \cdot 996{,}000 = 498{,}000\,\text{kN} \tag{5.299}$$

then the changes in the frame are

$$J(e) = -\Delta EA \int_0^{l_e} u_c' \, G' \, dy = -\Delta EA \int_0^{l_e} \frac{N_c}{EA_c} \frac{N_G}{EA} \, dy \tag{5.300}$$

where the integral extends over the length $(0, l_e)$ of the pier. To quantify these changes the derivative $u_c' = N_c/EA_c$ of the modified solution in the pier and the derivative G' of the Green's function must be known; the latter poses no problem, it can be extracted from $\boldsymbol{K}^{-1}$.

To determine N_c the idea is to apply (5.300) to N_c itself

$$N_c - N = -\Delta EA \int_0^{l_e} \frac{N_c}{EA_c} \frac{N_G}{EA} \, dy, \tag{5.301}$$

in which case G is the influence function for the normal force N in the pier and $N_G = EA\, G'$ is the EA-fold strain or normal force of G in the pier itself.

To calculate G the force

$$j = EA\, \varphi_1'(0) = -\frac{EA}{l_e} = -249{,}000\,\text{kN} \tag{5.302}$$

is applied at the upper node of the pier, see Fig. 5.27; it results in a normal force $N = 239{,}460$ kN in the pier. To this force must be added the normal force $N = -249{,}000$ kN when the pier, spread by one unit apart, presses against the fixed ends, see Fig. 5.27, so that the total force in the pier is

$$N_G = -249{,}000\,\text{kN} + 239{,}460\,\text{kN} = -9{,}540\,\text{kN} \tag{5.303}$$

and consequently the change in the normal force, see (5.301), should come out as

$$N_c - N = 0.13 = -498{,}000 \cdot \frac{3.47}{498{,}000} \cdot \frac{-9,540}{996{,}000} \cdot 4 = 0.13 \tag{5.304}$$

which it does. The general form of this equation is

$$N_c = \frac{EA\, EA_c}{EA\, EA_c - \Delta EA\, N_G\, l_e} N. \tag{5.305}$$

5.13.1 Computational Aspects

The nodal vector $\boldsymbol{g}$ of the Green's function is the solution of the system

$$\boldsymbol{K}\, \boldsymbol{g} = \boldsymbol{j} \tag{5.306}$$

where $j_i = EA\, \varphi_i'(x),\, x \in (0, l_e)$. If the global degrees of freedom of the two end nodes are i and k then—with $j_i = -EA/l_e$, $j_k = EA/l_e$ and all other j_i being zero—$\boldsymbol{g}$ is a linear combination of the two columns $\boldsymbol{c}_i$ and $\boldsymbol{c}_k$ of $\boldsymbol{K}^{-1}$

$$\boldsymbol{g} = -\frac{EA}{l_e} \boldsymbol{c}_i + \frac{EA}{l_e} \boldsymbol{c}_k \tag{5.307}$$

and g_i and g_k are the corresponding entries in $\boldsymbol{g}$. The Green's function is linear on $(0, l_e)$

$$G_h(y, x) = g_i\,\varphi_1(y) + g_k\,\varphi_2(y) = g_i\,\frac{l_e - y}{l_e} + g_k\,\frac{y}{l_e} \tag{5.308}$$

so that its normal force is $EA\,(g_i - g_k)$ and this force plus the compressional force $-EA/l$ is the normal force

$$N_G = -\frac{EA}{l} + EA\,(g_i - g_k). \tag{5.309}$$

5.13.1.1 Beam

This technique can be applied to any expression $d(u_c, G)$. In the case of a beam the linear bending moment distribution

$$M_c(y) = M_1^c\,\psi_1(y) + M_2^c\,\psi_2(y) = M_1^c\,\frac{l_e - y}{l_e} + M_2^c\,\frac{y}{l_e} \tag{5.310}$$

on the element must be known to apply the formula

$$J(e) = -d(u_c, G) = -\Delta EI \int_0^{l_e} \frac{M_c}{EI_c}\,\frac{M_G}{EI}\,dy. \tag{5.311}$$

Following the same line of reasoning as before a system of two equations for the two nodal values

$$M_c(x_i) - M(x_i) = -\frac{\Delta EI}{EI} \int_0^{l} \frac{M_c}{EI_c}\,M_G(y, x_i)\,dy \qquad i = 1, 2 \tag{5.312}$$

can be formulated where M_G is the bending moment of the influence function for $M(x_i)$

$$M_G(y, x_i) = -EI \sum_{j=1}^{4} g_j(x_i)\,\varphi_j''(y) + M_0(y, x_i) \tag{5.313}$$

and where

$$M_0(y, x_i) = f_2(x_i)\,\frac{l_e - y}{l_e} + f_4(x_i)\,\frac{y}{l_e} \tag{5.314}$$

is the bending moment distribution due to the fixed end forces in the frame element

$$f_2(x_1) = f_2(0) = 4\,\frac{EI}{l_e} \qquad f_4(x_1) = f_4(0) = -2\,\frac{EI}{l_e} \tag{5.315}$$

$$f_2(x_2) = f_2(l_e) = -2\,\frac{EI}{l_e} \qquad f_4(x_2) = f_4(l_e) = 4\,\frac{EI}{l_e}. \tag{5.316}$$

This provides a system of two equations for $M_c(x_1)$ and $M_c(x_2)$

$$\begin{bmatrix} a_{11} & a_{12} \\ a_{21} & a_{22} \end{bmatrix} \begin{bmatrix} M_c(x_1) \\ M_c(x_2) \end{bmatrix} = \begin{bmatrix} M(x_1) \\ M(x_2) \end{bmatrix} \tag{5.317}$$

where

$$a_{11} = -EI\Big(\sum_{j=1}^{4} g_j(x_1)\,b_{1j} - 1\Big) \qquad a_{12} = -EI\sum_{j=1}^{4} g_j(x_1)\,b_{2j} \tag{5.318}$$

$$a_{21} = -EI\sum_{j=1}^{4} g_j(x_2)\,b_{1j} \qquad a_{22} = -EI\Big(\sum_{j=1}^{4} g_j(x_2)\,b_{2j} - 1\Big) \tag{5.319}$$

with

$$b_{1j} = \int_0^{l_e} \psi_1(y)\,\varphi_j''(y)\,dy = \frac{1}{l_e}(1,\ -l_e,\ -1,\ 0) \tag{5.320}$$

$$b_{2j} = \int_0^{l_e} \psi_2(y)\,\varphi_j''(y)\,dy = \frac{1}{l_e}(-1,\ 0,\ 1,\ l_e). \tag{5.321}$$

The $2 \cdot 4$ eight nodal values $g_j(x_i)$, $j = 1, 2, 3, 4$; $i = 1, 2$ of the two influence functions come from the solutions of the two systems

$$\boldsymbol{K}\,\boldsymbol{g}^{(1)} = \boldsymbol{f}(x_1) \qquad \boldsymbol{K}\,\boldsymbol{g}^{(2)} = \boldsymbol{f}(x_2). \tag{5.322}$$

The two vectors $\boldsymbol{f}(x_1)$ and $\boldsymbol{f}(x_2)$ have only four non-zero components

$$f_i(x_1) = -EI\,\varphi_i''(x_1) = \frac{EI}{l_e^2}\,(-6,\ 4\,l_e,\ 6,\ 2\,l_e) \tag{5.323}$$

$$f_i(x_2) = -EI\,\varphi_i''(x_2) = \frac{EI}{l_e^2}\,(6,\ 2\,l_e,\ 6,\ -4\,l_e) \tag{5.324}$$

where $i = 1, 2, 3, 4$ are the local degrees of freedom which, when the system is assembled, are mapped to global degrees of freedom j_1, j_2, j_3, j_4.

Let us assume for simplicity that the local and global degrees of freedom of the end nodes of the beam element coincide. Under these circumstances the solution vectors

$$g^{(1)} = \sum_{i=1}^{4} f_i(x_1)\, \boldsymbol{c}_i \qquad g^{(2)} = \sum_{i=1}^{4} f_i(x_2)\, \boldsymbol{c}_i \tag{5.325}$$

are a linear combination of the first four columns $\boldsymbol{c}_i$ of the inverse stiffness matrix $\boldsymbol{K}^{-1}$ and the first four entries in these two columns are the $g_j(x_i)$ required to evaluate (5.313)

$$\boldsymbol{g}^{(1)} = \{g_1(x_1), g_2(x_1), g_3(x_1), g_4(x_1), *, *, \ldots, *\}^T \tag{5.326}$$

and for $\boldsymbol{g}^{(2)}$ analogously.

Remark 5.9 As a side effect these equations imply that if N or M respectively are zero before the element cracks then they are also zero afterwards. This applies of course also in the other direction: no increase in stiffness can revive a "dormant" element.

5.13.1.2 2-D Problem

To trace changes in the solution of a Poisson equation, $\kappa \to \kappa + \kappa_\Delta$,

$$J(e) = -\kappa_\Delta \int_{\Omega_e} \nabla u_c \cdot \nabla G \, d\Omega_y \tag{5.327}$$

the gradient $\nabla u_c = \{u_{c,x}, u_{c,y}\}$ of the modified solution u_c on Ω_e is needed. To this end (5.327) is used to formulate a set of equations for the gradient on Ω_e itself

$$u_{c,x_i}(\boldsymbol{x}) - u_{,x_i}(\boldsymbol{x}) = -\kappa_\Delta \int_{\Omega_e} \nabla u_c(\boldsymbol{y}) \cdot \nabla G^{(i)}(\boldsymbol{y}, \boldsymbol{x})\, d\Omega_y \quad \boldsymbol{x} \in \Omega_e \quad i = 1, 2 \tag{5.328}$$

where $G^{(i)}(\boldsymbol{y}, \boldsymbol{x})$ is the Green's function for $u_{,x_i}$ on Ω_e.

Assume the mesh consist of triangles with linear shape functions φ_i. So to calculate $G^{(i)}$ we would apply the nodal forces

$$j_j^{(i)} = \varphi_{j,x_i} \qquad j = 1, 2, 3 \tag{5.329}$$

at the three nodes of Ω_e, solve the system $\boldsymbol{K}\, \boldsymbol{g}^{(i)} = \boldsymbol{j}^{(i)}$, construct

$$\bar{G}^{(i)} = \sum_{j=1}^{3} g_j^{(i)}\, \varphi_j(\boldsymbol{x}) \tag{5.330}$$

(we again assume for simplicity that on Ω_e local and global degrees of freedom coincide) and add to its gradient the local gradient field generated when the center point of the element is spread by one unit apart in x_i-direction while the nodes are fixed. These two latter fields can be expressed in terms of the equivalent nodal forces $j_j^{(i)}$ so that

$$\nabla G^{(i)} = \nabla \bar{G}^{(i)} + \sum_{j=1}^{3} j_j^{(i)} \nabla \varphi_j = \sum_{j=1}^{3} (g_j^{(i)} + j_j^{(i)}) \nabla \varphi_j \tag{5.331}$$

which is a constant vector.

The further treatment of the linear system (5.328) contains no new aspects so that we can skip it here.

5.14 Adjoint Method of Sensitivity Analysis

Sensitivity analysis and Green's functions are closely related [3, 5]. We only want to show that the *adjoint method* of sensitivity analysis, is identical with the d-form in its *simplified* form

$$J(e) \approx -d(u, G). \tag{5.332}$$

Let $\boldsymbol{u}$ (n-components) be the displacement vector and let $\boldsymbol{p}$ (m-components) be the vector of *design variables* or model parameters. The functional $J(\boldsymbol{u}, \boldsymbol{p})$ depends on $\boldsymbol{u}$ and (possibly) directly on $\boldsymbol{p}$. The displacement vector $\boldsymbol{u}$ must satisfy a system of equations $\boldsymbol{K}\,\boldsymbol{u} = \boldsymbol{f}$ so that we can speak of the evaluation of a functional J under certain side conditions:

$$functional \quad J(\boldsymbol{u}, \boldsymbol{p}) \quad side\ condition \quad \boldsymbol{K}(\boldsymbol{p})\,\boldsymbol{u} = \boldsymbol{f}\,. \tag{5.333}$$

We are interested in the total differential of $J(\cdot, \cdot)$ with respect to an increment $\boldsymbol{d\,p}$ of the model parameter

$$dJ = \frac{dJ}{d\boldsymbol{p}}\,\boldsymbol{d\,p} = \frac{dJ}{dp_1} dp_1 + \cdots \frac{dJ}{dp_n} dp_n. \tag{5.334}$$

The gradient of J (here as a row vector)

$$\frac{dJ}{d\boldsymbol{p}} = \{J,_{p_1}, J,_{p_2}, \ldots, J,_{p_n}\} \tag{5.335}$$

has the components

$$J,_{p_i} = \frac{\partial J}{\partial p_i} + \frac{\partial J}{\partial u_k}\frac{\partial u_k}{\partial p_i} \tag{5.336}$$

or in matrix notation

$$\begin{gathered}\frac{dJ}{d\boldsymbol{p}} = \frac{\partial J}{\partial \boldsymbol{p}} + \frac{\partial J}{\partial \boldsymbol{u}}\boldsymbol{U}_{\boldsymbol{p}} \\ (1\times m) = (1\times m) + (1\times n)\,(n\times m)\end{gathered} \tag{5.337}$$

where

$$\frac{\partial J}{\partial \boldsymbol{p}} = \left\{\frac{\partial J}{\partial p_1}, \frac{\partial J}{\partial p_2}, \ldots, \frac{\partial J}{\partial p_n}\right\} \qquad \frac{\partial J}{\partial \boldsymbol{u}} = \left\{\frac{\partial J}{\partial u_1}, \frac{\partial J}{\partial u_2}, \ldots, \frac{\partial J}{\partial u_n}\right\} \tag{5.338}$$

and

$$\boldsymbol{U}_{\boldsymbol{p}} = \left[\frac{\partial u_i}{\partial p_j}\right] = (n\times m). \tag{5.339}$$

Because $\boldsymbol{K}\,\boldsymbol{u} - \boldsymbol{f} = \boldsymbol{0}$ holds true for any choice of $\boldsymbol{p}$ the side condition is invariant with respect to the single p_i

$$\frac{\partial}{\partial p_i}(\boldsymbol{K}\,\boldsymbol{u} - \boldsymbol{f}) = \boldsymbol{K},_{p_i}\boldsymbol{u} + \boldsymbol{K}\,\boldsymbol{u},_{p_i} - \boldsymbol{f},_{p_i} = \boldsymbol{0} \qquad (n\times 1) \tag{5.340}$$

or if this equation is solved for column i of $\boldsymbol{U}_{\boldsymbol{p}}$

$$\boldsymbol{u},_{p_i} = -\boldsymbol{K}^{-1}(\boldsymbol{K},_{p_i}\boldsymbol{u} - \boldsymbol{f},_{p_i}) =: -\boldsymbol{K}^{-1}\boldsymbol{d}_i \qquad (n\times 1) \tag{5.341}$$

so that (the columns of the matrix $\boldsymbol{D}$ are the vectors $\boldsymbol{d}_i$)

$$\boldsymbol{U}_{\boldsymbol{p}} = -\boldsymbol{K}^{-1}\boldsymbol{D} \qquad (n\times m) \tag{5.342}$$

and we finally have

$$\frac{dJ}{d\boldsymbol{p}} = \frac{\partial J}{\partial \boldsymbol{p}} + \frac{\partial J}{\partial \boldsymbol{u}}\boldsymbol{U}_{\boldsymbol{p}} = \frac{\partial J}{\partial \boldsymbol{p}} - \frac{\partial J}{\partial \boldsymbol{u}}\boldsymbol{K}^{-1}\boldsymbol{D} \qquad (1\times m)\,. \tag{5.343}$$

Let the (column) vector $\boldsymbol{\lambda}$ the solution of the system

$$\boldsymbol{K}\,\boldsymbol{\lambda} = \left(\frac{\partial J}{\partial \boldsymbol{u}}\right)^T \qquad \Rightarrow \qquad \boldsymbol{\lambda}^T = \frac{\partial J}{\partial \boldsymbol{u}}\boldsymbol{K}^{-1T} = \frac{\partial J}{\partial \boldsymbol{u}}\boldsymbol{K}^{-1} \tag{5.344}$$

then follows

$$\frac{dJ}{d\boldsymbol{p}} = \frac{\partial J}{\partial \boldsymbol{p}} - \boldsymbol{\lambda}^T\boldsymbol{D} \qquad (1\times m)\,. \tag{5.345}$$

If the functional J is linear

$$\frac{\partial J}{\partial \boldsymbol{u}} = \boldsymbol{j} \qquad (n \times 1) \qquad j_i = J(\varphi_i) \tag{5.346}$$

and does not directly depend on $\boldsymbol{p}$ so that $\partial J / \partial \boldsymbol{p} = \boldsymbol{0}$ then (5.345) reduces to

$$\frac{dJ}{d\boldsymbol{p}} = -\boldsymbol{\lambda}^T \boldsymbol{D} \qquad (1 \times m). \tag{5.347}$$

To keep it simple it is assumed that the single element matrix is of the type

$$\boldsymbol{K}_i = k_i \begin{bmatrix} 1 & -1 \\ -1 & 1 \end{bmatrix} \quad \Rightarrow \quad \frac{\partial \boldsymbol{K}_i}{\partial k_i} = \begin{bmatrix} 1 & -1 \\ -1 & 1 \end{bmatrix} = \frac{1}{k_i} \boldsymbol{K}_i \tag{5.348}$$

and that only one parameter, the stiffness in one element i changes, $dp_i = \Delta k_i$ and $\boldsymbol{f}$ does not depend on p_i then

$$\frac{dJ}{dp_i} = -\boldsymbol{\lambda}^T \boldsymbol{d}_i = -\boldsymbol{\lambda}^T \boldsymbol{K}_{,p_i} \boldsymbol{u} = -\boldsymbol{\lambda}^T \frac{1}{k_i} \boldsymbol{K}_i \boldsymbol{u} \tag{5.349}$$

and so it follows

$$J(\boldsymbol{e}) = J(\boldsymbol{u}_c) - J(\boldsymbol{u}) = \frac{dJ}{dp_i} dp_i = -\boldsymbol{\lambda}^T \frac{\Delta k_i}{k_i} \boldsymbol{K}_i \boldsymbol{u} \tag{5.350}$$

which is exactly the expression

$$J(e) \simeq -d(G, u) \tag{5.351}$$

in vector notation (finite elements) with $\boldsymbol{\lambda}$ being the nodal vector of the Green's function.

5.15 Linear Versus Nonlinear

What is the error of a linear analysis of a structure compared with a nonlinear analysis?

The governing equations of the linear problem are

$$\begin{aligned} \boldsymbol{E}_L(\boldsymbol{u}) - \boldsymbol{E} &= \boldsymbol{0} & \qquad \frac{1}{2}(u_{i,j} + u_{j,i}) - \varepsilon_{ij} &= 0 \\ \boldsymbol{C}[\boldsymbol{E}] - \boldsymbol{S} &= \boldsymbol{0} & C_{klij}\varepsilon_{kl} - \sigma_{ij} &= 0 \\ -\operatorname{div} \boldsymbol{S} &= \boldsymbol{p} & -(\sigma_{ij})_{,j} &= p_i \end{aligned} \tag{5.352}$$

and of the nonlinear problem

$$\begin{aligned} \boldsymbol{E}_{NL}(\boldsymbol{u}) - \boldsymbol{E} &= \boldsymbol{0} & \frac{1}{2}(u_{i},_{j} + u_{j},_{i} + u_{k},_{i}\, u_{k},_{j}) - \varepsilon_{ij} &= 0 \\ \boldsymbol{C}[\boldsymbol{E}] - \boldsymbol{S} &= \boldsymbol{0} & C_{klij}\varepsilon_{kl} - \sigma_{ij} &= 0 \\ -\mathrm{div}(\boldsymbol{S} + \nabla \boldsymbol{u}\, \boldsymbol{S}) &= \boldsymbol{p} & -(\sigma_{ij} + u_{i},_{k}\, \sigma_{kj}),_{j} &= p_i \end{aligned} \tag{5.353}$$

The variational formulation of the linear problem is

$$a(\boldsymbol{u}, \boldsymbol{v}) := \int_{\Omega} \boldsymbol{E}_L(\boldsymbol{v}) \cdot \boldsymbol{S}_L(\boldsymbol{u})\, d\Omega = (\boldsymbol{p}, \boldsymbol{v}) \tag{5.354}$$

where

$$\boldsymbol{S}_L(\boldsymbol{u}) := \boldsymbol{C}[\boldsymbol{E}_L(\boldsymbol{u})] \tag{5.355}$$

and of the nonlinear problem

$$a_{NL}(\boldsymbol{u}, \boldsymbol{v}) := \int_{\Omega} \boldsymbol{E}_{\boldsymbol{u}}(\boldsymbol{v}) \cdot \boldsymbol{S}_{NL}(\boldsymbol{u})\, d\Omega = (\boldsymbol{p}, \boldsymbol{v}) \tag{5.356}$$

where

$$\boldsymbol{E}_{\boldsymbol{u}}(\boldsymbol{v}) := \frac{1}{2}(\nabla \boldsymbol{v} + \nabla \boldsymbol{v}^T + \nabla \boldsymbol{u}^T\, \nabla \boldsymbol{v} + \nabla \boldsymbol{v}^T\, \nabla \boldsymbol{u}) \tag{5.357}$$

is the Gateaux derivative of the tensor $\boldsymbol{E}(\boldsymbol{u})$

$$\frac{d}{d\varepsilon}[\boldsymbol{E}(\boldsymbol{u} + \varepsilon\, \boldsymbol{v})]_{|\varepsilon=0} = \boldsymbol{E}_{\boldsymbol{u}}(\boldsymbol{v}) \tag{5.358}$$

and

$$\boldsymbol{S}_{NL}(\boldsymbol{u}) := \boldsymbol{C}[\boldsymbol{E}_{NL}(\boldsymbol{u})]. \tag{5.359}$$

It is evident that both tensors in the nonlinear formulation can be split into linear and nonlinear terms

$$\boldsymbol{E}_{\boldsymbol{u}}(\boldsymbol{v}) = \boldsymbol{E}_L(\boldsymbol{v}) + \boldsymbol{\Delta E}(\boldsymbol{u}, \boldsymbol{v}) \tag{5.360}$$

$$\boldsymbol{S}_{NL}(\boldsymbol{u}) = \boldsymbol{S}_L(\boldsymbol{u}) + \boldsymbol{\Delta S}(\boldsymbol{u}) \tag{5.361}$$

where

$$\boldsymbol{\Delta E}(\boldsymbol{u}, \boldsymbol{v}) : = \frac{1}{2}(\nabla \boldsymbol{u}^T\, \nabla \boldsymbol{v} + \nabla \boldsymbol{v}^T\, \nabla \boldsymbol{u}) \tag{5.362}$$

$$\boldsymbol{\Delta S}(\boldsymbol{u}) : = \boldsymbol{C}\left[\frac{1}{2}(\boldsymbol{\nabla}^T \boldsymbol{u}\, \boldsymbol{\nabla u})\right]. \tag{5.363}$$

On substituting these expressions into (5.356) it follows

$$a_{NL}(\boldsymbol{u}, \boldsymbol{v}) = a(\boldsymbol{u}, \boldsymbol{v}) + d(\boldsymbol{u}, \boldsymbol{v}) \tag{5.364}$$

where

$$d(\boldsymbol{u}, \boldsymbol{v}) = \int_{\Omega} (\boldsymbol{E}_L(\hat{\boldsymbol{u}})\, \boldsymbol{\Delta S}(\boldsymbol{u}) + \boldsymbol{\Delta E}(\boldsymbol{u}, \hat{\boldsymbol{u}})\, \boldsymbol{S}_L(\boldsymbol{u}) + \boldsymbol{\Delta E}(\boldsymbol{u}, \hat{\boldsymbol{u}})\, \boldsymbol{\Delta S}(\boldsymbol{u}))\, d\Omega \tag{5.365}$$

and so the error in a linear functional $J(\boldsymbol{u})$ amounts to

$$J(\boldsymbol{e}) = -d(\boldsymbol{u}_c, \boldsymbol{G}) \tag{5.366}$$

where $\boldsymbol{u}_c$ is the nonlinear solution and $\boldsymbol{G}$ the Green's function of the functional.

For an approximate analysis one could try to replace $\boldsymbol{u}_c$ by the linear solution so that

$$J(\boldsymbol{e}) \simeq -d(\boldsymbol{u}, \boldsymbol{G}) \tag{5.367}$$

though experience shows that this produces an acceptable solution only in mildly nonlinear problems [6].

There is an additional problem. In nonlinear problems some functionals as for example the horizontal stresses

$$J(\boldsymbol{u})(\boldsymbol{x}) = \sigma_{xx}(\boldsymbol{x}) \tag{5.368}$$

are defined differently with respect to linear problems. In such cases one needs to think about how to modify the Green's function of the linear problem to predict the nonlinear $J(\boldsymbol{u})$.

References

1. Novartis Pharma AG Campus, Structural Engineers: Schlaich Bergermann and Partner, Software: SOFiSTiK
2. Deif A (1986) Sensitivity analysis in linear systems. Springer, Berlin
3. Choi KK, Kim NH (2005) Structural sensitivity analysis and optimization 1—linear systems. Springer, Berlin
4. Choi KK, Kim NH (2005) Structural sensitivity analysis and optimization 2—nonlinear systems and applications. Springer, Berlin
5. Strang G (2007) Computational science and engineering. Wellesley-Cambridge Press, Wellesley
6. Nikulla S (2012) Quality assesment of kinematical models by means of global and goal-oriented error estimation techniques. PhD thesis, Bauhaus University Weimar

Chapter 6
Appendix

6.1 Nonlinear Elasticity

In the triple $\{\boldsymbol{u}, \boldsymbol{E}, \boldsymbol{S}\}$ the tensor $\boldsymbol{E}$ is the Green-Lagrangian strain tensor and $\boldsymbol{S}$ the second Piola-Kirchhoff stress tensor, and we assume the material to be hyperelastic, i.e., there exists a strain-energy function $\boldsymbol{W}$ such that $\boldsymbol{S} = \partial W/\partial \boldsymbol{E}$. Given volume forces $\boldsymbol{p}$ the elastic state $\mathcal{S} = \{\boldsymbol{u}, \boldsymbol{E}, \boldsymbol{S}\}$ satisfies at every point $\boldsymbol{x}$ of the undeformed body the system

$$
\begin{aligned}
\boldsymbol{E}(\boldsymbol{u}) - \boldsymbol{E} &= \boldsymbol{0} & \qquad \frac{1}{2}(u_{i,j} + u_{j,i} + u_{k,i}\, u_{k,j}) - \varepsilon_{ij} &= 0 \\
\boldsymbol{W}'(\boldsymbol{E}) - \boldsymbol{S} &= \boldsymbol{0} & \frac{\partial W}{\partial \varepsilon_{ij}} - \sigma_{ij} &= 0 \\
-\mathrm{div}(\boldsymbol{S} + \nabla\, \boldsymbol{u}\, \boldsymbol{S}) &= \boldsymbol{p} & -(\sigma_{ij} + u_{i,k}\, \sigma_{kj}),_j &= p_i
\end{aligned}
\tag{6.1}
$$

and satisfies displacement boundary conditions $\boldsymbol{u} = \bar{\boldsymbol{u}}$ on a part Γ_D of the boundary and stress boundary conditions $\boldsymbol{t}(\boldsymbol{S}, \boldsymbol{u}) = \bar{\boldsymbol{t}}$ on the complementary part Γ_N where

$$
\boldsymbol{t}(\boldsymbol{S}, \boldsymbol{u}) := (\boldsymbol{S} + \nabla \boldsymbol{u}\, \boldsymbol{S})\, \boldsymbol{n} \tag{6.2}
$$

is the traction vector at a boundary point with outward normal vector $\boldsymbol{n}$.

With symmetric stress tensors $\boldsymbol{S}$ we have the identity

$$
\begin{aligned}
&\int_\Omega -\mathrm{div}(\boldsymbol{S} + \nabla\, \boldsymbol{u}\, \boldsymbol{S}) \cdot \hat{\boldsymbol{u}}\, d\Omega \\
&\qquad = -\int_\Gamma \boldsymbol{t}(\boldsymbol{S}, \boldsymbol{u}) \cdot \hat{\boldsymbol{u}}\, ds + \int_\Omega \boldsymbol{E}_{\boldsymbol{u}}(\hat{\boldsymbol{u}}) \cdot \boldsymbol{S}\, d\Omega
\end{aligned}
\tag{6.3}
$$

F. Hartmann, *Green's Functions and Finite Elements*,
DOI: 10.1007/978-3-642-29523-2_6,

where

$$\boldsymbol{E}_{\boldsymbol{u}}(\hat{\boldsymbol{u}}) := \frac{1}{2}(\nabla \hat{\boldsymbol{u}} + \nabla \hat{\boldsymbol{u}}^T + \nabla \boldsymbol{u}^T\, \nabla\, \hat{\boldsymbol{u}} + \nabla\, \hat{\boldsymbol{u}}^T\, \nabla\, \boldsymbol{u}) \tag{6.4}$$

is the Gateaux derivative of the matrix $\boldsymbol{E}(\boldsymbol{u})$

$$\frac{d}{d\varepsilon}[\boldsymbol{E}(\boldsymbol{u} + \varepsilon\, \hat{\boldsymbol{u}})]_{|\varepsilon=0} = \boldsymbol{E}_{\boldsymbol{u}}(\hat{\boldsymbol{u}}). \tag{6.5}$$

Collecting terms we can formulate Green's first identity of the operator $\boldsymbol{A}(\mathcal{S})$, that is the system (6.1)

$$G(\mathcal{S}, \hat{\mathcal{S}}) = \underbrace{\langle \boldsymbol{A}(\mathcal{S}), \hat{\mathcal{S}}\rangle + \int_{\Gamma} \boldsymbol{t}(\mathbf{S}, \boldsymbol{u}) \cdot \hat{\boldsymbol{u}} ds}_{\delta W_e} - \underbrace{a(\mathcal{S}, \hat{\mathcal{S}})}_{\delta W_i} = 0 \tag{6.6}$$

where

$$\langle \boldsymbol{A}(\mathcal{S}), \hat{\mathcal{S}}\rangle := \int_0^l (\mathbf{E}(\boldsymbol{u}) - \mathbf{E}) \cdot \hat{\mathbf{S}}\, d\Omega + \int_{\Omega} (\mathbf{C}[\mathbf{E}] - \mathbf{S}) \cdot \hat{\mathbf{E}}\, d\Omega + \int_{\Omega} -\text{div}\, \mathbf{S} \cdot \hat{\boldsymbol{u}}\, d\Omega \tag{6.7}$$

and

$$a(\mathcal{S}, \hat{\mathcal{S}}) = \int_{\Omega} (\mathbf{E}(\boldsymbol{u}) - \mathbf{E}) \cdot \hat{\mathbf{S}}\, d\Omega + \int_{\Omega} (\mathbf{W}'(\mathbf{E}) - \mathbf{S}) \cdot \hat{\mathbf{E}}\, d\Omega + \int_{\Omega} \mathbf{E}_{\boldsymbol{u}}(\hat{\boldsymbol{u}}) \cdot \mathbf{S}\, d\Omega. \tag{6.8}$$

The identity (6.6) is the basis of many variational principles in nonlinear mechanics and can be formulated in the same way also for beams and slabs, [1].

In the case of a pure displacement formulation $\mathcal{S} = \{\boldsymbol{u}, \boldsymbol{E}(\boldsymbol{u}), \mathbf{W}'(\boldsymbol{E}(\boldsymbol{u}))\}$ and it is $\hat{\boldsymbol{u}} = \mathbf{0}$ on Γ_D, so that (6.6) reduces to

$$G(\boldsymbol{u}, \hat{\boldsymbol{u}}) = \int_{\Omega} \boldsymbol{p} \cdot \hat{\boldsymbol{u}}\, d\Omega + \int_{\Gamma_N} \bar{\boldsymbol{t}} \cdot \hat{\boldsymbol{u}}\, ds - \int_{\Omega} \mathbf{E}_{\boldsymbol{u}}(\hat{\boldsymbol{u}}) \cdot \mathbf{S}\, d\Omega = 0, \tag{6.9}$$

where $\boldsymbol{S} = \boldsymbol{W}'(\boldsymbol{E}(\boldsymbol{u}))$.

Next let $\boldsymbol{u}_h = \sum_j^n u_j\, \varphi_j(\boldsymbol{x})$ the FE solution and let $\hat{\boldsymbol{u}} = \varphi_i$ a virtual displacement then

$$\underbrace{\int_\Omega \boldsymbol{E}_{\boldsymbol{u}_h}(\boldsymbol{\varphi}_i) \cdot \boldsymbol{W}'(\boldsymbol{E}(\boldsymbol{u}_h))\, d\Omega}_{k_i} = \underbrace{\int_\Omega \boldsymbol{p} \cdot \boldsymbol{\varphi}_i \, d\Omega + \int_{\Gamma_N} \bar{\boldsymbol{t}} \cdot \boldsymbol{\varphi}_i \, ds}_{f_i} \tag{6.10}$$

or

$$\boldsymbol{k}(\boldsymbol{u}) = \boldsymbol{f} \tag{6.11}$$

where $\boldsymbol{u}$ is the vector of nodal coordinates.

6.1.1 Linearization

For computational purposes a linearization of (6.11) is necessary. Let $\boldsymbol{p}_\Delta$ and $\bar{\boldsymbol{t}}_\Delta$ be load increments, and let $\boldsymbol{u} + \boldsymbol{u}_\Delta$ be the displacement field corresponding to $\boldsymbol{p} + \boldsymbol{p}_\Delta$ and $\bar{\boldsymbol{t}} + \bar{\boldsymbol{t}}_\Delta$ then

$$\begin{aligned} G(\boldsymbol{u} + \boldsymbol{u}_\Delta, \hat{\boldsymbol{u}}) = & \int_\Omega (\boldsymbol{p} + \boldsymbol{p}_\Delta) \cdot \hat{\boldsymbol{u}} \, d\Omega + \int_{\Gamma_N} (\bar{\boldsymbol{t}} + \bar{\boldsymbol{t}}_\Delta) \cdot \hat{\boldsymbol{u}} \, ds \\ & - a(\boldsymbol{u} + \boldsymbol{u}_\Delta, \hat{\boldsymbol{u}}) = 0 \,, \end{aligned} \tag{6.12}$$

where

$$a(\boldsymbol{u} + \boldsymbol{u}_\Delta, \hat{\boldsymbol{u}}) := \int_\Omega \boldsymbol{E}_{\boldsymbol{u}+\boldsymbol{u}_\Delta}(\hat{\boldsymbol{u}}) \cdot \boldsymbol{W}'(\boldsymbol{E}(\boldsymbol{u} + \boldsymbol{u}_\Delta))\, d\Omega \,. \tag{6.13}$$

The Gateaux derivative of the strain energy product

$$a(\boldsymbol{u}, \hat{\boldsymbol{u}}) := \int_\Omega \boldsymbol{E}_{\boldsymbol{u}}(\hat{\boldsymbol{u}}) \cdot \boldsymbol{W}'(\boldsymbol{E}(\boldsymbol{u}))\, d\Omega \tag{6.14}$$

with respect to a displacement increment $\boldsymbol{u}_\Delta$ is

$$\begin{aligned} a_T(\boldsymbol{u}_\Delta, \hat{\boldsymbol{u}}) : &= \left[\frac{d}{d\varepsilon} a(\boldsymbol{u} + \varepsilon \, \boldsymbol{u}_\Delta, \hat{\boldsymbol{u}}) \right]_{\varepsilon=0} \\ &= \int_\Omega [\nabla \boldsymbol{u}_\Delta \, \boldsymbol{W}'(\boldsymbol{E}(\boldsymbol{u})) \cdot \nabla \hat{\boldsymbol{u}} + \boldsymbol{E}_{\boldsymbol{u}}(\hat{\boldsymbol{u}}) \cdot \boldsymbol{C}[\boldsymbol{E}_{\boldsymbol{u}}(\boldsymbol{u}_\Delta)]]\, d\Omega \,, \end{aligned} \tag{6.15}$$

where the tensor

$$\boldsymbol{C} = \frac{\partial^2 \boldsymbol{W}}{\partial \boldsymbol{E} \, \partial \boldsymbol{E}} = \frac{\partial}{\partial \boldsymbol{E}} \boldsymbol{S} \tag{6.16}$$

is the second derivative of $\boldsymbol{W}$, which is to be evaluated at $\boldsymbol{u}$. Note that $a_T(\boldsymbol{u}, \boldsymbol{u}_\Delta, \hat{\boldsymbol{u}})$ is linear in the second and third argument, $\boldsymbol{u}_\Delta$ and $\hat{\boldsymbol{u}}$.

We then let

$$a(\boldsymbol{u} + \boldsymbol{u}_\Delta, \hat{\boldsymbol{u}}) \simeq a(\boldsymbol{u}, \hat{\boldsymbol{u}}) + a_T(\boldsymbol{u}, \boldsymbol{u}_\Delta, \hat{\boldsymbol{u}}), \tag{6.17}$$

so that (6.11) becomes

$$\boldsymbol{K}_T(\boldsymbol{u})\,\boldsymbol{u}_\Delta = \boldsymbol{f} - \boldsymbol{k}(\boldsymbol{u}), \tag{6.18}$$

where now $\boldsymbol{u}_\Delta$ is the vector of nodal displacements of the field $\boldsymbol{u}_\Delta$ and $\boldsymbol{K}_T$ is the tangent stiffness matrix:

$$(\boldsymbol{K}_T)_{ij} = a_T(\boldsymbol{u}, \boldsymbol{\varphi}_j, \boldsymbol{\varphi}_i). \tag{6.19}$$

6.2 Software

The following software

- BE-PLATES
- BE-SLABS
- BE-LAPLACE
- MATHEMATICA™
- MATLAB™
- SOFiSTiK™
- WINFEM

was used to generate the plots and to do the calculations.

BE-PLATES: Figs. 2.17, 3.20, 3.21, 3.22, 3.34, 3.35, 3.37, 3.44, 3.46, 3.47, 3.48, 5.20, 5.22

BE-SLABS: Figs. 1.6, 1.7, 2.15, 2.22, 4.14, 4.15, 4.16, 5.21

BE-LAPLACE: Figs. 2.3, 5.19

MATHEMATICA: Fig. 2.13

MATLAB: Fig. 3.1

SOFiSTiK: Figs. 1.10, 5.1, 5.10, 5.25

WINFEM: Figs. 1.3, 1.4, 1.5, 1.8, 1.9, 1.19, 1.20, 3.4, 3.8, 3.9, 3.13, 3.17, 3.38, 3.39, 4.1, 4.4, 4.5, 4.8, 4.9, 4.11, 4.18

The programs BE-PLATES, BE-SLABS, BE-LAPLACE and WINFEM are public domain, see www.winfem.de or www.be-statik.de. For an educational version of SOFiSTiK contact www.sofistik.de.

Acknowledgments Parts of the techniques described in this book were developed in the context of the research project: "Preventive quality assurance in computer based structural design—Failure

Mode and Effects Analysis (FMEA)" which was supported by the German Federal Office for Building and Regional Planning under document number Z 6-10.08.18.7-08.26/II 2-F20-08-112, the SOFiSTiK AG and the Society of Inspection Engineers for Structural Analysis (VPI). Their support is gratefully acknowledged.

Reference

1. Hartmann F (1985) The mathematical foundation of structural mechanics. Springer, Berlin

Index

F. Hartmann, *Green's Functions and Finite Elements*,
DOI: 10.1007/978-3-642-29523-2, © Springer-Verlag Berlin Heidelberg 2013

FSC
www.fsc.org
MIX
Papier aus verantwortungsvollen Quellen
Paper from responsible sources
FSC® C105338

If you have any concerns about our products,
you can contact us on
ProductSafety@springernature.com

In case Publisher is established outside the EU,
the EU authorized representative is:
Springer Nature Customer Service Center GmbH
Europaplatz 3, 69115 Heidelberg, Germany

Printed by Libri Plureos GmbH
in Hamburg, Germany